VEGETABLE SCIENCE

D. Singh
Programme Coordinator,
KVK,CAZRI,Pali

Published by :

NEW VISHAL PUBLICATIONS
New Delhi

First Edition : 2011
Reprinted : 2013
Reprint: 2019

Published by :

NEW VISHAL PUBLICATIONS
E-153, West Patel Nagar,
New Delhi-110008
Ph. : 9811251105, 9868437555
E-mail : nvpbooks@hotmail.com

Available in Delhi at :

Pusa Agricultural Books Service
29, Shishir Hostel,
IARI, Delhi-12
Ph. : 011-25841048
E-mail : pusa.agri@gmail.com

Graphics and laser typesetting by :
Oberoi Graphics, New Delhi

Printed at :
Giriraj Printers, Jhilmill Industrial Area, Delhi

PREFACE

Vegetables and spices form the backbone of Horticulture. As the spectrum covered by these crops is too wide, the students had to study different book, assimilate their contents and prepare suitable notes. Often this task is not easy and in case of less time and shortage of books it is really impossible. Keeping these facts in mind we had made an effort to make available a book suitable for preparation and revision for competitive examinations, at the same time maintaining academic rigour and faculty accuracy. All the chapters in the book had been properly covered according to their families and relative importance. Questions given in the books are very relevant to the competitive examinations and most of the previous year questions asked in different exams had been incorporated. The appendix given at the end carries vast information on different aspects of vegetables and spices in short tabular form .This book fulfill the need of a good reference-cum text book for the student community.

I express my sincere thanks to scientists from ICAR institutes/SAUs for their inputs in terms of material collection, compilation and correction. Every effort has been made to ensure that the information collected from various sources such as individuals, institutions, organizations and publications are accurate. We hope that this book will be a useful reference material for the teachers, practitioners and students searching their careers in ICAR Institutes, State Agricultural Universities, State Agriculture/Horticulture Departments and for competitions like State Public Services/UPSE/ARS/JRF/SRF/NET along with entrance examinations for B.Sc, M.Sc and Ph.D Degree in Horticulture .

We hope that this effort will help the readers in their endeavours. All the suggestions in making the book further informative are always welcome.

Author

CONTENTS

1

GENERAL

- The term Horticulture first appeared in written language in 17th century.
- The word Horticulture is derived from the Latin words 'hortus' and 'celere' meaning garden and to cultivation.

Branches of Horticulture –

1. **Pomology** – Derived from Latin word 'pomus' meaning fruit and Greek word 'logy' meaning science.

   Thus pomology is the science of parody of fruits.

2. **Olericulture** – Latin word

   Oleris – Pot herb.

   Culture – Raising of plants.

   Olericulture is the science of vegetable crops.

3. **Floriculture / Landscape gardening** – Science of flower production.

   Webster's Dictionary defines a vegetable as "a herbaceous plant cultivated for food, as the cabbage, potato, bean, etc; also, the edible part or parts of such plants, as prepared for market or table." Vegetables are generally classified according to the source of the edible plant part(s):

   Root (e.g. beet, carrot, turnip)

   Bulb (e.g. leek, onion, garlic)

Stem (e.g. asparagus, kohl rabi)

Flower (e.g. cauliflower, broccoli)

Tuber (e.g. potato)

Immature fruit (e.g. cucumbers, green peas, summer squash, sweet corn)

Mature fruit (e.g. muskmelon, watermelon, tomato, winter squash)

Leaf (e.g. cabbage, lettuce, spinach)

Petiole (e.g. celery)

Seed (e.g. bean)

- Vegetable production for processing (e.g. canning, freezing, or dehydration) is usually distinct from fresh market vegetable production. Whereas vegetables for processing are grown exclusively in the field, a significant quantity of the fresh market production can be in greenhouse structures (e.g. tomatoes, cucumbers, and lettuce).
- Other specialized areas of the vegetable industry include the production of vegetable transplants and vegetable seed for commercial and home garden vegetable growers.
- Vegetables are generally cultured as annuals. Exceptions include artichoke, asparagus, cardoon and horseradish, which are grown as perennials. Onions, celery and carrots are typically grown as annuals for consumption, but as biennials for seed production.
- Propagation of vegetables is done mostly by seed, although artichoke is propagated by divisions, Irish potato by tubers or tuber sections, and sweetpotato by rooted shoots. A number of vegetable crops (primarily those of the cucurbitaceae) are pollinated by bees (e.g. cucumber, muskmelon, pumpkin, squash and watermelon). Although

eggplant, lima bean, okra and pepper will set fruit without bees, bee activity has been shown to increase yields of these vegetable crops.

- Honey bees do not assist in the pollination of peas, snap bean, sweet corn and tomato, but collects pollen and nectar from these vegetables.
- Vegetables have highly variable climatic requirements. Temperature, water availability, and to a lesser extent day length, are the climatic factors which have the greatest influence on productivity.
- Cool season vegetables (including artichoke, green peas, endive, lettuce, cabbage, cauliflower, broccoli, onions, and spinach) grow best between 12° and 20° C. Warm season vegetable crops (including bean, cucumber, eggplant, muskmelon, okra, pepper, sweet corn, summer squash, watermelon and tomato) grow best between 18° and 28° C.
- Most cool season vegetables are generally also classified as "hardy" to temperatures as low as 1° C during the juvenile period. Such chilling hardy vegetables include asparagus, broccoli, brussels sprouts, cabbage, garlic, horseradish, kale, onion, pea, radish, rutabaga, spinach and turnip.
- Warm season crops are generally also classified as chilling, tender (sweet corn, snap bean, tomato) or very tender (cucumber, eggplant, lima bean, muskmelon, pepper, pumpkin, squash, sweetpotato, watermelon). The latter are not transplanted or seeded until all risk of frost is over and until soil and air temperature is conducive to optimum growth.
- Almost all commercially grown vegetables are now raised under irrigated conditions or with supplemental irrigation.
- In some areas rainfall during certain times of the growing season is detrimental because it interferes with field operations and / or promotes plant disease infection.

- Vegetables are generally grown on mineral or organic soils. Sandy loam and loam soils are the preferred types of mineral soils because of the desirable growth conditions and the ease of cultivation they provide, especially for bed preparation and precision seeding.
- Heavier loams and clay soils are generally avoided because they are not amenable to such cultural practices.
- The production of vegetables is highly intensive in comparison to the production of field crops such as corn, wheat and soybean.
- Vegetable production requires considerable hand labour because of the intensive nature of cropping, the variable maturities of plants in a field, and the perishable nature of the harvested crop, especially for fresh market.
- In recent years considerable advances have been made in both the development of more uniformly maturing plants and mechanical equipment for planting, cultivating, and harvesting vegetables. Perhaps the most notable example of this has been the development of new processing tomato cultivars with characteristics which make them amenable to mechanical harvesting by efficient field harvesters.
- Most vegetables are perishable and maintain peak quality for only a short period after harvest.
- Cooling (to restrict respiration) and storage in a high relative humidity environment (to prevent desiccation) are important post-harvest handling practices for most fresh market vegetables.
- A few vegetables (e.g. potatoes and onions) remain in excellent condition for many months with good storage.
- Vegetables are of considerable importance in human nutrition. Generally they are rich in vitamins and minerals, low in calories, and they supply bulk and fiber. Some (e.g. beans)

are excellent protein sources when combined with other foods.

- Vegetables are rich source of vitamins and minerals.

- **Vegatable Consumption Per Capita**

  1. Recommendation – 300 gm/day

     125 gm. Green leafy vegetables.

     100 gm. Root vegetables.

     075 gm. Other vegetables.

  2. Present availability– 145 gm. (In India)

- India is the **second largest producer** of vegetables after China (43%) producing 13.38% of total world's production.
- Vegetable crops in India occupy only 2.8% of the total cropped area.
- Productivity of vegetables in India – 14.9 tonnes/ha.
- State having largest area under vegetables – West Bengal.
- State having maximum produce under vegetables–West Bengal.
- State having maximum productivity of vegetables–Tamil Nadu.
- Most ancient type of vegetable garden – Kitchen Garden.
- Very intensive method of vegetable cultivation – Market Garden.
- Very extensive method of vegetable cultivation – Truck Garden.
- Green leafy vegetables are rich source of "Folic acid".
- Vegetables are also called as **"protective food".**

- **Daily Requirement –**

Cereals	-	475 g.			
Calories	-	2800 g.			
Pulses	-	80 g.			
Protein	-	55 g.	Fruits	-	120 g.
Calcium	-	450 mg.	Milk	-	200 ml.
Iron	-	20 mg.	Sugar	-	40 mg.
Vitamin C	-	50 mg.	Oil	-	40 ml.
Thiamin	-	1.4 mg.	B-Carotene	-	3000 mg.
Riboflavin	-	1.5 mg	Folic Acid	-	100 mg.
Niacin	-	19 mg.	Vita B_{12}	-	1.0 mg.
Vitamin D	-	5 mg.			

- Major mineral present in vegetables – K.
- Fat percent in vegetable – < 0.1 %.
- Highest consumer of vegetables (per capita)
  1. Greece – 377 kg. / Year.
  2. Turkey – 327 kg. / Year.
  3. India – 62.79 kg./ Year.

- **Nutritive Value of vegetables**

1.	Vitamin A	Bathua leaves	11300 IU/100g.
		Calocassia leaves	10278 IU/100g.
		Beet leaves	9770 IU/100g.
2.	Vitamin B_1 (Thiamine)	Chillies	0.55 mg./100gm.
		Calocasia leaves	0.22 mg./100gm.
		Tomato (Red)	0.12 mg./100gm.

3.	Vitamin B_2 (Riboflavin)	Fenugreek leaves	0.31 mg./100gm.
		Amaranthus	0.30 mg./100gm.
4.	Vitamin C (Ascorbic Acid)	Drumstick leaves	220 mg./100gm.
		Coriander leaves	135 mg./100gm.
		Chillies	111 mg./100gm.
		Broccoli	109 mg./100gm.
		Tomato	31 mg./100gm.
5.	Carbohydrate	Tapioca	38.1%
		Sweet Potato	28.2%
		Potato	22.6%
6.	Protein	Lima Bean	7.9g./100gm.
		Peas	7.2g./100gm.
		Cowpea	4.3g./100gm.
7.	Fat	Potato	118mg./100gm.
8.	Fiber	Potato	752mg./100gm.
		Chilli	6.8g./100gm.
		Amaranthus	1.0g./100gm.
		Beet leaf	0.7g./100gm.
		Spinach	0.6g/100gm.
9.	Calcium	Agathi	1130mg/100g
		Curry leaves	813 mg/100g
		Amaranthus & Fenugreek leaves	395 mg/100g
10.	Phosphorus	Amaranthus	800 mg/100g
		Garlic	187 mg/100g
11.	Iron	Agathi	83.9 mg/100g
		Amaranthus	22.5 mg/100g
12.	Potassium	Spinach	605 mg/100g
		Amaranthus	230 mg/100g

All India area and production of vegetables.

- Vegetable area in India – 7.2 M. Ha (2005-06)
- Vegetable production in India – 113.5 M.T. (2005-06)

Crops	% Share in Area	% Share in Production
1. Potato	19.4%	23.7%
2. Brinjal	7.5%	8.2%
3. Onion	7.1%	5.0%
4. Tomato	7.3%	7.7%
5. Okra	5.6%	3.6%

- **Five major nutrients –**
  1. Carbohydrates, protein and fat – All three are required in large quality for supplying energy.
  2. Minerals and vitamin – These are required in small quality for physiological process and metabolic activity of body.
- Major mineral elements obtained from vegetables are Ca, P and Iron.
- Most of leafy vegetables are rich in Carotene, riboflavin and minerals.
- Flavour compound in vegetables–
  1. Volatiles – Aldehydes, ketones, alcohols, sulphar compound and aromatic hydrocarbons.
  2. Non Volatiles – Organic acids, sugar, fatty acids and amino acids.
- Characteristic flavour of cole crop is due to volatile compound – Dimenthyl trisulphide.

- Pungency and flavours of Onion – Allyl - Propyl –Disulphide.
- Pungency and flavours of Garlic – Diallyl - Disulphide.

Toxic Substance – Trypsin inhibitors present in beans particularly in raw soybean which inhibit proteolytic enzyme activity.

- Phytic acid (Inositol Hexaphosphoric acid) – Peas and beans.
- Colocasia – Oxalic acid (also present in elephant foot which is responsible for bitter taste).

Vegetable in disease prevention –

1. Bitter gourd – Treating diabetes.
2. Ash guard – For weak nervous systems.
3. Bioflavonoids – Compound closely associated with Vitamin - C and found in several Vegetables eg. Onion and Garlic (quercetin – protection against cancer and heart disease).
4. Other compounds

   Onion and Garlic – 1. Allicin

   2. Allistalin

   3. Garlicin

   4. Dially Disulphide

   5. Dially Trisulphide

   6. Allyl Propyl - disulphide

   (All are sulphar compounds effective in reducing harmful blood cholesterol)
5. Cheratin – Isolated from bitter gourd (effective against diabetes disease).

6. Diphenyl amino – Onion (used in curing Diabetes).
7. 3 - n - Butyl Pthalide – Celery (effective against Hypertension).
8. Diosgenin – Dioscorea (Manufacture of cortisone and contraceptive drugs).

- By the end of 2030, according to an estimate we will need 151 - 193 M.T. of vegetables to meet our requirement.
- The Research and development programme in Horticulture have received an impressive support from the Eight Five year plan onwards.
- At present 10 % of ICAR budget and 17% of the Department of Agriculture and Co - operation budget is earmarked for horticulture section in Ninth Plan.
- Horticulture sector covers only 8 % of the total crop area in the country, contributing 24.5 % to the GDP and 54.55 % to export earning in the agriculture sectors.
- In 1945 Horticulture section was established in division of Botany, IARI, New Delhi.
- A full fledged division of Horticulture was created at the ICAR Headquarter, headed by DDG during 1987.
- Division of Horticulture was carved out of crop division in 1981 and in 1985; a position of Horticulture Commissioner was also created.
- National Horticulture Board (1984) : Gurgaon (Haryana).
- 17 Centres work on Horticulture.
- Spice board works under Ministry of Commerce.
- India occupies Ist position in Cauliflower production.
- India occupies IInd position in Onion production.
- India occupies IIIrd position in Cabbage production.
- Trypsin inhibitor is also present in sweet potato.
- Rutabaga is a man made vegetable (*Brassica napus* var - *napobrassica*)

- The Salem belt of Tamil Nadu and Smalkot (Andhra Pradesh) are known for Cassava production as industrial crop.
- ISTA (International Seed Testing Association) was established in 1924 at Norway (Switzerland).

Major Acids present in Vegetables.

1. Citric acid – Tomato, Potato, Legumes, Beetroot and Leafy Vegetables.
2. Malic acid – Carrot, Celery, Lettuce, Onion and Brocoli.
3. Oxalic acid – Amaranthus, Portulaca, Celosia and Basella.

Aroma containing compounds

1. Non adienal	–	Cucumber.
2. Allyl - iso - thiocynate	–	Raw Cabbage.
3. Dimethyl disulphide	–	Cooked Cabbage.
4. Dimethyl pyarzin	–	Potato.
5. Iso-thiocynate	–	Radish.

Toxic Substance in Vegetables

1. Haemaglutine	French bean.
2. Saponine	Spinach, Tomato, Asparagus.
3. Serotonine	Water Melon.
4. Salosodine	Brinjal.
5. Apiin	Celery.
6. Cholic ertrass	Pumpkin.
7. Zea - Xanthin	Corn.
8. Sinigrin	Cole crops.

9. HCN (Dhurin) Cassava.
10. Dioscorine Yams.

- Highest production of vegetables was obtained in 2004-05 (101433 thousand tons) and productivity (15.2 M.T./Ha) in 1999-2000.

Indian Vegetable productivity compared to other country.

Crop	India Productivity (Mt./ha.)	Other Country (Mt./ha.)
Potato	18 Mt / ha.	45.7 Mt. (Netherlands)
Brinjal	16.4 Mt / ha.	34 Mt (Japan)
Tomato	17.3 Mt / ha.	63.5 Mt (Spain)
Cabbage	21.2 Mt / ha	63 Mt (Korea Republic)
Onion	12.7 Mt / ha	54.4 Mt / ha
Cauliflower	17.1 Mt / ha	27 Mt / ha (Germany)
Okra	9.8 Mt / ha	14.2 Mt / ha (Egypt)
Peas	7.1 Mt / ha	18.2 Mt / ha (Belgium)

- Of the ten top vegetable crops of India, Okra is only crop where India figures among top five countries in order of their productivity.
- AICRP for improvement of Vegetable crops was started in 1970 - 71 with headquarters at IARI, New Delhi
- Indian Institute of Vegetable Research, Varanasi (U.P.) & Project Directorate of Vegetable Research was started in 1987.
- All India Vegetable Improvement Project was shifted from IARI, New Delhi to Varanasi in 1992.

Major Vegatable producing country –

China	48%
India	13%
Turkey	3%
USA	3%
Italy, Egypt, Russian	2%
Japan, Spain, Iran, Korea	1%
Other Countries	22%

Largest Producer in World

1. Potato	22.83% (India IIIrd)
2. Brinjal	7.85% (India IInd)
3. Tomato	7.37% (China ranks Ist)
4. Cabbage	4.30% (India IInd)
5. Onion	8.79% (India IInd)
6. Cauliflower	3.53% (India IInd)
7. Peas	4.10% (India IInd)
8. Sweet Potato	2.02%
9. Tapioca	4.16% (Nigeria ranks Ist)
10. Okra	5.30% (India ranks Ist)
11. Yam	Africa (Ranks 1st)
12. Sprout	USA (Ranks 1st)

China has 1st position in production of all vegetables mentioned above except Tapioca (Nigeria Ist), Yam (Africa Ist), Okra (India 1st) and Sprout Brocoli (USA 1st).

Vegetables	Major Producing States	National Productivity (MT/ha)
Tomato	Karnataka (26.9%), Maharastra (26.4%)	17.4
Brinjal	Kartnatak (22.9%)	16.4
Okra	Bihar (13.0%), Andhra Pradesh (12.5%)	9.8
Onion	Gujarat (22.9%)	12.7
Cabbage	Uttar Pradesh (30%)	21.2
Cauliflower	West Bengal (26.1%)	—
Potato	Haryana (24%)	18.9
Sweet Potato	Maharashtra (14%) Bihar (13.9%)	8.9
Tapioca	Tamil Nadu (41.9%)	28.1

- Among the Vegetables tapioca has the highest productivity i.e. 28.1 Mt/Ha.
- In Tamil Nadu productivity of tapioca is 41.9% which is highest amongst all vegetables.

Classification of vegetables

- **Monocot Vegetables –**

Family

(a) Alliaceae Onion, Garlic, Leek.

Multiplier Onion - *Allium cepa var – aggregatum.*

Top Onion - *Allium cepa var – viviparum.*

Welsh Onion - *Allium cepa var – fistulosum.*

Shallat - *Allium cepa var – ascaloricum.*

Chive - *Allium cepa var – schoenoprasum.*

(b) Liliaceae Asparagus (*Asparagus officinalis*).

(c) Araceae Elephant foot yam (*Amorphosphorus companulatum*).

(d) Dioscoreaceae Yam – *Dioscorea alata.*

- **Dicot Vegetables :-**

(a) Aizoaceae
1. New Zealand spinach (*Tetragouia expansa*).

(b) Chenopodiaceae
1. Beetroot or Sugarbeet Palak.
2. Swiss Chard (*Beta vulgaris var cicla*).
3. Spinach.

(c) Compositae
1. Lettuce.
2. Chicory - *Chichorium intybus.*
3. Endive – *Chichorium endive.*
4. Artichoke - *Cynara scolymus.*

(d) Convonlvulaceae
1. Sweet potato (*Ipomea botatas*).

(e) Cruciferae
1. Cole crops.
2. Radish.

(f) Cucurbitaceae	1. All cucurbitaceous vegetables.
(g) Euphorbiaceae	1. Tapioca (*Manihot esculenta*).
(h) Leguminaceae	1. Pea. 2. Beans.
	3. Winged Beans (*Pisophocarpus tetragonolobus*).
(i) Malvaceae	1. Okra (*Abelmoschus esculentum*).
(j) Polygonaceae	1. Rhubarb – (*Rheum rhaponticum*).
	2. Sorrel – (*Rumex vesricarius*).
(k) Solanaceae	1. Potato. 2. Tomato.
	3. Brinjal. 4. Chilli.
(l) Umbeliferae	1. Carrot. 2. Parsley.
	3. Celery.

■ Pollination mechanism in Vegetables –

1. Self pollination – Tomato, Pea, Cowpea, Fenugreek, Lettuce, French bean, Lablab bean, Broad bean
2. Often Cross pollination – Okra, Brinjal, Chilli, Limabean, Sweet pepper
3. Cross Pollinated – Cole Crops, Root Crops, Cucurbits, Onion — All these Entomophillus (Insect Pollination)

Spinach, Amaranthus, Beet leaf (Palak), Garden Beet, Beet root — All these Anemophillus (Wind Pollination)

Sex Forms in vegetables –

1. Monoecious – Cucurbits, Amaranthus, Sweet corn, Cassava.
2. Dioecious – Pointed gourd (JRF, 2008), Scarlet gourd, Spine gourd, Spinach, Yams, Beet root.

- Male Sterility – Tomato, Brinjal, Chilli, Onion, Carrot, Sweet potato.
- Protandry – Onion, Carrot, Beet, Leek, Celery.
- Sporophytic Self incompatibility – Cole crops (JRF, 2008), Radish, Turnip, Beet.
- Gametophytic Self incompatability – Tomato.

Photosynthesis based classification

C4 plant – Amaranthus, Globe amaranthus (Artichoke).

C3 plant – Lettuce, Sugarbeet, Carrot, Tomato, Potato, Sweet potato.

Photoperiod based classification

1. Long Day Plant: Cabbage, Cauliflower, Radish, Carrot, Beet, Turnip, Palak, Spinach, Lettuce, Onion, Potato.
2. Short Day Plant: Sweet Potato, Indian Spinach, Dolichus bean, Cluster bean, Winged bean.
3. Day Neutral – Tomato, Brinjal, Chilli, Okra, Cucurbits, Amaranthus, French bean, Cow pea, Sweet Pepper.

Temperature requirement based classification

1. 18-25° C - Cabbage, Cauliflower, Radish, Carrot, Lettuce, Potato.

2. 20-25° C - Tomato, Brinjal, Pepper, Onion, Garlic, Cucumber, Pumpkin.
3. 25-27° C - Okra, Chilli, Sweet Potato, Water Melon, Musk Melon.

Respiration rate based classification

1. Very low – Potato, Onion.
2. Low – Sweet Potato, Cabbage
3. Moderate – Tomato, Chilli, Carrot.
4. High – Peas, Beans.
5. Very High – Leafy greens, Green onion, Musk Melon, Water Melon, Cauliflower, Asparagus.

Respiration based classification

(A) Climacteric

(1) Tomato, (2) Muskmelon, (3) Watermelon,

(B) Non climacteric

(1) Cucumber.

Plant reproduction

- For employment of suitable breeding method for genetic improvement of any plant species, knowledge of mode of reproduction is essential.
- Self fertilizing plants are genetically homozygous while cross-fertilizing plants are characterized with high degree of heterozygosity and heterogeneity so breeding method for genetic improvement of any plant species should maintain such a natural state for better fitness which ultimately leads to wider adaptability.

- Plants (crop or vegetable) can be divided into two main categories depending on mode of reproduction.
- The sexual reproduction involves the union of male and female gametes, and seeds are used for propagation.
- Asexual reproduction does not involve fusion of male and female gametes and new plants develop either through vegetative parts of plants (vegetative reproduction), e.g., plant cutting, grafting, layering, or they may arise from embryos that develop without fertilization (apomixes).

Brief description of plant reproduction system

SN Terminology	Characteristics
1 Asexual reproduction	Multiplication of plants through vegetative means bypassing sexual process
1a. Vegetative reproduction	Multiplication by vegetative plant parts
(i) Natural	Multiplication by rhizome, tuber, corm, bulb, runner, suckers and stolon
(ii) Artificial	Multiplication by stem and root cuttings, grafting, layering and budding
1b. Apomixis	Development of embryo without fertilization (sexual fusion)
(i) Parthenogenesis	Development of embryo from egg cell without fertilization
(ii) Apogamy	Development of embryo either from synergids or from antipodal cells
(iii) Apospory	Origin of embryo from diploid egg cell of another embryosac developed from diploid tissues

(iv) Adventive embryony	Origin of embryosac directly from diploid cells of either nucellus or integument
2 Sexual reproduction	Multiplication of plants through the union of male and female gametes (fertilized embryos)
a Autogamy	Fertilization of ovules by the pollens of same flower
b Allogamy	Fertilization of ovules by the pollen of another plants

■ Asexual Reproduction :

Potato :

Potato often flowers only in cooler regions in the hills where it is grown in summer under long-day condition. In normal growing season in the plains under short-day condition it generally fails to flower. Even if flowers emerge, fruit set does not occur. In this short - day condition in plains, it can be forced to flower and set fruits with the arrangement of artificial light to lengthen the photoperiod for at least 4 hours. At present, potato is being cultivated commercially, though to a very lesser extent, from true seeds.

Cassava :

It is a highly heterozygous and cross-polllinated crop and monoecy i.e., production of both male and female flowers on the same inflorescence render to cross-pollination. Male sterility, which is commonly found in cassava, also favours cross-pollination. Sexually propagated seedlings are highly variable in character due to heterozygosity and at the same time it takes long time to give potential yield. However, hybridization following clonal selection is an important method to develop a variety.

Sweet Potato :

Flowering itself is a major constraint in this crop and a number of elite clones do not flower. Flowers are bisexual but sterility of different intensities is found because of hexaploidy and accumulation of recessive sublethal mutations through long clonal propagation. It is highly heterozygous due to cross-pollination. Hence, the sexual seedlings are highly variable in nature. Sexual reproduction is only encouraged for breeding purpose.

Yam :

All the members of yams are dioecious. Hence; the male and female flowers are borne on separate plants. Flowering and seed production is greatly limited due to shy flowering and poor seed set.

Taro :

Generally, flowering is rare in most of the taro types. Inflorescence, the spadix, is cylindrical with female flowers at the base and a longer protruding male section of staminate flowers with aborted flowers in-between.

Elephant-foot-yam :

The crop cycle of one to three years causes the emergence of inflorescence, although very rarely. The spadix is extended. bearing male flowers at the top and female flowers at the bottom and the neuters are absent. There is complete absence of seed under natural condition due to extreme protogyny coupled with the delayed opening of the spadix.

Pointed gourd :

Seeds are recalcitrant in nature, which lose viability very soon. The seedlings normally do not flower in the first year. The seedlings from the open pollinated seeds are highly variable due to heterozygosity and generally low yielder than the mother clone. The seedling population gives 50 per cent each of male (unproductive) and female (productive) plants due to involvement of a distinct pair of chromosomes for determination of dioecious sex - mechanism. Hence, seed propagation is normally avoided.

Garlic :

Most of the garlic types are non - bolting where inflorescence does not emerge. Very few cultivars bear fertile flowers and viable seeds that germinate. In most of the cases, the inflorescence, if emerge, contain sterile flowers and bulbils (aerial bulbs).

- Each flower has four basic organs viz., sepals, petals, stamens and pistil.
- A flower having all the basic organs is called as complete flower (pea, tomato, cabbage) while incomplete flower lack one or more basic organs.
- Stamens and pistils are the indispensable organs as these organs produce male and female gametes and flowers having both these organs are called perfect flowers e.g. pea, tomato etc
- Imperfect flower lacks one of the sex organs (either staminate or pistillate).
- Apparently, the perfect flowers are hermaphrodite as these include both male and female organs in the same flower and practically all complete flowers are perfect flowers contrary the imperfect flowers are unisexual as these are either staminate (male) or pistillate (female).

- Plants in which staminate and pistillate flowers are borne on the same individual are called "monoecious"; while plants in which the staminate and pistillate flowers are borne on different individuals are called "dioecious".
- **Pollination** is the process of transfer of pollen grains from the anther to the stigma of the female reproductive organ.
- Transfer of pollen from an anther to a stigma within the same flower or to a stigma of another flower on the same plant is referred to as ***self pollination*** or ***autogamy***.
- Hermaphrodite or bisexual flowers and homogamy (simultaneous maturation of anthers and stigma of a hermaphrodite flower) are pre-requisites for self-pollination in self-pollinated plant species. But practically, self-pollination is not complete and cross-pollination may take place up to certain extent.
- The degree of cross-pollination in self-pollinated species is affected by several factors, *e.g.*, variety, environmental conditions like temperature and humidity, and location.
- Self pollination leads to very rapid increase in homozygosity and these species do not show inbreeding depression but may exhibit considerable heterosis. There are certain mechanisms in the flowering plants that encourage self-pollination which are as under:

    1. ***Cleistogamy***: When pollination and fertilization take place in an unopened flower bud or flower do not open at all this condition ensures self-pollination.
    2. ***Chasmogamy:*** In some crop plants flower opens but only after completion of pollination and this condition favours the self-pollination.
    3. ***Position of anthers:*** In some vegetable crops (tomato, brinjal), stigma is enclosed by anthers in such a manner that self-pollination is ensured.

■ Cross-pollination :

The process of transfer of pollen grains from anthers of one plant to stigma of another plant through wind (anemophily), water (hydrophily) or insects (entomophily) is referred as cross-pollination or allogamy.

- Cross-pollination is very common in almost all vegetable crops which preserve and promote heterozygosity in any population. These plant species are highly heterozygous and exhibits considerable heterosis in certain combinations and severe inbreeding depression on selfing. Therefore, breeding methods should be devised for exploiting heterosis and development of hybrids and synthetics.

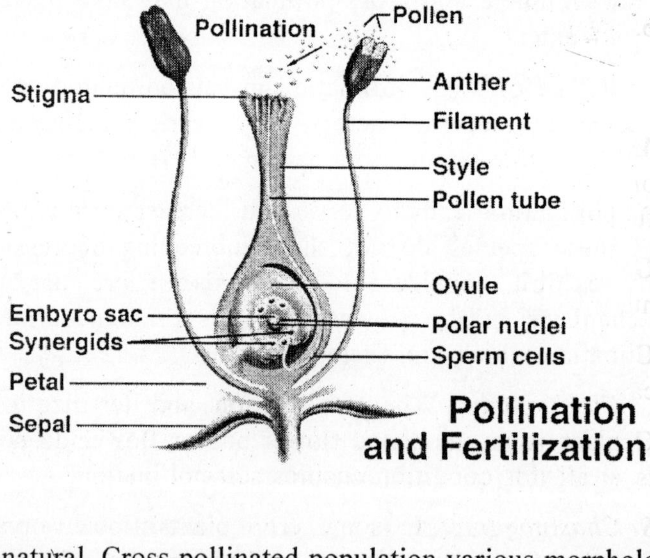

In natural, Cross-pollinated population various morphological, genetical, physiological and biochemical mechanisms are operative individually or jointly that promotes cross-pollination and restricts self-pollination in different manner which are as follows:

1. ***Dicliny or unisexuality***: In such condition flowers are either staminate (male) or pistillate (female). It can be further divided in to two classes first, when male or female flowers borne separately on the same inflorescence or on the same plant this condition is termed as monoecy. In some vegetable crops like gourds and cucumber, staminate (male) and pistillate (female) flowers are borne separately on the same plant. Second, when staminate and pistillate flowers are present on separate individuals, it is referred as dioecy.

2. ***Dichogamy***: The male and female reproductive organs mature at different times even in a perfect flower. Dichogamy is of two types:

   a. ***Protogyny:*** Stigma matures earlier than anthers of the same flower so that allogamy is facilitated.

   b. ***Protandry:*** Anthers mature earlier than the stigma of the same flower thus promoting allogamy as reported in carrot and onion.

3. ***Male sterility*** **:** A condition in which either pollen is absent or non - functional in flowering plants while female gametes functions normally.

(i) **Genetic Male Sterility :** Pollen sterility which is caused by nuclear genes.

(ii) **Cytoplasmic Male Sterility :** Pollen sterility which is caused by cytoplasmic genes.

(iii) **Cytoplasmic Genetic Male Sterility :** Pollen sterility that is controlled by both cytoplasmic and nuclear genes.

4. ***Self - incompatibility*** **:** The inability of a plant with functional pollen to set seeds when self pollinated. It is mainly of two types: (i) Heteromorphic and (ii) Homomorphic.

(i) **Heteromorphic system :** Self incompatibility due to differences in floral morphology. The incomapatibility reaction of pollen is determined by the plant producing them.

(ii) **Homomorphic system :** Presence of self incompatibility due to physiological or genetic causes rather than differences in floral morphology. The incompatibility reaction of pollen may be controlled by the genotype of plant on which it is produced (sporophytiv system) or by the genotype of pollen itself (gametophytic system).

1. **Cytoplasmic** Genetic male sterility was first discovered in onion by Jones and Davis in 1944.

2. ***Watermelon*** is a cross-pollinated species with monoecious or andromonoecious flowering habit. Watermelon will not cross with any other cucurbits except for species within the genus **Citrullus**.

3. *Carrot*: Carrots are protandrous. Two distinct types of male sterile flowers are reported in carrot.

2

SALONECEOUS VEGETABLES

"TOMATO" (No. 1 Processing vegetable)

- B. N. *Lycopersicon esculentum* (Wolf apple, Vilayti Baingan).
- Family – Solananceae.
- 2n = 24.
- Origin - Peru - Ecuador - Bolivia - (South America) Mexico.
- Type of Fruit – Berry.
- Edible part – Fruit.
- **Largest producer state in India – Karnataka.**
- Tomato is universally treated as "Protective Food".
- Total sugar content is 2.5% in ripe fruit.

  Ascorbic acid : 16-65 Mg/100g.

  Amino acid : 100-300 Mg/100g.
- Processed items – Paste, puree, syrup, sauce, ketchup, chutney and soup.
- Export quality of tomato producing area is Nasik and Pune in Maharashtra and Bengaluru in Karnataka.

- Processed product of tomato especially "puree" and "paste" have great demand in export.
- First known record of tomato is mentioned in the year 1554 in South America.
- It was introduced in Europe in the early 16th century. Since 1800 tomatoes are being used as food all over the world.
- Earlier the botanical name of tomato was *Solanum lycopersicon*; later Miller (1978) gave the name *Lycopersicon esculentum* to cultivated tomato and *L. pimpinellifolium* to wild form of tomato.
- Tomato has tap root system.
- For early crop – Sandy to loam soil is best.
- Tomato is moderately tolerant to acid soil (Ph 5.5)
- *L. esculentum* has five forms –
    1. *Cerasiforme* – Probable ancestor of cultivated tomato, also known as small fruited or cherry tomato.
    2. *Pyriformae* – Pear tomato.
    3. *Commune* – Common tomato.
    4. *Grandifolium* – Potato leaf type.
    5. *Validium* – Upright tomato.
- Muller (1940) divided the genus *lycopersicon* into sub genus

  Eulycopersicon – Red Fruited type.

  Eriopersicon – Green Fruited type.

Tomato has broadly two types of varieties.

1. Determinate (restricted with appearance of terminal flower)	2. Indeterminate
1. Inflorescence occurs at every internode.	1. Inflorescence occurs at every third internode.
2. Suitable for North Indian plains.	2. Suitable for Hills & South India.
3. Suitable for mechanized . harvesting	3. Suitable for polyhouses.

- Red colour of tomato is due to lycopene (20-50 Mg/100g).
- Lycopene is highest at 21-24° C temperature.
- Production of this pigment drops off above 27° C temperature.

Varieties –

S. N.	Variety	Features
1	Pusa Ruby	Indeterminate, Maturity 60-85 days. Yield 300 q/ha.
2	Pusa Early Dwarf	Determinate, Maturity 70-80 days. Yield 300 q/ha.
3	Pusa 120	Semi determinate, highly suitable for long distance transportation and processing, Maturity 110 days, Yield 300-400 q/ha.
4	Pusa Gaurav	Determinate, firm fruits highly suitable for long distance transportation and processing. Maturity 110 days, Yield 300-400 q/ha.

5	Pusa Sheetal	Semi determinate, capable of fruit setting at 8° C night temperature, Maturity 90 days, Yield 360 q / ha.
6	Pusa Uphar	Indeterminate upright plant, fruits smooth round to flattish round, Maturity 120 days. Yield 370 q / ha.
7	Pusa Sadabahar	Determinate, dwarf, accommodates more plants per unit area, suitable for growing under wide range (8 - 30° C) of night temperatures. Maturity 60 days, Yield 300-400 q / ha (main season).
8	Pusa Rohini	Determinate fruits, suitable for long distance transportation. Maturity 120 days, Yield 410 q / ha.
9	Roma	Determinate plants, medium sized fruits suitable for transportation and hilly areas. Maturity 100 - 120 days. Yield 300 - 350 q / ha.
10	Italian Red Pear	Determinate, fruits suitable for transportation. Maturity 90 - 110 days. Yield 200 - 300 q / ha.
11	Punjab Chhuhara	Determinate high yielding elongated fruits, low juice content suitable for processing and transportation. Maturity 100 - 110 days. Yield 325 q / ha.
12	La Bonita	Determinate bushy plants, fruits suitable for long distance transportation, maturity 100 - 110 days. Yield 350 q / ha.
13	S - 12	Determinate bushy plants, fruits round, medium size, uniform red colour at maturity, juicy highly acidic. Maturity 90 - 110 days. Yield 175 - 280 q / ha.
14	Pant T-3	Semi determinate, dark green foliage, fruits round smooth, medium size, suitable for processing, maturity 70 days. Yield 190 q / ha.

15	Pusa Hybrid-1	Determinate, can set fruit upto 28° C night temperatures, fruits from June to mid July the lean period for tomato in North India. Maturity 95 days. Yield 320 q/ha.
16	Pusa Hybrid-2	Determinate tolerant to root knot nematode, Maturity 110-120 days. Yield 591 q/ha.
17	Pusa Hybrid-4	Determinate highly tolerant to root knot nematode. Maturity 120 days. Yield 600 q / ha.
18	Pusa Hybrid-8	Determinate, fruits suitable for long distance transportation. Maturity 105 days. Yield 557 q / ha. (main season)
19	**Pusa Divya (F1 hybrid)**	**Only hybrid developed so far by utilizing male sterile line in India,** Indeterminate fruits suitable for long distance transportation. Maturity 90 - 100 days. Yield 350 - 450 q / ha.
20	Kashi Vishesh	Plants are determinate and fruits spherical. Recommended for J&K, HP, UP, Bihar, Uttarakhand, Punjab, Jharkhand, Chattishgarh, Orissa and Southern states. It is resistant to TLCV, Duration 70 - 75 days after transplanting, yield : 450 - 550 q / ha.
21	Kashi Amrit	Plants determinate, Duration 75 - 80 days and yield 500 - 550 q / ha. Suitable for Rajasthan, Gujarat, Haryana and Delhi.
22	Kashi Hemant	Suitable for MP, Chattishgarh, AP. Orissa, Maharastra. Yield 400 - 420 q / ha.
23	Kashi Sharad	Plants indeterminate, suitable for longer shelf life. Average Yield 400 - 500 q / ha. Suitable for J&K, HP and Uttarakhand.
24	Arka Saurabh	Developed by IIHR, Bangalore, High yielding, (40 tons/ha.) It is semi determinate, Very good keeping quality and suitable for transport. Kept for 8-10 days at room temperature (TSS 3.9%) and acidity (0.39%), suitable both for fresh and processing.

25	Arka Vikas	High yielding variety (35 - 40 tons / ha) developed by IIHR, suitable for fresh market.
26	Arka Abhijit	For Karnataka, Maharashtra and Madhya Pradesh. Plants semi determinate, fruits medium (65 - 70g), suitable for fresh market, resistant to bacterial wilt. Duration 140 days, yield 65 t / ha.
27	Arka Abha	Plants semi - determinate, average weight 75 g. Resistant to bacterial wilt, suitable for kharif and rabi. Duration 140 days .
28	Arka Alok	Plants determinate, fruits resistant to bacterial wilt, Both for kharif and rabi, duration 130 days, yield 40 t / ha.
29	Arka Ashish	Plants determinate, excellent fruit colour (lycopene 10 mg / 100 g) with TSS 4.8 %, Suitable for processing. Tolerant to powdery mildew and fruit cracking, for both kharif and rabi. Duration 130 days, yield 38 t / ha.
30	Arka Ahuti	Plants semi determinate. Fruits oblong. TSS 5.2 %, suitable for processing, 140 days, yield 40 t / ha.
31	Arka Meghali	Plants semi determinate, Suitable for rainfed cultivation and kharif season. Duration 125 days, yield 18 t / ha.
32	Arka Vishal	Plants indeterminate, fruits large (140 g), tolerant to cracking. Requires staking, suitable for kharif and rabi. Duration 165 days, yield 75 t / ha. Recommended for Karnataka, Bihar and UP.
33	Arka Vardan	Plants indeterminate, fruits large (140 g), tolerant to fruit cracking. Resistant to nematodes. Requires staking for both kharif and rabi. 160 days duration, yield 75 t / ha.
34	Arka Shresta	Plants semi determinate. For fresh market and processing. Resistant to bacterial wilt and suitable for rabi season. Yield 70 t / ha. Duration 140 days. Recommended for Karnataka.

35	Co-1	Dwarf, semi spreading habit with dark green foliage. The fruits are borne in clusters of 6 to 8, each weighting about 125g. Yield 80 q/ha.
36	Sweet 72	Cross between Pusa Red Palm and Sioux. Yield 200 q/ha. Developed at the Regional Agricultural Research Institute, Gwalior.

■ Crop Improvement :

(1) Introduction –

a) Roma (Pear shaped suitable for long distant transport).

b) Sioux.

c) Best of all.

d) Tip Top.

e) Labonita.

f) Marvel.

g) Monkey Marker.

h) Agethe.

i) Marglobe.

j) Keckruth.

All these are introductions from USA.

(2) Selection –

a) Improved Meeruti - Suitable for autumn, winter and spring summer.

b) HS 110 – Determinate and potato leaf type.

c) Co1-

d) Pant Bahar – Resistant to verticillium and fusarium wilt.

e) Arka Saurabh-

f) Arka Vikash – Seletion from American variety Tip Top.

g) Sonali-

(3) Hybridization –

a) Pusa Ruby – Sioux × Improved Meeruti, best combiner variety.

b) Pusa Early Dwarf – Developed at NBPGR, New Delhi (Improved Meeruti × Red cloud), Suitable for rainy season.

c) Pusa Gaurav – Glamour × Watch.

d) Marglobe – Marvel × Globe.

e) Pusa Red Plum – *L. esculentum* × *L. pimpinellifolium.*

f) Punjab Chhuhara – EC 55005 × Punjab Tropic (Pear Shaped).

g) Sweet – Pusa Red Plum × Sioux.

h) Pusa Sheetal – Known as "Cold set" variety.

i) Hisar Anmol – Hisar Arun × *L. hirsutum.*

j) Hisar Arun – Pusa Early Dwarf × K1.

k) Hisar Lalit – HS 101 × Resistance Bangalore.

l) Hisar Lalima – Pusa Early Dwarf × H.S. 101.

m) Pusa Uphar – Suitable for processing.

(4) Mutation –

a) S12 – By X - rays.

b) Co3

c) PKM 1

d) Pusa Lal Meeruti – By gamma rays.

e) Maruthan

Other varieties

a) Arka Vishal
b) Arka Vardan – Nematode and bacterial wilt resistant.
c) Arka Abhijit – Bacterial wilt resistant.
d) Rajshree
e) Pusa Divya – F1 hybrid developed using male sterile line (functional male sterility).

- **Bacterial wilt resistant varieties –**

  a) Shakti.
  b) Arka Alok.
  c) Arka Abha.
  d) Sonali.
  e) BT – 1.
  f) BT – 10.
  g) Arka Shreshta.
  h) Arka Abhijit.
  i) VC 48-1.

- Sel 120 – Ist Root knot nematode resistant variety developed in India.
- Hisar Gaurav – Leaf curl resistant variety.
- Hisar Anmol – Leaf curl resistant variety.
- **Indeterminate variety –**

  a) Pusa Ruby – Suitable variety for processing & fresh market.
  b) Pusa Divya.
  c) Arka Saurabh.
  d) Pant Bahar.
  e) Best of All – Suitable variety for distant transport.
  f) Sioux.
  g) Marglobe.
  h) DT - 10.

- Pusa Sheetal – For low temperature (8° C or low) region (Cold set variety).
- Pusa Hybrid 1– For high temperature region (Hot set variety).
- Pusa Sadabahar – For high and low temperature region (8-28° C temperature, fruit sets at night).
- Hisar Arun – Extreme early variety.
- Pusa Rohini – Recently released variety.
- Angur Lata – Tolerant to fruit borer.
- Mangla and Karnataka hybrid varieties are resistant to root knot nematode.
- PWR – 7 – Resistant to both root knot nematode and fusarium wilt.
- Jawahar Tamatar – Fruits produced at a time.
- Pusa Red Plum (Interspecific hybrid).
- Hisar Anmol (Interspecific Hybrid).
- Arka Vikash – Suitable for drought conditions.
- Arka Meghali – Suitable for rainfed conditions.
- Variety developed by use of biotechnology- Flaursaur.
- To increase fruit set and yield faliar spray of PCPA (20 ppm) is very effective.
- "B" and "Zn" - Important micronutrient required for tomato cultivation.
- Breaker stage (10% lycopene) – Suitable for long distance transport.
- **Private sector Hybrid –**

  (1) Naveen (Resistant to fusarium and Verticilium wilt).

(2) Avinash (Tolerant to TLCV).

(3) Meenakshi.

(4) Manisha.

(5) Krishna.

- In private sector first F1 hybrid of tomato Karnataka Hybrid was released for commercial cultivation in 1973 by private seed company "Indo - American Hybrid Seed Company" – Bengaluru (Karnataka).
- Sel - 32 – Determinate, Tolerant to leaf curl virus.
- Punjab Tropic – Late maturing with luxirant growth.
- S - 12 - Dwarf type variety.
- Punjab Kesari - Dwarf, foliage heavy, protects fruit from sun burn.
- NDT – 108 - Fruit chhuhara sized.

Semi-determinate varieties –

1. Pusa 120 – Resistant to root knot nematode.
2. Pusa Hybrid - 2 – Highly tolerant to root knot nematode.
3. Arka Saurabh – Suitable variety for fresh market and processing.
4. Arka Vikash – Suitable for rainfed condition.
5. Arka Shreshta – Resistant to bacterial wilt.
6. Arka Abhijit
7. Arka Abha
8. Arka Vishal
9. Pant T - 3

Other varieties –

Ratna, Rupali, Rashmi gives better yield.

- Solarization is done to reduce incidence of damping off disease in nursery.
- Seed treatment with 2, 4-D @ 2-5 ppm increase early fruit set and parthenocorpy.
- Tomato is moderately tolerant to acidic soil (Ph 5.5).
- Tomato has tap root system.
- Training and Prunning – Required in Indeterminate varieties (Single stem system).
- Staking – It is done for indeterminate types.
- About 33% of total area (Tomato) is covered by F1 hybrid which is highest among vegetables.
- In Punjab only spring summer crop is taken due to heavy incidence of tomato leaf curl virus (TLCV) in autumn.
- Tomato is susceptible to cold winds and frost.
- Temperature for flowering and fruit setting : 25° C.
- Temperature for flowering & fruit setting at night : 15 - 20° C (18° C is best).
- Temperature for pollination : 21° C.
- Below 10° C fruits are unable to develop red colour.
- Above 40° C lycopene is completely destroyed.
- Seed rate : 400 - 500 g – Open pollinated varieties.

  125 - 175 g – Hybrid varieties.
- Sowing Time –

(A) In Northern plain -

(a) Autumn – July - August.

(b) Spring Summer – November.

(B) Southern part -

1. June - July
2. October - November
3. January - February.

- Close spacing is kept– For mechanical harvesting and processing.
- When excessive growth occurs in indeterminate varieties, pinching or pruning of side shoots improves size and uniformity of fruit.
- Methyl ester of long chain acid and alcohol suppress the side shoot formation.
- "B" Borax aaplication @ 20 - 25 Kg / ha or spray of 0.2 - 0.4 % borax at fruit formation stage improves shape, size and colour.
- "Zn" – Zinc sulphite @ 25 Kg / ha is applied for better growth and development.
- To enhance ripening ethrel @ 1000 - 2000 ml / lit is sprayed.
- High temperature and humidity and oxygen pressure increases spoilage.
- High CO_2 (5%) and low O_2 (2.5%) increases shelf life and are best atmospheric storage condition.
- After harvesting tomatoes are cooled rapidly at 13°C which increases its shelf life.
- Yield : 250 Quintal / ha – Open pollinated.

  400 Quintal / ha – Hybrid.

Physiological disorders –

1. BER (Blossom end rot)
   (a) Due to "Ca" deficiency in greenhouse.
   (b) Reduced soil moisture.

2. Fruit cracking – Four types
   (a) Radial – (Very severe).
   (b) Concentric (very severe).
   (c) Burst.
   (d) Cuticular.

   Causes –

   (a) Effect of soil moisture, rain, dew when rain follows a long dry spell.

   (b) High day temperature followed by low day temperature with high relative humidity.

   (c) Deficiency of "Boron" (main cause).

3. Blotchy ripening –

   Symptoms – Greenish-yellow to whitish patch on tomato, mostly on stem end portion.

   (a) Mainly by "K" deficiency.

   (b) Imbalance of "N" and "K".

   (c) High level of N, Ca and Na are associated with blotchy ripening.

4. Puffiness –

   Symptoms –
   1. Fruit surface flattened.
   2. Locules unfilled with pulp and seed.
   3. Affected fruit becomes light in weight or less hollow.
   4. Parenchymatous tissue retart in growth.

Causes – Puffiness is associated with poor pollination and abortion of ovules due to adverse environmental conditions (Low and High temperature).

5. Sun Scald – Intensive sunlight (40°C or above) causes sun scald.

6. Cat face – Distortion of the blossom end rot and development of ridges, furrow, identation and blotches. Cause – Unfavourable climatic conditions during flowering.

7. Unfruitfulness –

   Causes -

   - Day Temperature above 32° C.
   - Night Temperature above 20° C and < 13° C adversely affecting the fruit setting of tomato.

Source of resistance in Tomato –

1. *L. pimpinellifolium* - Fruits rich in vitamin C and resistant to

   (a) Fusarium wilt.

   (b) Early blight.

   (c) LCV (Leaf curl virus).

   (d) Low temperature.

2. *L. pissisi* – Perennial, fruits green and free from bitterness.

3. *L. peruvianum* – Perennial, resistant to

   (1) TMV (Tomato Mosaic Virus).

   (2) TLCV.

   (3) Septoria blight.

   (4) Cladosporium leaf mould.

   (5) Late blight.

4. *L. hirsutum* – Perennial, fruit green, bitter and resistant to fruit borer.
5. *L. gladulosum* – Resistance to curly top disease and leaf spot.
6. *L.cheesmani* – Perennial, tolerant to salinity and crossable with *L. esculentum.*
7. *L. chilense* - Resistant to curl top, TMV, verticillium wilt, powdery mildew, leaf curl, Gametophytic self incompatibility is present in *L. chilense.*
8. *L. pennellii* – Drought tolerant, gametophytic self incompatibility is present.
9. *L. parviforum* – Fruit small (1.0-1.4 cm in diameter).
10. *L. chmielewskii* – Plants similar to *L. parviforum*, flowers large and yellow.

Disease	**Mechanism of disease resistance**
Fusarium wilt	Tomatin content of root tissue increased activity of peraoxidase and polyphenoloxidase.
Root knot nematode	High content of carbohydrate in root and shoot.
Late blight	Tomatin content of leaves.
Verticilium	Antifungal compound rishitin and tomatin content.

- **Nature of resistance of disease in tomato –**

Disease	Nature of resistance
1. Fusarium wilt 2. Root knot nematode 3. TMV 4. Rhizoctonia root rot 5. Spotted wilt virus	Monogenic dominant
1. Bacterial Canker 2. Early blight 3. Rhizoctonia root rot 4. Spotted wilt virus	Oligogenic factor
1. Bacterial Wilt 2. Bacterial canker 3. Leaf curl viruses 4. Late blight	Polygenic factor
1. Early Blight 2. Fruit rot	Additive Factor
1. Bacterial wilt 2. Early Blight 3. Buckeye rot	Polygenically partially dominant Two or more recessive gene Br gene

Virus	Resistant gene
1. Tomato mosaic	TM-2, Single dominant.
2. Tomato spotted wilt	SW-2, SW-3, SW-4, Recessive gene.

Insect	Resistant source
Fruit borer	1. Angurlata
	2. HT 64
	3. Hybrid No-37- tolerant.
	4. Sabour Prabha
	5. Atkinson
	6. Punjab Chuhara

- Varieties suitable for fresh market- Pusa Early Dwarf (P.E.D.), Pusa Ruby, Arka Vikash, Arka Saurabh, Punjab Kesari, Pant Bahar .
- Varieties suitable for processing -Pusa Gaurav, Roma, Punjab Chuhara, Pusa Uphar, Arka Saurabh.
- In 1992 Greshof and Doy produced haploid plant of tomato by anther culture.
- Tomato is preferred for export to European countries.

"BRINJAL" (Egg plant, Aubergine)

- Botanical Name – *Solanum melongena.*
- Family – Solanaceae.
- 2n = 24.
- Origin - India (Indo–Burma), China (Secondary Origin).
- Fruit type – Berry.
- Edible part – Fruit.
- Brinjal originated from a wild species *Solanum sysimbrifolium.*
- De Candel (1886) wrote that *Solanum melongena* has been known in India since long.
- Brinjal has long bearing period when grown under mild climate of southern state but its bearing is shortened under hot summer and cold winter season of Sutlej - Ganga.
- In hills – Only summer season crop is taken.
- India stands IInd in brinjal production after China.
- Brinjal covers 8.14% of total vegetable area and 9.00% of total vegetable production.
- Brinjal fruits are good source of Vitamin - B (particularly) and are also good source of Ca, P and Iron.
- Pigmented, dark-purple brinjal has more Vitamin - C than those with white skin.
- Roasting of fruit raises the nicotinic acid at expense of trigonellin.
- The de-cholestrolizing action is due to presence of poly-unsaturated fatty acid (linoleic and lemolenic) which are present in flesh and seed (65.1%).

- Bitter taste of fruit is due to presence of glycoalkaloids.
- Dry fruit is reported to contain goitrogenic principal.
- Anthocyanin pigment present in fruit is responsible for purple colour of fruit skin.
- Heterostyly is common feature in multi flower cymes except in bunchy top cultivar.
- .Brinjal is a self-pollinated crop but cross-pollination also occurs. This is because it has pronounced hetersostyly which favours cross-pollination.
- 60-70% fruit setting happens through pollination by insect while 30-40% is by selfing.
- Insects (1) Bumble bees (*Bambus sp.*) and (2) Honey bees (*Apis sp.)* are most suitable for pollination in Brinjal.
- Lack of suitable pollinators favours selfing in crop.
- Another bee sp. *Exomalopsis biliottii* has been noted in solanaceous crops especially in brinjal for pollination.
- *Mimosa pudica* plant should be planted in vicinity of brinjal to encourage pollination and visit of these bees.
- Brinjal is a day-neutral plant.
- Optimum temperature –

  21 - 30° C for growth and development

  20 - 25° C with 50% relative humidity is required for seed germination.
- The effect of maximum temperature and precipitation on fruit setting is separate from each other and governed by additive factor.
- Flowering or flower formation are affected by nutritional factor.

- *S. melongena* is more closely related to *S. incanum* than that to any other species of solanum.
- KKM – Milky white coloured cultivar of Brinjal.
- Flower opening 6 - 7:30 hr. in summer.
- Flower opening 11 : 15 hr. in winter.
- High temperature and high humidity in morning hasten opening and dehiscence of flower.
- Brinjal flower – Divided into four types (pentamerous, hermaphrodite and solitary.)
  1. Long styled – Maximum fruit setting (70 - 86.7%).
  2. Medium style – Fruit setting (12.5 - 55.6%).
  3. Pseudo short styled – No fruit setting.
  4. Short styled – No fruit setting.
- All the present day cultivar belongs to three species –
  1. *Esculentum* - Round and egg shaped cultivar.
  2. *Serpentinum* - Long, slender type.
  3. *Depressum* - All early and dwarf brinjal cultivars.
- Filov (1940) classified both cultivated and wild forms of brinjal on agro-ecological basis.
- Five sub species are :
  1. *Agrestis* – Extremely bitter and inedible fruit.
  2. *Occidentale* – Grows in Asia Minor, adequate humidity and high temperature.
  3. *Orientale* – Origin in Japan.
  4. *Palestinicum*– Hydrophytic type can withstand high temperature.
  5. *Arabico–italicum*–Bitter fruit & only of ornamental value.
- White brinjal is best suited for the patient suffering from diabetes.

- 14 % area under brinjal is covered by hybrid varieties.
- Orobanche (root parasite) is serious weed in solanaceous vegetables.
- 2, 4 - D is used to control this weed.

Recommended varieties of Brinjal

S. N.	Variety	Features
1	Pusa Purple Long	Early fruiting with purple colour, 20 to 25 cm long fruits, first picking 75 - 80 days after sowing. Yield 300 q / ha.
2	Pusa Kranti	Fruits stocky, uniformly long, attractive dark purple. First picking 80 - 85 days after sowing, Yield 350 q / ha.
3	Pusa Bhairav	Resistant to phomopsis blight, fruits dark purple, glossy, attractive and 12 - 15 cm long. First picking 80 - 85 days after sowing. Yield 300 to 350 q / ha.
4	Pusa Purple Cluster	Plants erect tall, compact sturdy with purple pigmentation all over. Fruits borne in clusters of 4 - 6 fruits, moderately resistant to bacterial wilt. First picking 85 - 90 days after sowing. Yield 250 q / ha.
5	Pusa Anupam	Plants medium tall, fruits in cluster of 3 to 5, cylindrical, tender, purple colour and resistant to wilt. First picking 95 - 100 days after sowing. Yield 370 q / ha.
6	Pusa Uttam	Fruits oval round, glossy, dark purple and medium sized, 4 - 5 fruits in one kg. First picking 85 days after sowing. Yield 410 q / ha.

7	Pusa Upkar	Fruits round, glossy, dark purple and medium sized, 4 - 5 fruits in one kg. Fruits picking 90 days after sowing. Yield 400 q / ha.
8	Pusa Bindu	Plants pigmented all over, fruits small, oval round, glossy, dark violet purple, about 14 fruits in one kg. First picking 85 - 90 days after sowing. Yield 450 to 650 q / ha.
9	Pusa Ankur	Plants semi erect, fruits small, oval round, glossy, purple, about 15 fruits per cluster, each fruit weight 100g. First picking 80 - 85 days after sowing. Yield 450 to 650 q / ha.
10	Pusa hybrid-6	Plants semi-erect, vigorous, fruit round, glossy, attractive purple colour, each weighing 200 - 250 g. First picking 85 - 90 davs after sowing. Yield 400 - 600 q / ha.
11	Pusa Hybrid-9	Plants strong, upright branches, fruit oval round, glossy, purple, each fruit weights about 250g. First picking 85 - 90 days after sowing. Yield 500 - 600 q / ha.
12	Azad Kranti	Local cultivar used for selection, oblong thick and dark purple.
13	Shyamala	Local Selection, round and dark purple.
14	Arka Neelkanth	Tall compact plants bearing small fruits on clusters, purple fruits, resistant to bacterial wilt. 150 days duration, yield 43 t / ha.
15	Arka Keshav	Long fruits in clusters, green leaves with purple glossy fruits, resistant to bacterial wilt. 150 days duration, yield 45 t / ha. Recommended for Karnataka.
16	KashiSandesh	Round and purple fruits medium sized, maturity 76 DAT, yield 780 q / ha.

17	Kashi Komal	Long fruited, prolific bearing, fruits light purple, maturity 65 - 70 DAT, Yield 700 - 720 q/ha. Suitable for all brinjal growing areas.
18	Kashi Prakash	Open pollinated, plants semi upright, maturity 80-85 DAT and yield 680 - 700 q / ha. Recommended for UP, Bihar and J & K.
19	Kashi Tarun	Open pollinated, tall and erect plants, Maturity 75 - 80 DAT, Yield 700 - 750 q / ha.
20	Arka Anand	Hybrid variety suitable for summer and rainy seasons. Resistant to bacterial wilt. Fruits are green, medium long, borne in clusters. 140-145 days duration and 60-65 t/ha.
21	Pusa Anmol	It is a new early hybrid variety with attractive dark purple oblong fruits. Fruits are ready to harvest after 70-75 days of transplanting. The average yield is about 300-350 q/ha.
22	Krishnanagar Green Long	The fruits are long (25-30 cm). Green fleshy with scanty seeds. The fruits are ready to harvest 75-80 days after transplanting. The average yield varies from 250-300 q/ha. It is a very popular variety cultivated in South India.
23	H 4	Fruits are long (15-20 cm) and thick, average fruit weight is 125-150 g, ready to harvest in 75-80 days after transplanting. Yield is about 250 q/ha. Released from HAU, Hissar.
24	Arka Sheel	The fruits are medium long, thick with deep shining purple colour. The yield varies from 350 - 400 q / ha. Released from IIHR Bangalore.
25	Punjab Barsati	Fruits are 18 - 20 cm long and 4 - 6 cm in diameter. The fruits are dark purple in colour soft and shining. Released from PAU, Ludhiana.

26 Arka Nidhi	Fruits are 20 - 25 cm long purple in colour. Released from IIHR, Bangalore.
27 Hisar Jamuni	Fruits are light purple in colour, 10 - 15 cm long with 3 - 4 cm diameter. Released from HAU, Hisar.
28 NDB-25	Fruits are 15 - 20 cm long and dark red in colour. Released from ND University, Faizabad.
29 Arka Shirish	The fruits are green, extra long with very few seeds. It is an early variety and is ready to harvest after 60 - 65 days. If transplanting the average yield varies from 350 - 400 q / ha. Released from the IIHR, Bangalore.
30 Arka Kusumakar	The finger shaped fruits are borne in clusters, light green. Commonly cultivated variety in South India. Yield varies from 350-400 quintals per hectare, fruits are ready to harvest 70 - 80 days after transplanting. Released by the IIHR Bangalore.
31 Vijay Hybrid	The fruits are long, thick and deep purple in colour. They are ready to harvest in 90-100 days after the transplanting. The average yield varies from 270 - 280 q / ha. Released by Vegetable Research Station, Kalyanpur, Kanpur.
32 Pant Samrat	Fruits are ready to harvest in about 70 days after transplanting. Fruits are borne in clusters. It has resistance to bacterial wilt and phomopsis blight diseases and is also resistant to shoot and fruit borer and jassids. The average yield varies from 300 - 50 q /ha. Released by GBPUAT, Pantnagar.

33	Pant Rituraj	Plants are semi-dwarf and semi-spreading. Fruits are round, medium in size with dark purple shining colour. Harvest after 60-80 days of transplanting, yield 300 q/ha. Developed at GBPUAT, Pantnagar.
34	Pusa Purple Round	Fruits are round, purple coloured, 400 - 500 g. Highly resistant to the 'little leaf' virus disease. Ready to harvest 80 - 90 days after transplanting, yield 250 - 300 q / ha. Does not fruit in summer.
35	Manjri	Fruits are medium sized round, Resistant to wilt, yield is 300 - 350 q / ha. Recommended for Maharashtra State.
36	Surti Gola	Fruits are medium sized, round, yield 245-300 q / ha.
37	Krishnanagar Pruple Round	Fruits are large, round, slightly oval, dark purple, flesh with very less seeds, yield is 250 - 300 q / ha. Suitable for West Bengal.
38	Type-3	Fruits are round purple colour 300g - 400g. Can be grown in spring summer also, Yield 250-300 q/ha. Developed at Kanpur.
39	Punjab-8	Fruits are round and medium in size with light purple colour. Yield is 250-300 q/ha. Developed by PAU, Ludhiana.
40	Punjab Bahar	Fruits are round with deep shining purple colour, weight 200g - 300g, yield 350 - 400 q / ha. Developed by PAU, Ludhiana.
41	Arka Navneet	Fruits are round, shining and dark purple in colour. Recommended from IIHR, Bangalore.
42	Hisar Shayamal	Fruits are round, short with dark colour. It is released from HAU, Hissar.

■ Crop Improvement –

1. Selection –

(i) Pusa Purple Long – Extra early maturing variety.

(ii) Pusa Purple Cluster – Tolerant to bacterial wilt.

(iii) Pusa Purple Round.

(iv) Arka Shirish – Green type fruits are produced by this variety.

(v) Arka Kusumkar- Green type fruits are produced by this variety.

(vi) Arka Sheel.

(vii) Arka Nidhi.

(viii) Punjab Bahar – Round type fruits are produced by this variety.

(ix) Punjab Chamkila.

(x) Punjab Neelum.

(xi) Pant Samrat – Tolerant to bacterial wilt, less infected by fruit and shoot borer, it is a pure line selection variety.

(xii) Azad Kranti – Less seeded variety.

(xiii) Manjri Gota – Bicolour round type.

(xiv) Bhagyamati.

(xv) Annamalai – Resistant to aphid.

(xvi) Jamuni Gole.

(xvii) BR 112.

(xviii) Type 3 – Round Type.

(xix) Co1.

(xx) Co2.

(xxi) MDV – 1 : Round type.

(xxii) PKM – 1.

2. Hybridization –

1. Pusa Kranti – Pusa Purple Long × Hyderpur × Wynad local – Long type.
2. Pusa Bhairav – Resistant to phomopsis blight.
3. Pusa Anupam - Resistant to phomopsis blight (Pusa Purple Long × Pusa Kranti), fruits long.
4. Pusa Uttam – GR × Pant Rituraj.
5. Pusa Upkar.
6. Pusa Bindu – GR × Pant Rituraj.
7. Punjab Neelum – Jamuni Globe × Pant Rituraj.
8. Punjab Barsati – Pusa Purple Cluster × PH - 4, Long type.
9. Punjab sadabahar – Tolerant to fruit borer.
10. Hisar Shaymal.
11. Hisar Jamuni – Suitable for both kharif and rabi season.
12. Pant Rituraj – T3 × Pusa Purple Cluster, round type.
13. Vaishali – Arka Kusumkar × Manjri Gota, Bicolour.
14. Pragati – Vaishali × Manjri Gota.

Public sector Hybrid –

(a) Pusa Anmol.

(b) Pusa Hybrid 5 – Long type, early cultivar.

(c) Pusa Hybrid 6 – Round type, early cultivar.

(d) Pusa Hybrid 9 – Round type, early cultivar.

(e) Arka Navneet - IIHR – 22 - 1 × Supreme – Highest yielding variety.

(f) Vijay.

(g) Azad.

(h) ABH 1, 2 – Oval, Early, purple type.

- Annamalai – Aphid resistant variety.
- Pusa Kranti – Can grow both in spring and autumn season.
- Black Beauty – Introduced variety.
- Private sector Hybrid – Vardan, Nisha, Shiva, Vaishali, Bicolour, Suphal.

- **Bacterial wilt resistant** -

1. Pusa Anupam	2. Pusa Purple Long
3. Arka Keshav	4. Arka Nidhi
5. Pant Rituraj	6. Neelkantha

- **Phomopsis blight resistant** -
  1. Pusa Bhairav (Highly resistant)
  2. Pusa Anupam
  3. Florida Market
- Phomopsis blight and bacterial wilt resistant variety - Pant Samrat.
- NDB – 25 – It has soft joint and is very easy to harvest.
- Pusa Ankur – New variety of brinjal.
- Little leaf resistant varieties -
  1. Arka Sheel 2. Manjarigota
- Application of "Cu" favours increased yield in Brinjal crop.
- Parachlorophenoxy acetic acid (4-CPA) and 2, 4 dichlorophenoxy oxalic acid improves fruit set, induces parthenocarpy, increases fruit set and advances maturity in Brinjal.

- Application of 2, 4 – D (2 ppm) at flowering stage increases total yield of Brinjal.
- Seed rate – 200 g / ha is sufficient.
- Approximately 30,000 – 45,000 seedling / ha is required.
- In North India sowing is done in June - July for autumn winter (Takes 4 week for seeding to become ready) and November for spring summer (Takes 6 week).
- Spraying of B, Mn, Cu and Zn stimulate photosynthesis and increases sugar level.
- Yield – 25 - 30 tonnes / ha.
- Storage – At 10 - 11° C and 92 % relative humidity for 2 - 3 weeks.
- Resistance source –

  (1) *S. xanthocarpum* – Resistant to phomopsis blight.

  (2) S. *indicum* – Resistant to root knot nematode.

  (3) *S.sisymbrifolium* – Resistant to root knot nematode.
- Sandy loam soil – Early crop production can be taken.
- Silt loam or Clay loam soil – Most suitable for heavy crop production.
- Brinjal green fruit have longest shelf life (4 weeks).
- Brinjal have bi - directional root system.
- Arka Keshaw, Punjab Moti, Bharamashi are other improved varieties of Brinjal.

"CHILLI" (Hotpepper)

- B.N. – *Capsicum annum.*
- Family – Solanaceae or nightshade family.
- 2n = 24.
- Origin – Mexico.
- Fruit type – Berry.
- Edible Part – Fruit.
- Chilli was introduced into India from Brazil in the 16th century by Portuguese (1584).
- India is a major producer, consumer and exporter of chilli in the world.
- Andhra Pradesh had been leading both in area and production, contributing an average of 25 % of the total area and over 40 - 50 % of total production.
- Andhra Pradesh, Karnataka, Maharashtra, Orissa, Tamil Nadu, Madhya Pradesh, West Bengal and Rajasthan account for more than 80 % of the total area and production.
- Capsicum also known as Bell pepper is restricted to cooler regions of the country like Bengaluru, Mysore, Nilgiri, Tamil Nadu, Pune, Ranchi, Darjeeling, Himachal Pradesh and Jammu and Kashmir.
- Major Bell Pepper (Capsicum) producing country is China.
- Chilli is mainly used in culinary, adding flavour, colour, Vitamin - A, C, E, P, and pungency.
- Curry powder, curry pastes are also prepared from Chilli.
- In food and beverage industries, Chilli is being used in the form of oleoresin which permits better distribution of colour and flavour in food.

- Chilli is used in a dish called bhazzi.
- Capsicin is pungent principle of chilli, which is used in preparation of balms.
- Capsanthin is principle pigment of chilli, which is responsible for red colour.
- A non conventional use of chilli is in self defence spray which is gaining popularity in USA.
- Indian chillies are mainly exported to Sri Lanka, USA, Canada, UK and Saudi Arabia etc.

• Chilli Species -

a. ***C. frutescens*** – Grown where extreme pungency is required such as oleoresin extraction, large type of fruits of frutescens include the Tabasco Pepper. In India grown in Kerala, Karnataka, Maharashtra, Vizianagar district of Andhra Pradesh and North Eastern states.

b. ***C. chinense*** – Most variable of domesticated species and most widespread within Latin America. In India grown in backyard of Kerala and North Eastern states.

c. ***C. baccatum*** - Grown on garden and field scale in South America – Peru, Bolivia and Brazil. It has useful gene mainly for powdery mildew and anthracnose fruit rot resistance. In India this species is not known.

d. ***C. pubescens*** – High Land pepper does not set fruit satisfactorily, pungent, fruits, thick pericarp and c□ntains lot of moisture.

- Genus capsicum is divided into three sections by Hunziker (1956).

  1. Monotypic 2. Tubocapsicum 3. Pseudoacnistus

- **White Flower** – (1) *C. annum.*
  (2) *C. chinensis.*
  (3) *C. frutescens.*
  (4) *C. baccatum. var – baccatum.*
  (5) *C. baccatum var – pendulum.*
- **Purple Flower** - *C. pubescens.*
  C. eximinum.
  C. cardenasii.

Recommended varieties of Chilli

S. N.	Variety	Features
1	Pusa Jwala	Plants dwarf, light green fruits, 9-10 cm long, thin, turning red at ripening, dry fruits wrinkled, suitable for both the seasons in plains and March-April in the hills. Mildly pungent. Maturity 130-150 days. Yield 75 - 80 q / ha fresh and 7 - 8 q/ha (Dry).
2	Pusa Sadabahar	Plants erect dark green foliage, Fruits 6-8 cm long, erect bearing habit in clusters of 6 - 14 fruits / cluster. Highly pungent, resistance to leaf curl virus, Suitable for both the seasons. Maturity 150 - 170 days. Yield 110 - 125 q / ha. (Fresh) and 15 - 20 q / ha (Dry).
3	Bhagyalakshami (G-4)	Leaves narrow, dark green, fruits olive green turning bright red on ripening. Fairly tolerant to diseases and insects. Suitable for sowing in both the seasons.

		Maturity 150 - 170 days. Yield 110 - 130 q/ha. (fresh) and 10 - 13 q / ha (dry).
4	Sindhur(CA-960)	Fruits long, 7 to 8 cms, light green turning deep red on ripening, surface smooth, midly pungent, suitable for both the seasons. Maturity 130-180 days. Yield 120-140 q/ha. (fresh) & 9-11 q/ha. (dry).
5	Pant C - 1	Plant short statures, more primary branches, fruit erect bearing small in size, highly pungent, 5.5 cm long. Light green fruits turn red at maturity. Tolerant to leaf curl virus. Suitable for both the seasons. Maturity 90-130 days. Yield 100 - 120 q/ha. fresh and 9-11 q/ha. dry fruits.
6	Punjab Lal	Plant dwarf, bushy with dark green foliage, fruits erect bearing medium sized, dark green turning red at ripening. Rich in capsaicin content suitable for both the seasons. Maturity 120-180 days. Yield 100-120 q/ha. fresh and 9-10 q/ha. (dry).
7	Kashi Anmol	Plants determinate, fruits are compact, dark green, pungent (0.5%) capsicin, suitable for long distance transport, harvesting 50-55 DAT. Yield 250 q/ha. (Green fruits) recommended for UP, Karnataka, MP, Uttranchal, Jharkhand, Chatthisgarh.
8	Kashi Vishwanath	Red ripe fruits in 220 days and yield 200 q/ha. It is a cytoplasmic male sterile based hybrid sutiable for dry fruit production in UP, Bihar, MP, Jharkhand, Uttranchal, Karnatakam AP, Delhi and Haryana.

9	Kashi Early	For both green dry fruits, yield 300 q/ha. in 240 days. Suitable for AP, MP Jharkhand, Uttranchal, Karnataka, Delhi, Punjab and Haryana etc.
10	Arka Lohit	Plant tall, fruits dark green, turn deep red on maturity (capasanthin 0.205%). Highly pungent suitable for irrigated and rain fed cultivation. Duration 180 days yield 25 t/ha. (Green chillies) and 37 t/ha. (Dry chillies), Suitable for Orissa, Maharashtra, Karnataka.
11	Arka Suphal (PMR 57/88 K)	Plants tall and erect. Fruits green, medium long, pungent, suitable for irriaged and rain fed cultivations. Tolerant to powdery mildew, Duration 180 days. Yield 25 t/ha green chilli and 3 t/ha dry chilli. Suitable for Karnataka, Kerala, AP.
12	Arka Meghna (F1 hybrid)	Plants tall and spreading, fruit length 10 cm. very early type (24 days for 50% flowering after transplanting). Fruits green turn deep red on maturity. Duration 120-140 days. Tolerant to powdery mildew and viruses. Yield 33.5 t/ha. green and 5 t/ha. dry chilli.
13	Arka Harita (F1 hybrid)	Plants tall, spreading, fruits dark green (10 cm long.) Duration 150-160 days, resistant to powdery mildew and viruses. Yield 33.5 t/ha. green and 5 t/ha. (dry chillies)
14	Arka Abhir	Plants tall, fruits light green, winkled, has low pungency (0.05%), suitable for Oleoresin extraction. Duration 160-180 days, yield 2 t/ha (Dry).

[A] Crop improvement in Chilli -

- Chilli is often cross – pollinated crop. Cross pollination is reported upto 63%.
- Flower opens at 5 AM.
- For making cross, a bud one day prior to anthesis is ideal for emasculation.

1. Selection –

(i) NP 46-A-Pure line selection tolerant to virus.

(ii) G1, G2, G3, G4- Pure line selection.

(iii) NP – 41.

(iv) K1 – Tolerant to thrips.

(v) CO1.

(vi) Patna Red.

(vii) NP 34.

(viii) Sindhur- Mild pungent variety.

(ix) Bhagyalaxmi (G4)-Fairly tolerant to disease and pest.

(x) Arka Lohit.

(xi) Sabour Angar.

(xii) Kalyanpur Yellow.

2. Hybridization –

(i) Andhra Jyoti - Cherry type fruit.

(ii) Pusa Jawala - NP 46 A × Puri Red – Tolerant to thrips, mites and aphid.

(iii) Pusa Sadabahar – (Pusa Jwala × IC 31339), Virus resistant.

(iv) Pant – C1 (NP46 - A × Kandhari), Tolerant to viruses.

(v) Pant – C2 (NP46A × Kandhari), Tolerant to virus.

(vi) Punjab Lal – (Perennial × Long Red).

(vii) Bhaskar – (G4 × Yellow Anther mutant).

(viii) X 235 – (G4 × Anther mutant).

3. Mutation –

Alberto Pirovano (1922) - First man to use X - rays for inducing mutation in capsicum.

I. MDU – 1 Through gamma - rays.

II. Ankur – 1 Gamma rays.

4. Private Sector hybrid –

i. Tejashwini	ii. Champion
iii. Gayatri	iv. Gni
v. Delhi hot	vi. Skyline

- CH – 1 : A public sector hybrid, related by Punjab Agriculture University, Ludhiana by using Genetic Male Sterile Line (GMS).
- In private sector first – F1 hybrid of tomato "Karnataka Hybrid" and of sweet pepper "Bharat" was released for commercial cultivation in 1973 by a private seed company "INDO-AMRICAN" Hybrid seed company, Bengaluru (Karnataka).

[B] Recommended Varieties of Green Pepper (Capsicum)

S. N.	Variety	Features
1	California Wonder	Upright Plants, medium green foliage, good cover, bearing fruits continuously. Fruits smooth, blocky, 3-4 lobed, dark green turning red on ripening, 6.5-8.5 cm long. 6.5-7.5 cm in diameter, erect bearing habit thick flesh. Suitable for sowing in plains from September to October and March to April in hills. Maturity 90-120 days. Yield 150-180 q/ha.
2	Yolo Wonder	Dwarf bushy plant, prolific bearer, fruits blocky, smooth 3-4 lobes pendent, 6.5-8.5 cm long and 4.5 to 6.5 cm in diameter, thick flesh, deep green turning red on ripening. Suitable for sowing in both the seasons. Maturity 110-150 days. Yield 130-180 q/ha.
3	Arka Mohini	Medium to large, drooping fruits, dark green turning red on ripening, 3-4 lobs suitable for both the seasons in plains and March-April in hills, Maturity 125-140 days. Yield 180-250 q/ha.
4	Arka Gaurav	Fruits dark green, blocky erect bearing, 3 to 4 lobed. Thick fleshed medium size, fruits turn to orange yellow on ripeining. Foilage covers the fruits to avoid sunscald. Good for both seasons. Maturity 125-150 days. Yield 180-250 q/ha.
5	Arka basant	Fruits waxy (creamish white), conical, erect bearing, thick fleshed prolific continous bearing suitable for both the seasons. Maturity 125-140 days. Yield 150-200 q/ha.

6	Bharat	Plant tall, prolific bearing, fruits dark green, blocky, 3-4 lobed, thick fleshed, large size, suitable for both the seasons. Maturity 125-140 days. Yield 150-200 q/ha.
7	Pusa Deepti	Plants erect, medium tall, bushy, fruits smooth, erect bearing, conical, light green turning dark red on ripening, thick fleshed, 9-11 cm long and 3-5 cm in diameter, very early fruiting, suitable for both the seasons. Maturity 90-105 days. Yield 300-350 q/ha.

■ Crop Improvement :

1. Introduction –

(i) California Wonder. (ii) Yolo Wonder.

(iii) Golden Wonder. (iv) Bullnose.

(v) Sweet Banana. (vi) World Beater.

(vii) Chinese Giant.

2. Selection –

(i) Arka Mohini.

(ii) Arka Gaurav.

(iii) Arka Basant.

3. Public Sector Hybrid -

(i) Pusa Deepti-IARI Katrain, F1 Hybrid, Light green in colour.

(ii) Green Gold.

4. Private sector hybrid –

(i) Bharat. (ii) Indira.

(iii) Early Bounty. (iv) Hira.

(v) Lario.

- Fruit rot resistant variety – K2.
- **Leaf Curl resistant variety –**

  (i) Pusa Jawala. (ii) Pusa Sadabahar.

  (iii) Pant C1.
- Leaf curl, TMV, wilt and dieback resistant variety – Punjab Lal.
- Multiple disease resistant variety – Punjab Surakh.
- Bacterial wilt resistant variety – Utkal Rashmi, Arka Gaurav.
- Jawalamukhi – Suitable for high density plantation, low pungency and resistant to viruses.
- Arka Lohit – Tolerant to powdery mildew.
- Bhaskar – Yellow anther type, resistant to thrips and mites.
- Pusa Sadabahar- Variety of *C. fruitescence* which is perennial in nature.
- Arka Abir – Paprica (*C. fruitscence*), suitable for colour extraction.
- Punjab Lal – Used for colour extraction.
- Himachal Pradesh is leading state in supplying capsicum in off season.
- Green to dry chilli ratio – 10 : 1.
- Drying is done for 3-10 days under sun, lowering moisture to 8 – 10%.
- 15 irrigations are required in capsicum.
- India's share in total export of chilli in the world is 4 %.
- Fruit drop – NAA and triconitol is used to control fruit drop.
- 7g. good quality seed provides 100 plants.

- Most extensive cultivation of chilli can be seen on vertisols covering the state Andhra Pradesh, Karnataka, Maharashtra and Tamil Nadu.
- Sowing time – (1) June to Oct. (Kharif season)

  (2) South India (March-April)
- **Seed rate –**

  Chilli – 1.5 kg./ha.

  Capsicum – 250-300 g./ha.
- 50-60 thousand seedlings are sufficient to cover 1 ha.
- Yield –

  7.5-10 tons/ha. Green chilli.

  12-20 tons/ha. Bell pepper.
- Long dry period – Flower shedding and fruit fall occurs.
- Dry chilli contains about 6% stalks, 40% pericarp and 54% seed.
- Excess of 'N' with water stress leads to BER of capsicum (Blossom and Rot).
- Variety having thin pericarp, less seed and strong spike is suitable for drying.
- Punjab Guchedar – Highly resistant to mosaic and tolerant to LCV.
- Punjab Sadabahar – Resistant to Chilli mosaic virus, Tomato mosaic virus, Leaf curl virus.
- Ujjwala – Resistant to bacterial wilt.

"OKRA" (Lady Finger)

- B.N. – *Abelmoschus esculentus.*
- Family – Malvaceae.
- 2n = 130.
- Origin – Africa
- Type of fruit – Capsule.
- Edible part – Fruit.
- India is the largest producer of Okra in the world.
- Major growing states – Bihar and Andhra Pradesh.
- Production is declining due to yellow vein mosaic virus (YVMV).
- The root and stem are used for clearing cane juice in preparation of "Gur" and "Jaggary".
- High Iodine content of fruit helps to control goitre while leaves are used in curing inflammation and dysentery.
- The fruit also helps in cure of renal colic, leucorrhoea and general weakness.
- Dry seed – Contains 13-22% good edible oil & 20-24% protein.
- Okra accounts for 70% of the (30%) exchange earning other than onion from export of vegetables.
- For fruit export Okra fruit should be tender and 6-9 cm long.

(1) Pusa Sawani
(2) Prabhani Kranti (Mainly for export)
(3) Versha Uphara
(4) Pusa A – 4

All these are suitable for export

- Before the release of Prabhani Kranti, Selection-2 released earlier, was the major exported variety.
- Major okra exporting area – Nasik, Ozar, Saikheda, Dindori, Kolhar, Sholapur (Maharashtra).
- Dutta and Naug (1968) suggested that the chromosome number for *A esculentum* was 2n = 72, 108, 120, 132, 144 are an indication of a series of polyploids of x = 12.
- The polyploidy nature of *A .esculentus* was also suggested by Charrier (1984).
- Okra variety Pusa Sawani resistant to YVMV was the first example of successful breeding for disease and pest resistance at IARI, New Delhi.
- Pusa Swami has revolutionized the production of okra in all okra growing areas but now it is susceptible to YVMV.
- Major breeding objective in Okra is to develop variety resistant to YVMV.

Recommended varieties of Okra

S. N.	Variety	Features
1	Pusa Sawani	Early, plants generally single stemmed, fruits 12-15 cm long, 5 edged, smooth and dark green when fully developed, suitable for spring summer cultivation. Less sensitive to temperature fluctuations but sensitive to virus. Maturity 50 days (first harvest). Yield 120 q/ha.
2	Pusa A-4	Plants with single stem, sparse pigmentation and short inter-nodes. Fruits dark green, 12-15 cm long. Excellence shelf life, resistant to yellow

		vein mosaic virus, tolerant to aphids, jassids and borer. Responds to pruning to extend the crop period for additional harvests. Maturity 45 days (first harvest). Yield 140 q/ha.
3	Punjab Padmini	Plants tall (180-200 cm) plant parts pigmented and hairy fruits dark green. 15-20 cm long, 5 ridges, remain tender for a long period (3-4 days). Resistant to yellow vein mosaic virus and tolerant to jassid and cotton boll warm. Suitable for kharif and summer of North and winter season of Southern states. Maturity 50-55 days. Yield 110-125 q/ha.
4	Parbhani Kranti	Plants single stemmed, tall, deeply lobed green leaves. Fruits smooth, dark green, tender, 5 ridged with long narrow up. Resistant to yellow vein mosaic virus. Maturity 50 days. Yield 85-90 q/ha. during summer and 110-115 q/ha. during kharif season.
5	Punjab – 7	Plants medium tall (105 cm) partial pigmentation. Fruits medium long, green, tender, 5 ridged, slightly furrowed. Resistant to yellow vein mosaic virus. Maturity about 54 days. Yield 80-100 q/ha.
6	Pusa Makhamali	Excellent fruit quality, susceptible to virus, suitable for cultivation in hills and virus free areas and season. Plants hairy, erect, 180-200 cm tall. Fruits light green, tapered, attractive, 12-15 cm long. First harvest in 50-55 days. Yield 80-100 q/ha.
7	Co-1	Plants 90-120 cm tall, scarlet red, medium sized leaves. Fruits glossy, slender, 5 ridged, scarlet red (colour not persistent on cooking) Maturity 50 days. Yield 85-90 q/ha. cultivation year round in Tamil Nadu.

8	Arka Anamika	Released by Central Variety Release committee. Excellent yield in southern but lower in north India, Resistant to YVMV. Plants open, 100 cm tall partial pigmentation. Fruits dark green, 5 ridged. Furrowed. Maturity 55 days. Yield 120 q/ha. Developed by IIHR Bangalore.
9	Punjab 8	Plants 150 cm tall, hairy partial pigmentation. Fruits medium long, green, tender, 5 ridged. Plants resistant to YVMV and tolerant of fruit borer. Maturity 50 days. Yield 105 q./ha. Variety developed by mutation at PAU, Ludhiana.
10	Hisar Unnat	Plants are medium tall with 3-4 branches, green foliage. Fruits green, attractive, 5 ridged, 15-16 cm long. Resistant to YVMV. Maturity 55 days. Yield 120 q./ha. Variety developed at HAU, Hissar.
11	Varsha Uphar	It was released in 1992 & notified in 1995 by Central sub committee on crop standard. Plants 90-120 cm tall, 2.3 branches and pigmented petioles. Fruits attractive dark green. 18-20 cm long, long tapering up. Resistant to YVMV, field tolerant to borers. Maturity 45-50 days. Yield 100 q./ha. Variety developed by HAU, Hissar following pedgee selection..
12	Azad Kranti	Plants fast growing occasionaly branched, spares pigmentation. Fruits 5 ridged, smooth, shiny, green, long beak. Highly tolerant to YVMV. Maturity 50-55 days. Yield 125 q/ha. Variety developed by CSAUAT, Kanpur.
13	Arka Abahay	It is a sister line of Arka Anamika, Resistant to YVMV, tolerance to fruit borer and may suit pruning for a ratoon crop. The plant

		resemble Arka Anamika in appearance as well as YVMV resistance. Plants tall, well branched, fruits lush green, tender and long, fruits borne in clusters, green stem with purple shade. Fruits free from spines having 5-6 ridges and delicate aroma. Maturity: 120-130 days. 18 t/ha. Recommended for Karnataka, WB and Assam.
14	Kashi Vibhuti	Resistant to YVMV, OLCV, yield 170-180 q/ha. A plant bears 18-22 fruits having 8-10 cm length.
15	Kashi Pragati	Fruits 23-25 plants 8-10 long. Yield 180-190 q/ha. rainy season, 130-140 q/ha. in summer season. Recommended to Chattisgarh, Orissa and AP.
16	Kashi Satdhari	Resistant to YVMV, OLCV, 18-15 fruits / plant and 110-140 q/ha. Recommended to Punjab, Bihar and Jharkhand,
17	Shitta Jyothi	Suitable for warm, humid climate. Yield 180-200 q/ha. Resistant to YVMV and OLCV. Recommended for Rajasthan, Gujarat, Haryana, Delhi, Chattisgarh, Orissa and AP.
18	Kashi Bhairav	Fruit length 10-12 cm. Yield 200-220 q/ha. Suitable for all the states.
19	Kashi Mahima	Fruit length 12-14 cm. Yield 20-220 q/ha. UP, Punjab, Bihar, Karnataka, Chattisgarh, Orissa, AP and Maharastra.

■ Crop Improvement

1. Introduction –

a) Velvet Green.

b) Ghana Red.

c) Perkins Long Green.

d) Clemon's Spineless.

2. Selection -

(a) Pusa Makhmali -Pure line selections susceptible to YVMV.

(b) Gujarat Bhindi No. 1 Pure line selection.

3. Hybridization –

i. Pusa Sawani – (Pusa Makhmali × IC 1542)-Developed by Dr. Harbhajan Singh through selection from a cross. Earlier (1962), it was resistant to YVMV but now resistance had been broken down. It is a day neutral variety which is spineless.

ii. Punjab Padmini – (*A esculentus* × *A. manihot sp. manihot*).

iii. Prabhani Kranti – Interspecific cross (1955), (*A exculentus* cv.-Pusa Sawani × *A. manihot*).

iv. Arka Anamika – (*A esculentus* × *A tetraphyllus*).

v. Arka Abhay.

vi. Varsha Uphar – (Lam sel. 1 × Parbhani Kranti)

4. Mutation –

i. MDU-1 – Evolved at TNAU, Coimbatore in 1978 as an induced mutation from Pusa Sawani.

ii. Punjab-8 (EMS8) Induced mutant from Pusa Sawani by treating seed with EMS 1%, resistant to YVMV and tolerant to fruit borer.

5. Hybridization –

1. Punjab 7 – *A. esculentus* cv. (Pusa Sawani) × *A. manihot* sp. *manihot*, resistant to YVMV.

2. Hissar Unnat (HRR55) – (Selection 2-2 × Prabhani Kranti), Released by Central Variety Release committe and notified in 1996, resistant to YVMV.

3. Selection 2-5 – (Pusa Sawani × Best one) × (Pusa Sawani × IC 7194), It was once very popular in Maharashtra for fresh fruit export.

6. **Other Varieties –**

- Perkins Long Green – Suitable for slightly cooler area.
- Hissar Barsati – Resistant to YVMV.
- Pusa A-4 – Released in 1994 by IARI as a substitute for Pusa Sawani, resistant to YVMV and tolerant to jassid and shoot and fruit borer, it has excellent shelf life.
- Co. 1 -Developed in 1976 as single plant selection from a heterogeneous population of Red Wonder, tolerant to YVMV, resistant to fruit borer and powdery mildew.
- Harbhajan Bhindi – The variety Perkins long Green was released in 1983-84 as Harbhajan Bhindi from College of Agri. Solan to commemorate the memory of Dr. Harbhajan Singh, resistant to YVMV.
- Azad Kranti – Tolerant to YVMV.
- Punjab no. 13– (*A. esculentus* (Pusa Sawani) × *A. Manihot* var. Manihot) Susceptible to YVMV.
- Higher temperature – 40-42°C flower may desiccate and drop, causing yield loss.
- Pusa Sawami has some tolerance to salt and also to large variation in PH range.
- **Seed rate –**

  18-28 kg/ha   Spring (Summer).

  8-10 kg/ha   Rainy (Kharif).
- Export market requires tender, dark green, straight, short 6-8 cm fruits.
- Vector of YVMV is white fly (*Bemicia tabaci*).
- Jassid (Erias sp.) – Serious pest of Okra.

- Arka Abhay and Pusa A-4 give quick branching after pruning.
- Private sector hybrid – Resistant to YVMV.
  (1) Vash.
  (2) Vijay – Indo American Hybrid Seed Company.
  (3) Variety Adhunik and Panchali – Century Seed.
  (4) Variety Hybrid No. 6 7 - Mahyco seed company.
  (5) Variety Nath Sobha - Nath seed company.
  (6) Variety Sungro 35 - Sungro seed company.
- Okra gives 300-350% crop land use efficiency as an intercrop in cassava and cucurbits.
- Dr. H. Singh started systematic work on okra.
- *A manihot* is also used as leaf vegetable.
- Okra flowers are self – fertile but usually upto 10% cross-pollination occurs by insect, thus it is classified as an often cross – pollinated crop.
- 5000 plants/ha are recommended for branching type.
- 66000 plants/ha are recommended for non-branching type.
- Resistant source –
  (1) *A. manihot* sp- *manihot*
  (2) *A. tuberculatus*
  (3) *A. pungens*

  All species are resistant to YVMV.

Okra Amphidiploids 2x =192

A. esculentus × *A. manihot*

(2x = 124) (2x = 68)

Resistant to YVMV

3

COLE CROPS

Origin – Mediterranean region.

- Cole crops consist of a large group and have originated from a common parent wild cabbage or colewart (*Brassica pleracea var sylvestris*).
- The term cauliflower is derived from two Latin words caulis means cabbage and floris means flower.
- All species of genus Brassica contain gluconsinolates which in crushed leaves is broken down by myrosinase enzyme to give bitter taste and goitrogenic substance.
- Cole Crop Family – Crucifereae/Brassicaceae.

S.N	Common Name	Botanical Name	Ch. (2x=)	Edible Part
1.	Cabbage	*B.O. var-capitata*	18	Head
2.	Cauliflower	*B.O. var-botrytis*	18	Curd
3.	Knol-Khol	*B.O. var-gongylodes* *(Syn. B. Caulorapa)*	18	Knob/tuber/Flower bud
4.	Sprout brocoli	*B.O. var-italica*	18	Unopened flower bud
5.	Brussels sprout	*B.O. var-gemmifera*	18	Sprout/bottons/ minicabbage
6.	Chinese Cabbage	*B.C. var-pekinensis* *B. compestris Sp. Chinensis*	20	Leafy head

7.	Kale/Collard/ tree Kale	*B.O. var-acephula*	18	Leaves and Shoot
8.	Rutabaga	*B. napobrassica (syn. B. napus var - napobrassica)*	38	Root
9.	Red cabbage	*B.O. var-capitata f. rubra*	18	Leafy head
10.	Savoy cabbage	*B.O. var-sabauda*	18	Leafy head

■ Seed rate of Cole Crops

S.N.	Crop Name	Seed rate/ha.
1.	Cauliflower	500-700 g.Early type 300-500 g. Mid and Late type
2.	Cabbage	500- 600 g. Early type 400g-Mid and Late type
3.	Knol-Khol	1-1.5 kg
4.	Sprout Brocoli	400-500 g.
5.	Brussel's Sprout	500 g.
6.	Kale	350-400 g

■ Sowing Time

Cauliflower	In N Indian Plains	May – July – Early July-Aug – Mid Sep. – October - Snowball type
Cabbage	In Hills	August – Sept
	In N Indian Plains	July-August – Early September-Mid October-November Late

Knol-Khol (Best time October)		August – Early August - September - Late
Sprout Broccoli	North India	August – October
	Hills	March and July-August
Brussel's Sprout		August – October
Kale		October – November

- Genus brassica has three monagenic and three amphidiploids digenomic species.

Species	**Cultivated type**	**Ch. No. (n)**	**Genomic constitution**
1. Monogenomic –			
B. nigra	Black mustard	8	b
B. pleracea	Cole group	9	c
B. compestris	Turnip group	10	a
2. Digenomic –			
B. carinata	Abyssinian Mustard	17	bc
B. juncea	Brown mustard	18	ab
B. napus	Swede group	19	ac

- All the amphidiploids digenomic sp. have been developed from the crosses between the monogenomic sp. as reported by U.N. (1953) and illustrated by U.S. triangle.

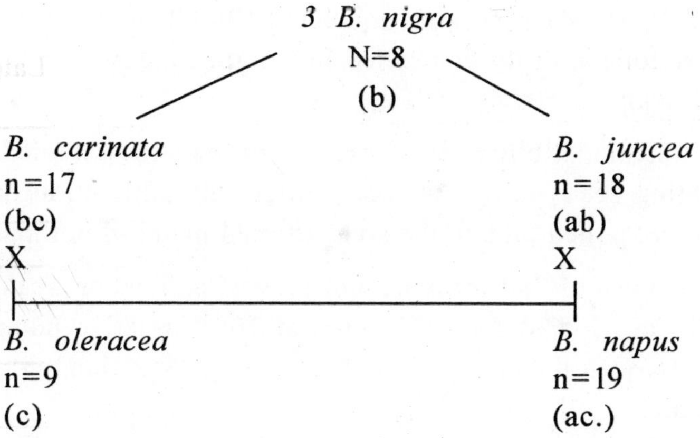

■ M.S. (Male sterility) -

1. **Genetic Male Sterility (GMS)** – Reported in calabrese and purple heading brocoli by Cole (1959) and Brocher (1966) and in brussels sprout by Johnson (1958).
   - GMS is governed by recessive male sterility gene (MS) mutated from male fertile gene (MF).
   - In brocoli several male sterile genes MS1, MS2, MS4, MS5 and MS6 have been identified which are not allelic.
2. **Cytoplasmic Male Sterility (CMS)** – Not found apparently in *B. oleracea.* It was transferred into *B. oleracea* from *B. nigra* by "Pearson" through repeated back crosses and embryo rescue technique.

Self incompatibility – Genetically controlled physiologically hindrance to self – fertilization.

(a) Gametophytic – Pollen pistil interaction is genetically controlled by haploid genome of each pollen grain and diploid genome of pistil tissue.

(b) Sporophytic – Pollen pistil interaction is genetically controlled by the genome of somatic tissue in which pollen grain is formed.

- Brocoli, Brussels sprouts and Knol-Khol have sporophytic self incompatibility controlled by single gene with multiple alleles known as "S" alleles.
- Self incompatibility in these crops can be studied by counting seed set a few weeks after self pollination or by counting pollen tube in the style after 24 hours of pollination.
- Break down of Self incompatibility - Bud pollination treatment with Co2 (3-5%) for 8-24 hours at 100% relative humidity and spraying the crop at full bloom with 5% sodium chloride solution.
- Self incompatibility is used to develop hybrid varieties.

"CAULIFLOWER"

- Cauliflower was introduced in India in 1822 from England by Dr. Jemson, Incharge Company Bagh Saharanpur (U.P.) from Kew Botanic garden.
- In India two separate groups of Cauliflower are commonly grown.
  1. Indian or Tropical type (Originated in India during the last 175 years).
  2. Annual temperature type known as "Erfurt" or "Alpha" or "Snowball type".
- Cauliflower is a thermosensitive crop i.e. temperature play an important role influencing vegetative curding and reproductive phase.
- India is the largest producer of Cauliflower in the world.
- Tropical type requires 20-25° C temperature for curd formation.

- Temperate type requires 10-16° C temperature for curd formation.
- Curd is a prefloral fleshy apical meristem.
- Snowball type sets seed only in the hill region.

Recommended varieties of Cauliflower

S. N.	Variety	Features
1	Early Kunwari	Curds are semi-spherical with even surface. Suitable for early July transplanting. Maturity 80-85 days. Yield 80-100 q/ha. A very early variety for growing in Punjab, Haryana, Himachal Pradesh and Delhi.
2	Pusa Early Sythetic	Erect plants with bluish green leaves, curds small to medium in size, flat, creamy white and compact. Suitable for transplanting in July beginning. Adaptable to non-traditional areas like Karnataka and Kerala, besides North Indian plains. Yield 110-120 q/ha.
3	Pusa Meghna	Plants dwarf with medium sized sessile green leaves, curds semi dome shaped, white, compact, medium sized, weighing about 350-450 g. suitable for planting in the beginning of July, Maturity 90-100 days. Yield 125 q/ha.
4	Pant Gobhi-3	A synthetic variety having long stem, semi-erect leaves and hemispherical creamy white, medium compact, non-ricey curds. Suitable for planting in the beginning of July, Maturity 80-85 days. Yield 120 q/ha.

5	Pusa Deepali	Plants medium tall, erect with bluish green and waxy leaves curds compact, retentive white and medium in size. Suitable for transplanting in the first fortnight of July. Maturity 100-120 days. Yield 200 q/ha. Recommended for entire north Indian plains.
6	Pant Gobhi – 2	Curds are medium compact and yellowish. Suitable for transplanting in the first fortnight of July. Maturity 95-100 days. Yield 100 q/ha. Recommended for the northern plains of India.

■ Varieties according to maturity time:

[A] Early maturing (September-October)

(a) Early kunwari – Punjab Agriculure University, Ludhiana, Sensitive to riceyness.

(b) Pusa Katki – Selection.

(c) Pusa Deepali – Cultivation in entire North Indian plains.

(d) Pusa Early Synthetic.

(e) Improved Japanese- Introduced from Israel.

(f) Pant Gobhi-1.

(g) Pant Gobhi-3 – A synthetic variety developed by crossing eight inbred lines of Katki, free from riceyness.

(h) Pant Gobhi – 2 – For North Indian plains.

[B] Mid season varieties (November – December)

- Pusa synthetic – Synthesized from seven lines with good general combining ability and resistant to curd blight.
- Pusa Hybrid 2 - First public sector hybrid of Cauliflower made by crossing self incompatible line and selection 1-3-18-19, resistant to downy mildew.
- Pusa Aghani – Selection.

- Pusa Subhra – Selection from a triple cross. Free from riceyness, resistant to black rot and inflorescence blight.
- Pusa Himjyoti – Only variety which can be grown in spring summer (April-July) in hills, it is a self blanched variety.
- Pant Shubhra – Selection.
- Pant Gobhi – 4 : Free from riceyness.
- Sel. 235 S.
- Hisar – 1.
- Punjab Giant – 26, 35.

[C] Late season variety (December- January)

- Snowball – 16 : Introduced from Holland.
- Pusa Snowball developed by crossing (EC 12012 × EC 12013) – 1, Self blanching and tightly covering the curd.
- Pusa Snowball – K1-Self blanched and hidened, resistant to black rot and inflorescence blight.
- Pusa Snowball – 2-Selection from an introduction EC 120-12.
- Lata Dania – Sensitive to fluctuation in environment.
- Most of late types (snowball) have self blanched habit.
- Light soil is suitable for early crop.
- Loamy and Clay Loam - Suitable for mid and late season crop.
- It can be grown in saline soil provided the C- Index is within tolerance limit.

  (1) C2 and C4 – Early cultivar (Nieuwhof 1969).

  (2) C6 – Late Cultivar.
- Cauliflower is sensitive to high acidic conditions.
- Blanching – Preventing the curds of Cauliflower from direct sunlight that make the curd white and improve flavour and tenderness.

- Cauliflower is only crop in group of cole crops in which the intermediate stage of curding lies between vegetative and reproductive stage.
- Scooping – Removal of central portion of curd for easier initiation of flower stalk.
- The first illustration and description of Cauliflower was presented by the Herbalist Dodens in 1544.
- Application of $GA^4 + GA7$ @ 80 ppm shorten the periods from transplanting to harvest.

Physiological disorder –

1. Riceyness – Curd surface loose and velvety appearance due to elongation of pedicel and formation of small white flower buds at curding stage. Such curds are called ricey.

   Causes -

   a. Fluctuation and unfavourable temperature.

   b. Heavy application of nitrogen and high humidity.

2. Buttoning – Development of small curd in young plant and fewer less developed leaves is known as buttoning.

   Causes –

   a. Deficiency of nitrogen.

   b. Water stagnation in field.

   c. Transplanting of old seedlings.

   d. Planting early type Cauliflower under lower temperature.

3. Browning – (Brown-rot or red-rot)-Water soaked, light brown to dark brown spot formed on stem and curds, which leads to the formation of hollow stem and brown colour of curd.

   Causes –

   a. "Boron" deficiency is main cause of browning.

4. Whiptail – Normal leaf blade development fails and only strap like savoyed leaves are formed. In extreme cases, only midrib develops hence the name whiptail is given to this disorder.

   Causes –

   a. Deficiency of "Molybdenum" especially in acidic soils.

5. Chlorosis – Interveinal yellow and mottling of lower older leaves.

   Causes –

   a. Deficiency of "Mg" when grown on high acidic soils.

6. Yellow Heart – Growing plant of Cauliflower develops hollow stem and curd.

   Causes –

   a. Heavy fertilized soil (Particularly "N")

7. Blindness –

   (1) Plants grow without terminal bud and fail to form any curd

   (2) Leaves of blind plant become thicker and leathery owing to accumulation of carbohydrate.

   Causes –

   a. In early stage damage by insect at growing point, low temperature, frost etc.

8. Fuzziness.

9. Leafy curd.

"CABBAGE"

1. In India cabbage ranks IInd in area and production.
2. Thick tightly packed overlapping leaves manifesting a large head is cabbage..

3. Cabbage covers 4% area under vegetable crops in our country.

- Hybrid – 22.2% area, Fruit type – Siliqua.
- Among Cole crops cabbage rank Ist in production in world as well as in India.
- Flavour in cabbage leaves is due to glucoside sinigrin.
- Cabbage has an anticancer property, it protects against bowel cancer due to the presence of Indole -3- Carbinol.
- Cabbage is a rich source of Vitamin A, B and C.
- Cabbage juice was used as remedy against poisonous mushroom and as a gargle against hoarseness.
- Cabbage helps in digestion and is rich in minerals and vitamins.
- Cabbage is highly cross-pollinated (as high as 73%) crop.
- Self incompatibility is the genetically controlled mechanism which expresses the inability of a plant having functional male and female gamets to set seed through self-pollination or cross-pollination with a plant having the same self incompatible alleles.
- Self incompatible alleles are controlled by one locus called "S" alleles which are active in the stigma in flowering stage but remain inactive in the bud stage i.e. 2-4 days before its opening.
- Glycoproteins presents in the stigma hinder the pollen germination and penetration. This phenomenon helps to maintain an individual plant or develop inbred lines for breeding work by bud pollination method.
- Spray of NaCl 3% half to one hour after pollination may also overcome self incompatibility.

■ Factors responsible for breakdown of Self incompatibility.

1) Genetical – Modifier gene, less active "S" alleles, competitive interaction between "S"- alleles.
2) Physiological (intrinsic) –Age and stage of flowering.
3) Environmental – High temperature, high humidity, high concentration of Co_2.

- Protogyny is the phenomenon where the stigma matures earlier then the anthers.
- In cabbage the stigma remains receptive for five days before anthesis due to protogynous nature of the flower and four days after anthesis.
- Cytoplasmic and genetic male sterility found in Cabbage is also helpful for the production of F1 hybrids.
- Cross-pollinated crops- Any suitable breeding method whether it is controlled by oligogens or polygenes.
- Myers (1915) classified cabbage depending on maturity size, shape and colour of head into eight district groups -
    1. Wakefield on Winningtadt group
    2. Copenhagen Market group
    3. Flat Dutch or Drumheaded group
    4. Savoy group.
    5. Danish Ballhead group
    6. Alpha group
    7. Volga group
    8. Red cabbage group
- Mass selection has been found to be effective in cabbage breeding for disease resistance and ascorbic acid content.

- Cabbage requires specific low temperature for chilling 4-8° C for 40-60 days so that plants grow to reproductive stage of flowering.
- Back cross method of breeding is used for transferring the disease resistance governed by one or two genes, from the donor parent to an otherwise desirable cultivar.
- About seven generations of back crossing are needed to transfer a dominant or recessive gene from the recurrent parent.
- A detail breeding programme has been given by Dickson and Wallace (1990) on back cross breeding.

Recommended varieties of Cabbage

S. N.	Variety	Features
1	Golden Acre	Head small, round, compact, average weight 1.2 kg. Fewer cup shaped outer leaves. Maturity 60-70 days. Yield 250-300 q/ha.
2	Pusa Mukta	Flattish round, medium sized compact head, average weight 1.4 kg. Resistant to black rot (*Xanthomonas compestris*). Puckered outer leaves. Maturity 70-80 days. Yield 275-325 q/ha.
3	Pride of India	Small, round, medium sized, compact head, average weight 1.2-1.5 kg. Maturity 70 days. Yield 275-375 q/ha.
4	Copenhagen Market	Large sized, round, compact head, average weight 1.5-3.0 kg. Maturity 75-85 days. Yield 350-400 q/ha.
5	Pusa Drum Head	Plants with larger frame, large size flat shapped, less compact head, weighing 3.5-

		5 kg. Resistant to black leg (*phoma lingam*) disease. Maturity 100 days. Yield 400-500 q/ha.
6	September	Bluish green foliage and large sized frame. Oblong large, very compact head weighing 4-6 kg, resistant to cracking, maturity 110 days. Yield 400-500 q/ha.
7	Late Large Drum Head	Wide frame, large sized, medium compact head, weighing 3.5kg, maturity 115-120 days. Yield 350-500 q/ha.
8	Pusa Ageti	First tropical variety developed for cultivation under high temperature conditions, can grow and produce marketable head at a temperature range of 15-30° C, Flattish round, compact, medium sized head, weighing 0.6-1.2 kg. Medium frame, green foliage, suitable for plating in March and August-September. Head weight 600-1200g, Yield 11-30 t/ha. Seed can be produce under subtropical condition of North Indian plains. Able to produce seed in northern plains of India. Maturity 70-90 days.
9	Sri Ganesh Gol	Round, compact, bluish green head, weighing 1.5-1.8 kg with better staying weight 1.5-1.7 kg. Maturity 85-95 days. Yield 500-700 q/ha.
10	Nath Laxmi 401	Round, very compact uniform sized heads, longer shelf life, average weight 1.5-1.7 kg. Maturity 85-95 days. Yield 500-700 q/ha.
11	BSS-32	Very compact, round large sized heads weighing 1.5-1.8 kg. Longer shelf life. Maturity 90 days. Yield 700-800 q/ha.
12	Quisto	Large sized frame, dark green foliage, very

		compact, large sized, round head weighing 1.6-1.8 kg. Suitable for early sowings. Maturity 110-120 days. Yield 700-800 q/ha.
13	Drumhead Savoy	It has a dark green foliage which is much blistered or wrinkled, the quality is superior but it is not commercially acceptable in our country.
14	Red Cabbage	The leaves are deep purplish red. The yield is usually low.

■ Varieties

(a) On the basis of type of Head –

Round head or ball head	Conical Head	Flat Head or Drum Head
(i) Golden Acre. (ii) Pride of India. (iii) Green express. (iv) Copenhagen market. (v) Mammoth Rock Red. (vi) Pusa Sambandh.	Jersey Wakefield	Pusa Drum Head Savoy – Chieftain

(b) On the basis of maturity group –

Early varieties – (50-70 days)

- Golden Acre (Introduced from Denmark) – Selection from Copenhagen market.
- Pride of India – Selection from Copenhagen Market, 1-1.5 Kg head weight.
- Pusa Synthetic – Synthetic variety.
- Pusa Ageti
- Pusa Sambandh – Suitable for close spacing (H.D.P.) (has wider adaptability). It is a synthetic variety.

- Copenhagen Market - Exotic variety, popular in West Bengal, Head 2-3 Kg in weight.
- Shri Ganesh Gol- Mahyco, Maharashtra Hybrid Seed company Ltd. It has much better staying ability.
- Green Boy – Introduced from Japan.
- Early Drum Head.

Mid varieties –

(a) Pusa Mukta – (EC 24855 × EC 10109), Resistant to black rot (*Xanthomonas compestris*).

(b) September – Introduce from Germany, popular in Nilgiri hills, better staying power, Weight of head 4-6 kg., highly susceptible to black rot disease.

Late Varieties (100 – 120 days)

(a) Pusa Drum Head – Pure line selection variety, Resistant to black leg.

(b) *Phomalingum* or dry rot common in hills, earliest of this group.

(c) Late K 1 – Resistant to black rot.

(d) Late Large.

(e) Drum Head – Introduced from Denmark.

Private sector Hybrids

(a) Questo. (b) Shree Ganesh Gol.

(c) Uttam. (d) Bajrang.

(e) Green Boy.

(f) Green Express – Can tolerate hot weather (30-35^0 C temperature).

(i) Sudha. (ii) Stone Head.
(iii) Green Challenger. (iv) Green Cornet.
(v) Savarna. (vi) Vishesh.

- In cabbage round variety matures first (earlier) followed by conical variety.
- Spray of 50 ppm boric acid at flowering enhances the seed yield.
- Sauerkraut – Value added product prepared from white cabbage used to cure scurvy, very popular in European countries.
- Spray of CCC (2500-5000 ppm) or SADH increase the low temperature resistance.
- Cabbage grown in saline soil is more prone to black leg disease.
- Cabbage produces seed in the temperate areas only.
- Cabbage bear seed in special kind of bicapillary pod called as "Siliqua".
- Popularity of F1 hybrid cultivars in cabbage is due to comparatively easy and cheaper method of hybrid seed production due to presence of genetic mechanism of self incompatibility.
- Speedling – Sowing of seed in theremacole plates, followed in cabbage.
- Growth of most cabbage varities is arrested when temperature rise above 25° C.
- Variety – Nath Laxmi – has better shelf life.
- Beejo Sheetal – 70-80 tons/ha yield.
- Optimum temperature for seed germination – 15-20° C.
- Storage temperature : 0-1.7° C with 90-95% relative humidity

for 4-6 weeks of early cabbage, where as late cabbage can be stored for 12 weeks at same temperature.

- Nieuwhof (1969) recognized the three cultivated forms of cabbage:
  (a) White Cabbage (b) Red Cabbage (Thell)
  (c) Savoy Cabbage

"KNOL-KHOL"/ Kholrabi, Biennial crop.

- Very popular in Kashmir, West Bengal and Karnataka.
- Edible part – Tuber/Knob formed by the thickening of stem tissue above the cotyledons.
- Tubers are produced above the ground, consumed mostly as salad.
- Excess use of cole crops results in swelling of thyroid glands, a disease known as goiter in human beings and other animals.
- Goitrin, thiocynates, isothiocynates and glucosinolates derived product interfere with the functioning of thyroid glands causing the disease.
- Knol-Khol was known to the Romans.
- Inflorescence – Racemose.
- The species of *B. oleracea* is a triple tetrasomic with six basic genome and genomic formula is AABBCCDEF.
- Highly cross pollinated crop (upto 91%).
- Main pollinating agent – Honey bees and flies.
- Male sterility – Absence of functional pollen grain (Male gametets).

Recommended varieties of Knol-khol

S. N.	Variety	Features
1	White Vienna	An early variety, plants dwarf with light green foliage and stem. Knobs are globular round and green with fewer leaves. Flesh creamy white, tender and crisp. Maturity 55-60 days. Yield 120-150 q/ha.
2	Early White Vienna	An early variety, knobs are globular-round and green with fewer leaves. Flesh creamy white, tender and crisp. Maturity 50-55 days. Yield 110-130 q/ha.
3	Large green	A late variety. Plants vigorous with dark green foliage, knob flattish round, dark green. Maturity 60-65 days. Yield 140-170 q/ha. Adapted well in Himachal Pradesh.
4	King of North	It is dark green, flattish round knobs. Dark green leaves are well spread over the knobs. Maturity 60-65 days. Yield 130-150 q/ha.
5	Early Purple Vienna (Verma Pocha)	Early variety. Foliage purple, knobs globular round, large with purple skin and green flesh. Maturity 55-60 days. Yield 120-140 q/ha.

Other varieties :

1. Golith White (Sadashiv).
2. Suttons Earliest purple.
3. Purple Vienna.
4. Purple speck-Earliest variety.
   - Knol-Khol
     For seed production-Annual
     For flowering and fruiting-Biennial.

- Early varieties are more prone to premature bolting.
- Purple varieties are more susceptible to bolting.

"SPROUT BROCOLI"

- Brocoli is an Italian word derived from Latin word brachium meaning "an arm or branch".
- Among cole crops it has the highest nutritive value.

Iron	:	205 mg/100 g.
Vitamin A	:	9000 IU/100 g.
Thiamin	:	300 mg/100 g.
Vitamin C	:	137 mg/100 g.

- It has 130 times more Vitamin A than Cauliflower and 22 times more than cabbage.
- On the basis of colour of varieties they are classified as white, green and purple.
- Brocoli is very popular in hotels in Mumbai, Calcutta, Delhi and Chennai.
- Mainly cultivated in hilly areas of Himachal Pradesh, Jammu and Kashmir, Uttarakhand and Uttar Pradesh.
- USA – Largest producer of sprout brocoli in world.
- Early type- Annual.
- Late type – Biennial.
- Rich source of sulphoraphane – compound associated with reducing risk of cancer.
- In dry condition crop become fibrous.
- Inflorescence – Cymose.
- Mass Selection – Most common method used in cross-pollinated crop for improvement of qualitative characters

governed by single or few genes.

- **Varieties** –
  - (1) Palam Samridhi – Mainly for Sub-Tropical conditions.
  - (2) Punjab Brocoli 1.
  - (3) Pusa Brocoli KTS 1.
  - (4) Italian Green – From University of Horticulture and Forestry (UHF), Solan.
  - (5) Green Head - (UHF, Solan).
  - (6) Green Bud.
  - (7) Green Mountain.
  - (8) Decico.
  - (9) Sparten Early.
  - (10) Costal Atlantic.
- Purple Sicilian brocoli is also known as purple cauliflower.
- Green type cultivars are more commonly cultivated than white type.
- Calabrese type brocoli developed from Italian green sprout brocoli are becoming more popular with the growers as well as consumers for both fresh market and processing.
- In India mainly early and mid type of brocoli are grown and these are susceptible to frost.
- Late type cultivar – Requires vernalization for the development of head and initiation of flower stalk.

"BRUSSEL'S SPROUTS"

- Brussels – Capital of Belgium.
- Genetic Male Sterility reported by – Johnson (1958).
- Edible part – Axiliary bud known as sprouts, buttons or mini-cabbage.

- Sprouts after cutting vertically into two halves or fried with besan (gram flour) and paste can be used for preparation of "Pakoras".
- Rich Source of Vitamin –A, C, B2, Niacin, Ca and Fe.
- The crop in the present form appeared in Brussels (Belgium) in 1750 and reached England and France in 1800.
- **Varieties** –

1. Hilds Ideal.
2. Rubine – F1 hybrid.
3. Amager Market.
4. Jade cross.
5. Danish Prize.
6. Long Island.
7. Dwarf Improved.
8. Kvik.
9. Frontier Zuerg.
10. Early Mom.
11. Catskill.
12. Red Vein.

Storage – 0° C for 16 weeks.
20° C for 11 days.

- Sprout in sealed polythene packs at 0-1° C in modified atmosphere, 2.5% CO_2 and 14.5-16.8% O_2 reduces yield losses and improves quality.

"CHINESE CABBAGE"

- It is a non heading strain of cabbage.
- Chinese cabbage – Turnip (*Brassica compestris sp.- rapifera*).

  Pak-choi (*Brassica compestris sp.- chinesis*).

■ **Varieties –**

1. Palampur Green.
2. Pusa Sag – Wongbok (Suttons) x Turnip.
3. Solan Selection –
4. Chinese sarson No. 1 – Non Heading type, resistant to alternaria leaf spot.

"KALE"

- Hardiest crop (as low as 0-15° C).
- Leaves are used for vegetable purpose.
- In India, it was introduced in 19th century.
- Very popular in Kashmir valley.

■ **Varieties –**

1. Karamsag – Tall variety, mostly grown in Jammu and Kashmir.
2. Dwarf Green.
3. Hamburger Market.
4. Scotish – For temperate region.
5. Dwarf Moss.
6. Siberian – For temperate region.

4

ROOT CROPS

S.	C.N.	B.N.	Family	2x	Edible part
1.	Radish	*Raphanus sativus*	Cruciferae	20	Fusiform
2.	Carrot	*Daucus carota*	Umbelliferae	18	Conical form
3.	Turnip	*Brassica rapa*	Cruciferae	20	Napiform
4.	Beetroot	*Beta vulgaris*	Chenopodiaceae	18	Fusiform

"RADISH"

- Origin : Indian (red tail) –India.
  Japanese radish : Japan.
  European radish : Mediterranean region.
- Radish is a cross-pollinated vegetable due to sporophytic self-incompatibility.
- Rat-tail *Raphanus sativus var-caudatus* (pods are used).
- Pollination – Entomophilous (Honey Bees).
- Radish show considerable inbreeding depression upon selfing.
- Red coloured roots are high in ascorbic content.
- Pungency – Due to Isothiocynates.
- Red colour – Anthocyanin is responsible.
- Roots are used in treating urinary complaints and piles.

Recommeded varieties of Radish

S. N.	Variety	Features
1	Pusa Desi	Roots 30-35 cm long, tapering, white with green stem end and mildly pungent in taste. Leaves medium size and green. Suitable for sowing from August to October. Maturity 50-55 days. Yield 300 q/ha.
2	Pusa Reshmi	Roots 30-35 cm long, tapering, white with green stem end. Mildly pungent in taste. Leaves small to medium size and green. Suitable for sowing from October to December. Maturity 60-65 days. Yield 350-400 q/ha.
3	Japanese White	Roots 35-40 cm long, stumpy in shape, pure white, less pungent or sweet in taste. Leaves medium size and greyish green. Suitable for sowing from October to December. Maturity 60-65 days. Yield 350-400 q/ha.
4	Pusa Himani	Roots 30-35 çm long, tapering white with green stem end, mildly pungent in taste. Leaves small in size and dark green. Suitable for sowing from December to February in plains and in hills it is only variety which can be grown throughout the year barring three winter months viz – November – January. Maturity 60 to 65 days. Yield 300-325 q/ha.
5	Pusa Chetki	Roots 15-20 cm long, stumpy in shape, pure white and mildly pungent or sweet in taste. Leaves medium in size, erect, soft and green. Suitable for sowing from the mid April to August. Maturity 40 days. Yield 200-250 q/ha.

6	White Icicle	Roots small (8-10 cm long), thin, pure white, tapering, sweet in taste. Suitable for sowing in kitchen garden or container garden from October to January. Maturity 30 days. Yield 50 q/ha.
7	Rapid Red White tipped	Roots small, round or globular, upper portion of the roots is red and lower portion is white, sweet in taste. Suitable for sowing in kitchen garden or container garden from October to January. Maturity 25 days. Yield 50 q/ha.
8	Arka Nishant	An Asiatic variety. Matures within 45-55 days. Roots are medium sized (25 cm × 3.4 cm), marble white, crisp texture, pleasant aroma, free from early bolting, pithiness, splitting & forking, each root weighing 300-400 g. Pungency is mild. It yields 200-300 q/ha. Developed at IIHR, Bangalore.
9	Punjab Safed	Roots are pure white, tapering, smooth, mild. In taste, 30-40 cm long and 3-5 cm thick. Developed at PAU, Ludhiana.

■ Varieties :

1. European or Temperate type (Biennial) - Short duration (25-30 days), produce good quality roots with less pungent, Small size, yield low (75-100 q/ha.), requires chilling temperature for bolting, seeds are not produced under North Indian plains.
   - (a) White Icicle.
   - (b) Pusa Himani – (Black × Japanese white).
   - (c) Rapid Red White Tipped – Globular shape.
   - (d) Scarlet Globe.
   - (e) Scarlet Long.

2. Asiatic or Tropical type – More pungent, long duration (45-55 days), large size roots, High yield (250-300q/ha.), do not require chilling temperature for bolting, sets seed freely under tropical condition, can be produced under North Indian plains also.
   (a) Pusa Desi – Selection variety.
   (b) Pusa Chetki – Suitable for hotter months, developed by selection. Introduction from Denmark.
   (c) Pusa Reshmi – Suitable for main season (September-October), tolerant to slightly high temperature.
   (d) Punjab Safed
   (e) Punjab Pasand
   (f) Punjab Ageti
   (g) Arka Nishant – By mass selection, Multiple resistance, resistant to white rust, pithiness, premature bolting, root branching, forking, splitting etc. root and shoot length ratio is 1:1.
   (h) Japanese White – Introduced from Japan, seed produced only in hills, less bolting.
   (i) Kalyanpur White
   (j) Kalyanpur No. 1
   (k) Col – Can be grown throughtout the year, highly damaged by rust and aphid.
   (l) Nadauni – Popular variety in Himachal Pradesh, top of root is green.
   (m) Jaunpuri Mooli – Less pungent and sweet.

- **Autopolyploids** –
  Raphanobrassica (Radish cabbage) – Cross done by Karpenchenko (An allotetraploid).
- Karpenchenko wanted to synthesis a plant with root like radish and shoot like cabbage but unfortunately "Raphanobrassica" has root like cabbage and shoot like radish, hence it was not applicable.

- Temperature : 10-15° C - Better root development in cooler areas.
- Seed rate –
  1. 8-10 kg - Tropical/ha.
  2. 10-12 kg - Temperate/ha.

■ Physiological disordes –

1. Akashin – Induced by high day and high night temperature 30° C and 20 ° C, as well as by low soil moisture. Caused by Boron deficiency in which root growth is checked.
2. Pithiness – More in summer crop than spring or autumn.
   - Caused due to high temperature three weeks before harvest.
   - Excess application of N, P, K causes pithiness.
   - Soil moisture stress.
3. Hollow root wart – High temperature during 16-30 days of sowing.
   - Soil moisture stress causes wart disorder.

   Storage – 0-1 ° C at 90-95% R.H. for 2 months.

   Yield - 15-20 tonnes/ha. (Viability of seed 4-5 years).
   - Brown heart in radish is due to Boron deficiency.
   - Application of fresh undecomposed manure will result in forking of roots in radish.
   - White rust is caused by *Albugo condida.*
   - Cultivated Radish – Originated from *Raphanus raphanistrum.*
   - Fodder type – *Raphanus sativus variety mongri.*
   - Phyllody – Seed disease of radish.

"CARROT"

- Origin – Southern Western Asia (especially Afghanistan).
- Annual – For root production.
- Biennial – For flowering and seed set.
- Taste of carrot is due to the presence of glutanic acid.
- Carrot juice is a rich source of carotene.
- A special type of beverage known as Kanji is prepared from black carrot which is used as appetizer.
- A sweet preparation called Gajar halwa is very famous dish in North.
- Outer core consist of a thin periderm (a layer of cork cell) and a relatively wide band of secondary phloem. In this region only the sugar is stored.
- Inner core has the secondary xylem and pith.
- Inflorescence - A compound "umbel".
- Flower – Protandrous.
- Orange coloured (Temperate) carrot contains more carotene (a precursor of Vitamin – A) than Asiatic type and is rich in thiamine and riboflavin.
- Asiatic – Contains more anthocyanin pigment but poor in nutritive value.
- Carrot green leaves – Contain appreciable amount of protein, minerals, and vitamins.
- Breeding Objective – To develop varieties with high root yield and carotene content, self coloured core, cylindrical, uniform tapering, non-branching.
- The Temperate types form root under both tropical and temperate conditions but set seed only under temperate condition. They need low temperature of 5-8° C for 40-60 days before throwing out the seed stalk (break dormancy).

- Asiatic type are high yielding and produce seed under tropical condition, but they are poor in carotene content and other quality attributes.
- Carrot juice is used for colouring butter and other foods.
- Highest accumulation of carotene occurs in older cells of the phloem.
- Cytoplasmic male sterlity (CMS) is reported in carrot.

Recommeded Tropical Varieties of Carrot

S. N.	Variety	Features
1	Pusa Kesar	Red coloured roots and self coloured core, early, rich in carotene 38 mg/100g (7753 IU/100g). Most desirable features of this variety is that roots stay for about a month longer in the field than local Red without bolting, sets seeds freely in plains. Suitable for early sowing (mid August to early October). Maturity 90-100 days. Yield 300 q/ha.
2	Pusa Meghali	Orange coloured roots, self coloured core and short top. Suitable for early sowing (mid August-early October). Maturity 100-110 days. Yield 250 q/ha.
3	Arka Suraj	The variety was developed by pedigree selection between Nantes and IIHR 253. Deep Orange roots with self coloured core, conical shape; root length is 15-18 cm with TSS 8-10%, tolerant to powdery mildew and nematode. It flowers and sets seeds under tropical condition. Developed at IIHR Bangalore.

Recommeded Temperate Varieties of Carrot

S. N.	Variety	Features
1	Nantes Half Long	Cylinderical, stumphy, orange coloured roots, self coloured core suitable for sowing from mid October to early December. Maturity 100-120 days. Yield 100-125 q/ha.
2	Pusa Yamdagni	Roots with orange flesh and self coloured core. Suitable for sowing from mid October to early December. Maturity 100-110 days. Yield 150-200 q/ha.
3	Early Nates	Roots with orange flesh and self coloured core. Almost cylindrical roots, terminating abruptly in small thin tall. Suitable for sowing from mid October to early December. Maturity 90-110 days. Yield 100-120 q/ha.
4	Chantaney	Roots with orange coloured flesh and self coloured core, thick at the shoulder with gradual tapering. Suitable for sowing from mid October to early December. Suitable for canning. Maturity 100-120 days. Yield 125-150 q/ha.
5	No : 29	Selection from alocal variety, Roots light red, long tapering& thin. Developed by IARI.
6	Sel. 233	Roots are orange, semi cylindrical, maturity duration 90-100 days.
7	Sel 21	Roots red with light orange core, quick growing, suitable for pickles, candy & juice.
8	Imperator	Deep orange coloured roots, self coloured core, suitable for sowing from mid October to early December. Maturity 100-120 days. Yield 120-150 q/ha.

9	Zeno	Introduction from Germany. Roots long orange. Medium tops maturity 110-120 days and suitable for Nilgiries.

■ Varieties –

I. European/Temperate Type –

1. Nantes Half Long – Self coloured core.
2. Early Nantes – Self coloured core.
3. Chantenay – Excellent cultivar for Canning and storage.
4. Imperator – (Nantes × Chantaney), mid to late variety.
5. Zeno – Introduced from Germany, suitable for Nillgiri Hills, self coloured core, exposed portion-pinkish green.
6. Ooty – Self coloured core.
7. Pusa Yamdagni – (EC 9981 × Nantes) (Earlier by a week /10 days than Nantes).
8. Early Horn.
9. Early Gem Imperator.
10. Danvers – Suitable for fresh market and processing.

II. Tropical / Asiatic Varieties –

1. Pusa Kesar – Selection Local red × Nantes half long.
2. Pusa Meghali – Selection Pusa Kesar × Nantes, self coloured core. Highest vitamin A (carotene) content (11574 mg/100g.).
3. Sel. No. – 29
4. Sel. No. – 233 (Nantes × No-22) free from spliting.
5. Sel. 21 – By Punjab Agriculture University, Ludhiana, retains edible quality for a longer period in field without bolting.
6. Golden Ball.

- Seed rate – 5-6 kg/ha.
- Excess "N" – Effects quality of root like sugar content, dry matter, carotene, vitamin-C, storage quality etc.

- Excess "N" – In field will increase the rotting of roots during storage.
- High temperature in carrot cause bolting.

■ Physiological disorder –

1. Forking – When hard soil does not allow straight growth
2. Splitting/Cracking-

   Causes –

   (a) Heavy application of "N", High soil concentration of ammonium compound cause more serious splitting than by other form of nitrogen.

   (b) Sudden change in soil moisture status.

   (c) Wider spacing.

   (d) Low chloride content.

3. Cavity spot – Characterized by appearance of cavity in the cortex, in most cases the subtending epidermis collapses to form a pitted lesion caused due to "Ca" deficiency and increase of "K" content.

   Yield – Asiatic : 25-30 tons/ha.

   European : 10-15 tons/ha.

   Storage : 0-4.4 °C temperature and 93-98% relative humidity for 6 months.

Crop	Botanical Name	Ch No
1. Parsnip	*Pastinaca sativa*	22
2. Chervil	*Anthiriscum cerebolium*	32
3. Ceraric	*Apium graveolum*	22
4. Skirret	*Sium sisaru*	12

- All four plants belong to Umbelliferae family.

- The free sugar, glucose, fructose and sucrose contribute to sweetness whereas volatile mono and sesquiterperiods contributes to harshness.
- Presence of polyacetylenes, contributes to the unpleasant odour in carrots.
- Carrot seed contains germination inhibitors known as carrotal.
- Propazine – Excellent residual and selective weed control in carrots.
- In India, carrot was introduced from Persia, probably carrot was first grown in Arne, the Salem gardens about 1620 AD.
- Dietary supplementations of a combination of carrot and orange juice have been found to control the oxidation of low density lipoproteins in habitual cigarette smokers.
- Carotene reduces the risk of cataract formation.
- Lygus bug – Serious pest for seed crop.

"TURNIP"

- Origin from Mediterranean region.
- Inflorescence is terminal raceme.
- Young leaves are rich source of vitamin–A (15660 IU/100g.), vitamin–C and Iron.
- It is a biennial in nature.
- The fleshy thickened underground portion of turnip is actually the hypocotyl.
- A distinct tap root and secondary roots arise from the lower part of the swollen hypocotyl.
- Mainly grown in Jammu and Kashmir, Himachal Pradesh and Haryana.

- It can be grown at an elevation of 1500 M. at MSL or above, but it is not suitable for growing in low lands of wet tropics.
- Mass Selection – Most adoped breeding method for crop improvement.
- Turnip has sporophytic self-incompatibility; hence there is no need for emasculation.
- Pollination – Entomophilous (flies) and Honey bees (main source).
- Sodium chloride ($NaCl_2$) and Carbon dioxide (CO_2) can also be used for overcoming self imcompatibility.
- Selfing can be achived by application of fresh pollen on decapitated stigma.
- Thinning –Important operation to maintain optimum plant to plant spacing (10-15 days after germination).

Recommended varieties of Turnip

S. N.	Variety	Features
1	Pusa Sweti	Roots pure white, round and slightly flat with rat tail habit. Flesh white, sweet and mildly flavoured. Leaves green and small to medium in size. Suitable for sowing from early August to September. Maturity 45-50 days. Yeild 250-300 q/ha.
2	Purple Top White Globe	Roots large, round, smooth upper part of skin purplish and lower part in creamy white. Flesh white, firm, crisp and mildly sweet. Leaves dark green and erect. Suitable for sowing from October to December. Maturity 60-65 days. Yield 350 q/ha.

3	Pusa Swarnima	Roots large, flattish round, pale amber coloured, rich in carotene content and sweet in taste. Leaves medium and green. Suitable for sowing from October to December. Maturity 65-70 days. Yield 350-375 q/ha.
4	PusaChandrima	Roots medium to large, smooth, pure white, sweet in taste, leaves medium & green. Suitable for sowing from October to December. Maturity 60-65 days. Yield 350 q/ha.
5	Golden ball	The roots are perfectly globe shaped, medium sized and smooth. Has bright creamy yellow skin and pale amber coloured flesh of fine texture and flavour. The top is small erect with cut leaves.
6	Snow ball	It is an early variety with medium sized roots. Roots are round, smooth with pure white skin, whereas the flesh is white, finely grained, sweet and tender. It has a small erect top with medium yellow green and cut leaves.
7	Early Milan Red Top	It is an extra early (45 day) variety. The roots are deep flat with purplish red tops and white underneath. The flesh is pure white, well grained, crisp and mildly pungent. The tops are very small with only 4 to 6 sessile leaves. It is a very high yielding variety.
8	Pusa Kanchan	Selection from a cross between Asiatic Red and Golden Ball, its roots look just like the Local Red round turnip from which it differs in having creamy yellow flesh. The roots stay longer in the field without getting spongy. The leaf top is

		shorter than the local. It produces seed satisfactorily in the plains, though the seed is harvested a fortnight later than the Local Red. It has excellent flavour and taste. Developed at IARI New Delhi.
9	Punjab Safed – 4	A selection from local cultivar. Roots pure white, round, medium size and mild taste. Developed by PAU, Ludhiana.

■ Varieties –

I. European or Temperate type –

1. Pusa Swarnima–Japanese white (Asiatic type) × Golden Ball (European Type).
2. Pusa Chandrima – Japanese White (A) × Snowball (E).
3. Purple Top White Globe – Best for dehydration, heavy yield.
4. Golden Ball.
5. Snow Ball.
6. Early Milan Red Top.

II. Asiatic Tropical Type

1. Pusa Kanchan – Seletion (Asiatic Local Red Round × Golden Ball).
2. Pusa Sweti – Very early maturing variety.
3. Punjab Safeda – Early maturing.

- **Sowing time –**

  July-September (Asiatic Types) : In North Indian Plains

  October-December: European Types : In North Indian Plains

- July-September in Hills.

- Root develops best flavour, texture and size at a temperature of 10-15 ° C.
- Short day length and cool weather favours proper root development and long day and high temperature induces early bolting.

Seed rate	:	3 - 4 kg/ha.
Seed sowing depth	:	1.5-2.5 cm.
Yield	:	200-400 q/ha.

■ Physiological disorder –

1. Whiptail – Due to "Mo" deficiency, more common in acidic soils.
2. Brown Heart – "B" deficiency, more common in acidic soils.

- Turnip yellow mosaic virus disease- transmitted through flea beetle, cabbage may serve as an alternate host for this virus.
- Crinkle – Viral disease.
- Seed viability – 5-6 years.

"BEET ROOT" (Chukandar)

- Origin from Mediterranean region.
- Fruit type – Capsule.
- Inflorescence – Spike (develops in 2nd year).
- Probable progenitor of garden beet is *Beta vulgaris sps. maritina.*
- After sugar cane it is the second most important crop for the preparation of sugar.

- Beet root is the rich source of protein, carbohydrate, Ca, P and vitamin C.
- The colour of root is due to the presence of red-violet pigment Beta-cyanins and yellow pigment Beta-xanthins.
- Upper portion of root develops from the hypocotyl and the lower portion from the tap root, from which the secondary roots arise.
- Thinning is practiced to maintain the intra-raw spacing and optimum population.
- Crops can tolerate freezing to some extent.
- Under high warm weather "zoning" is caused which is marked by appearance of alternating light and dark-red concentric circles in the roots.
- The seeds usually contain one or more seeds and are called "Seedball" which is viable for 5-6 years under normal storage condition.
- Seedball's rate – 7-9 kg./ha.
- **Spacing** –

  40-60 cm Row × Row

  8-10 cm Plant × Plant

  2.5 cm Depth of seed

  Sowing time-

March-July	In Hills
September-November	North Indian Plains
July-November	South India.

- During sowing, if temperature is high, the roots produced would be coarse with woody-flesh and dull colour.
- Beet root seed is multigerm which produce 3-4 seedlings per seed ball hence thining is an important operation.
- Temperature of 4.5-10^0 C for 15 days induces premature bolting.

Recommended varieties of Beetroot

S. N.	Variety	Features
1	Crimson Globe	Roots are globular to flattened, medium red skin, flesh medium dark, crimson red with indistinct zoning, top medium to tall, large bright green leaves with maroon shades. Heavy yielder. Maturity 80-90 days. Yield 180 q/ha.
2	Detroit Dark Red	Roots are round with smooth, uniform, deep red skin. Flesh dark blood red with light red zoning, tender, top small, leaves dark green tinged with maroon. Heavy yielder. Maturity 80-100 days. Yield 180 q/ha.
3	Crosby Egyptian	Roots are flat globe with a small taproot and smooth outer surface. The flesh colour is dark purplish red with some indistinct zoning. The top is medium tall and green with red veins. Maturity 55-60 days. Yield 170 q/ha.
4	Early Wonder	Roots are flattened globe with a smooth, dark red skin. Flesh dark red with some light red zoning. Top is heavy, green with red veins. Maturity 55-60 days. Yield 160 q/ha.

■ Other Varieties –

1. Ooty 1- Selection High yielding, flesh colour is blood red, suitable for growing in South Indian Hills.
2. Detroit Dark Red – Heavy yielder, mostly for salad making and storage.
3. Crimson Globe – Heavy yielder.

4. Crosby Egyptian.
5. Early Wonder.
6. Ohio Carner.

- Beetroot is sensitive to high acidity and low temperature.
- It can grow in saline and alkaline soil upto 9-10 pH.
- Raking is done to break the crust as seeds sprout poorly in crusted soil.
- Seed production is done in hills above 1200 meter elevation.
- Optimum temperature for growth is 18.3-21° C.
- The probable origin of beet root is mediterranean region for *B. vulgaris sp. maritima* possibly by hybridization with *B. patula*.
- Harvesting is done when they are about 3.5 cm in diameter.
- Beets for bulk marketing or canning are topped in the field when they attain a diameter of 6.5-8 cm.
- Yield : 250-300 q./ha.
- Storage : 0° C with 90% R.H.
- Beet mosaic yellow a viral disease is transmitted by aphid.
- Curly top disease is transmitted by hoppers.
- High temperature in beet causes poor colour development.

■ Physiological Disorder

1. Internal Black Spot / Brown Heart/Heart rot/Breakdown of beet – Hard and corky black spots are scattered throughout the root.
   - Caused due to "B" deficiency in neutral or alkaline soils.
   - "B" deficiency first affects young cells and tissue.

- Detroid Dark Red variety of beet show comparatively less symptoms of "B" deficiency.

2. Speckled yellow is caused due to deficiency of "Mn". "Mn" deficiency is very severe in sandy and very alkaline soils. Symptoms – leaves with yellow blotches between veins tend to curl up in summer.

   - "B" deficiency is common in alkaline soil or on soils relatively high in "Ca".

5

BULB CROPS

S.N.	B.N.	Family	2n	Origin	Edible Part
Onion	*Allium cepa*	Amaryllidaceae	16	Central Asia Primary Mediteranean Region Secondary	Bulb
Garlic	*Allium Sativum*	Amaryllidaceae	16	Central Asia	Cloves
Leek	*Allium Porrum*	Amaryllidaceae *Allium Ampeloprasum* Var-porrum	32	Central Asia	Stem and Leaves
Multiplier Potato/ Onion/ Under-ground Onion	*Allium Cepa* Var aggregatum	Amaryllidaceae	16	Central Asia Primary Mediteranean Region Secondary	Leaves and bulbs
Perennial onion Shallot	*Allium Cepa Var ascalonicum*	Amaryllidaceae	16	Central Asia	Leaves and bulbs
Tree Onion	*Allium Cepa* Var viviparum /proliform	Amaryllidaceae	16	Central Asia (Primary) Mediteranean Region (Secondary)	Leaves and bulbs
Japanese Onion	*Allium fistulosum*	Amaryllidaceae	16	Central Asia	Leaves and bulbs
Chive	*Allium Schoenopr--asum*	Amaryllidaceae	16, 24, 32	Central Asia	Leaves and bulbs

- All bulb crops belong to monocotyledonae vegetables.

"ONION"

- Economically most important bulb vegetable.
- Harvested organ: swollen base of the stem (bulb) with numerous fleshy leaves.
- India ranks **second** in area and production in the world after China and third in export (12% of total world export).
- In export Netherland (21% of world export) ranks first and Spain IInd.
- In India Maharashtra, Karnataka, Gujarat, Orissa, Andhra Pradesh, Uttar Pradesh, Tamilnadu, Rajasthan and Bihar are major onion producing states in India.
- Maharashtra is leading state with 23.4% area and 27.50% production followed by Karnataka (19.91% area and 10.80% production).
- Average yield per unit area is very low 10.6 tonnes/ha when compared with world (16.5 tonnes/ha.) (USA 43 t/ha.), (Japan 42.6 t/ha.).
- Onion accounts for 77% of total foreign exchange earning among fresh vegetables.
- In India, yield is highest in Gujarat (23.37 t/ha.) followed by Punjab (20.13 t/ha.).
- Per capita availability of onion in Netherlands (32.99 kg/year) is highest followed by Turkey (29.82 kg/year).
- In India, the availability of onion per capita is 3.94 kg/year.
- Onion is divided into 3 main groups:
    1. The common onion bulbs are formed as single plants, inflorescence does not form bulbets, biennial, often grown as an annual ,most important commercially

propagated in the main from true seeds, sometimes planted as transplants, small bulbs (sets) are produced from thickly planted seed when sets are replanted, these bulbs reach maturity rapidly, cultivars vary for bulb shapes, dry scale color (white, yellow, and red), pungency (ranging from mild and sweet to pungent), soluble solids, adaptation to different day lengths when harvested early they produce spring or bunching onions mature bulb onions are usually dried before marketing.

2. The aggregatum group with many lateral bulbs or shoots inflorescences lacking bulbets sterile seed production propagation by vegetative means includes the potato onion or multiplier onion, ever-ready onions, and shallots are sometimes called scallions, a source of confusion because *A. fistulosum* is also called scallions.

3. The proliferous (proliferum) group ground bulbs are sometimes poorly developed infloresence bears bulbets true seed is usually lacking reproduction is by vegetative infloresence and bulbets not commercially cultivated almost exclusively for home gardening includes tree onions, top-set onions, and Egyptian onions.

- Mostly the Indians prefer red and pungent type of onion while in Japan, Europe and America yellow coloured varieties are preferred.
- Alliums are characterized by the presence of remarkable sulphar containing compounds, which gives their distinctive smell and pungency.
- Oil yields methyl and propyl compounds 0.002-0.03g/100g. of fresh bulb by steam distillation.

- Onion is a rich source of carbohydrate, proteins, minerals (calcium) and vitamins (riboflavin).
- Pungency is due to Allyl propyl -di-sulphide
- Red colour is due to Anthocyanin
- Yellow colour is due to Quercetin
- Antifungel activity-Phenolic factor i.e. "Catechol"
- Inflorescence – Umbel
- Flower colour – White or blueish
- Number of seed stem – 1 to 20 or more.
- Onion is a highly cross pollinated crop due to its protandry.
- Pollination is chiefly by honey bees.
- Flies are excellent pollinators as they go from flower to flower.
- Onion (Allium) is known to benefit in the prevention and treatment of atherosclerosis and coronary heart disease.
- Onion is a cool season crop.
- For good vegetative growth 12.8 - 23° C temperature before bulbing is best.
- For bulb development 15.5-25° C temperature is best.
- In onion, very low temperature in early stage favours bolting.
- Sudden rise in temperature causes early maturity in Rabi crop resulting in small bulb size.
- Almost all cultivars grown in plains in India are short day cultivars.
- "Brown Spanish" – A long day variety require day length of more than 14 hours and low temperature for proper bulb development.

- Kharif onion varieties require day length of 10-11 hours.
- Rabi onion varieties require day length of 12-13 hours.
- Long day varieties do not bulb under short day condition where as short day varieties if planted under long day condition will develop early bulbs.
- Seed rate 10-12 kg/ha for Rabi season
  12-15 kg/ha for Kharif season
- Seedlings are ready for transplanting in 8-9 weeks in Rabi where as 6-7 weeks in Kharif season.

  Sowing Time

  August - September for Kharif season crop.

  October - November for Rabi season crop.

  March April for Hilly areas
- Spacing 4-5 cm apart in Rabi season
  5-7 cm apart in Kharif season
  2-3 cm deep
- Planting by bulb – To meet the demand of green onion for sale in early winter, planting by bulbs is done.
- 750 kg medium size bulbs planted in August-Sep, become ready in October - December
- If planting is done by sets, 1.5-2.0 cm of sets amounting 10 q/ha. are required.
- "Zn" application (1-3 PPM) Increase yield and improves quality.
- For direct sowing of the variety "Agrifound Rose", 22.5 kg/ha seed is required.
- In case of multiplier onion sets 1-1.2 tonnes of bulbils are required for planting in a hectare.
- "Cu" and "B" @ 13.4 kg and 1.8 kg/ha respectively are beneficial in increasing the yield.

■ Crop Improvement

- Jones and Clarke (1943 and 1963) were first to use Cytoplasmic Male Sterility (CMS) for producing hybrid seed in onion.
- Selfing and massing technique – Suggested by Jones and Mann (1963) Improvement in cultivar can be affected by selfing followed by massing.
- Onion seed production industry has moved away from open-pollinated types to hybrid production systems.
- Cytoplasmically inherited male sterility factor (*S*) and a nuclear encoded gene (*Ms*) which can normally override the male sterility or "restore" the plant to a normal fertile phenotype.
- When this "restorer" gene is in the homozygous recessive condition only in the *S* cytoplasm does male sterility result

  S msms (= sterile)

  S Msms (= fertile)

  S MsMs (= fertile)

  N msms (= fertile) (*N msms* plants are known as maintainers)

  N Msms (= fertile)

  N MsMs (= fertile)

- Onion is insect (bee) pollinated.
- In hybrid production important to have both parental lines flowering at the same time.
- Ratio of 8 female : 2 male rows commonly used.
- Three-way hybrids prepared by crossing an *S msms* F_1 hybrid (male sterile) to a third inbred parent.
- Similar male sterility system in chives (*A. schoenoprasum*)
- Five major genes determine onion bulb color:

I- inhibits the production of color and the bulbs are white (the expression of red, yellow or chartreuse pigments requires that the plant be *ii*) *C* locus conditions the production of colored bulbs all *iicc* bulbs are white plants with a dominant allele at the *C* locus and homozygous recessive *rr* at the *R* locus are yellow to brown in color red bulbs are conditioned by dominant alleles at both the *C* and *R* loci (*iiC-R-*) a dominant allele at the *G* locus conditions a golden color without pink chartreuse bulbs are homozygous recessive at the *G* locus *L* conditions a red pigment when the plant has the genotype *iiC-rrG-L-*.

Recommended Varieties of Onion

S. N.	Variety	Features
1	Pusa Red	Medium sized, flattish round, light red coloured bulbs. Bulb weight 90 g. TSS 12-13%. Very good storage. Maturity 140-145 days after transplanting. Yield 250-300 q/ha.
2	Early Grano	Large globular yellow bulbs, good for salad and for green onion purposes. Bulb weight 100-120g. TSS 6-7%, very poor in storage. Maturity 95-110 days. Yield 400-500 q/ha.
3	Pusa Madhvi	Light red, roundish flat bulbs. Bulbs weight 90-100 g. TSS 11-12%. Good in storage. Maturity 130-135 days. Yield 300-350 q/ha.
4	Pusa White Flat	Falttish round, white bulbs. Bulb weight 95 100g. TSS 11-12%. Drying ration 9: 1. Suitable for dehydration purposes and also good for green onion purpose. Good in storage, Maturity 130-135 days. Yield 325-350 q/ha.
5	Pusa White Round	Roundish flat, white coloured bulbs. Bulb weight 90-95g., TSS 12-13%, drying ratio 8:

		1. Suitable for dehydration and green onion purposes. Maturity 130-135 days. Yield 300-325 q/ha.
6	Pusa Ratnar	Roundish flat, deep red bulbs. Bulbs weight 100-110g. TSS 9-10%. Fair in storage. Maturity 125 days. Yield 300-400 q/ha.
7	N 2-4-1	Roundish flat, deep red coloured bulbs. Bulb weight 80 g. TSS 12%, good in storage. Maturity 130-140 days. Yield 200-250 q/ha.
8	N-53	Roundish flat, deep red coloured bulbs. Bulb weight 80 g. TSS 9-10%. Poor in storage. Suitable for Kharif season. Maturity 100-110 days. Yield 150-200 q/ha.
9	Agrifound round	Roundish flat, light red coloured bulbs. Bulb weight 90 g. TSS 12%, good in storage. Maturity 135-140 days. Yield 250-300 q/ha.
10	Agrifound Dark Red	Roundish flat, deep red coloured bulbs. Bulb weight 80 g. TSS 9-10%, poor in storage. Suitable for kharif season. Maturity 95-110 days. Yield 150-200 q/ha.
11	Arka Niketan	Developed through mass selection from IIHR-153 (sel.-13). Globe shaped medium large bulbs with an average weight of 100-180 g. TSS 12-14% Duration 145 days. Suitable for rabi season. Yield 37 t/ha. Recommended for UP, Maharastra, MP, Karnataka.
12	Arka Bindu	Developed by 5 generation of mass selection for IIHR-402 (local collection) (Rose onion selection). Bulbs small (2.5-3.5 cm diameter), deep red, highly pungent, flattish globe shape, high TSS (14.16%) with high dry matter content. Suitable for export market. Duration 90-100 days. Yield 25 t/ha. Recommended for Karnataka.

13	Arka Pitamber	Developed through pedigree selection from the cross U.D 100 x IIHR-396 (Yellow onion). Bulbs uniform yellow, medium sized (5.2-6.0 cm diameter) bulbs with globe shape and thin neck. Less pungent with TSS 11% and total sugar 9.81%. Tolerant to purple blotch; basal rot diseases and thrips. Suitable for export market, suitable for kharif and rabi seasons. Duration 140 days. Yield 35 t/ha. Recommended for Karnataka.
14	Arka Kirthiman	An F1 hybrid between MS-65 and Sel. 13-1 1 (Hybrid). Medium sized bulbs with globe shape and firm texture. Bulbs red in colour with bulb weight 120-130g. Tolerant to purple blotch, basal rot and thrips, long storage (4-5 months). Suitable for kharif and rabi seasons. Duration 125 days. Yield 47 t/ha. Recommended for Karnataka.
15	Arka Lalma	An F1 hybrid between MS-48 and sel. 14-1-1 (Hybrid 5). Medium sized bulbs with globe shape and firm texture. Bulbs red in colour with bulb weight 120-130g. Tolerant to purple blotch, basal rot and thrips, long storage (4-5 months). Suitable for kharif and rabi seasons. Duration 130-140 days. Yield 50 t/ha. Recommended for Karnataka.
16	BrownSpanish	Developed by IARI, Long day type, can grow in hills, maturity 160-180 days, Yield 25 t/ha.
17	Arka Kalyan	Globe shaped bulbs with deep-pink coloured outer scales, internal scales fleshy, succulent, concentric, high TSS (10.5 to 12%) suitable for Kharif season having a yield of about 470 q/ha. It takes 140 days from seed to bulb.

18	Arka Pragati	Globe shaped thin necked and deep pink bulb. It has fleshy scales with high pungency. TSS (10.5-12.5%). It has yield potential of about 450 q/ha. Suitable for both the seasons. It takes only 130 days from seed sowing to bulb harvesting.

Varieties

1. **Red coloured varieties – High yielding, Mild to high pungent, good in storage.**

a. Agrifound Dark Red – Suitable for Kharif season.

b. Agrifound Light Red – Recommendation for Rabi season.

c. Agrifound Red – Multiplier onion variety.

d. Agrifound Rose – Suitable for kharif, export purpose mainly grown in Kolar and Bengaluru district of Karnataka and Chuddaph district of Andhra Pradesh.

e. N.53- Most suitable for kharif, developed by MPKVP Rahuri (Maharashtra) yield 250-300 q/ha.

f. Arka Kalyan – Kharif season, resistant to purple blotch.

g. Arka Niketan – Rabi season, good for storage.

h. Arka Bindu – Recommendation for Rabi season, suitable for export to Malaysia and Singapore.

i. Arka Pragati – Suitable for both Rabi and Kharif season.

j. Pusa Red – Rabi season, very good keeping quality, less bolting and less pungent, short to intermediate day length type.

k. Pusa Ratnar – Rabi season, fair in storage quality.

l. Pusa Madhvi – Rabi season, good keeping quality.

m. Punjab Naroya – Tolerant to purple blotch disease and insect thrips (*Bemisia tabaci*)

n. Punjab Red Round – Rabi season, early maturing.

o. Punjab sel. – Rabi season, comparatively more tolerant to purple blotch and thrips, suitable for dehydration.

p. Kalyanpur Red Round – From Chandrasekhar Azad University of Agriculture and Technology, Kanpur, suitable for Rabi season.

q. Udaipur 101 – Rabi season, good for salad

r. Udaipur 103 – Rabi season

s. Hisar -2 – Rabi season

t. N 24-1 – Kharif season, tolerant to Alternaria blight and thrips

u. Banglore Rose – Suitable for export.

v. Basant -780 – Kharif season, good for storage, developed at MPKVP Rahuri (Maharashtra).

w. VL-3 – Suitable for Rabi in Hills.

2. White coloured varieties – Best suitable for dehydration purpose, TSS 11-12%

a. Pusa White Flat – Rabi season, suitable for dehydration, drying ratio – 9:1 good for green onion. Developed in 1975

b. Pusa white round – TSS 12-13% dry ratio 8:1 suitable for dehydration and green onion purpose, very good for storage.

c. Punjab White – High TSS (15%), suitable for dehydration

d. Punjab -48 (S48) – Suitable for dehydration, Rabi sowing, Identified in 1975 for sub humid plains of Sutlaj Ganga, Good keeping quality.

e. Udaipur 102 - Suitable for dehydration and rabi season, TSS 12%.

f. N 257 9-1 – Suitable for dehydration and Rabi season, Developed at MPKVP Rahuri (Mh.)

3. Yellow coloured varieties

a. Arka Pitamber – Short day, developed by pedigree selection by crossing between two short day local genotypes UD 102 (White) x IIHR -396 (Red)

b. Phule Swarna – Developed by MPKV Rahuri (Maharashtra), less pungent, suitable for export to Europe, Australia and America.

c. Early Grano – Good for salad, green onion purpose, Introduced from USA.

d. Brown Spanish - Long day type variety, excellent storability.

4. Multiplier onion varieties :

1. CO1 : 7-8 bulblets / plant.
2. CO2 : 7-9 bulblets / plant, Resistant to purple blotch.
3. CO3 : 8-10 bulblets / plant, Resistant to thrips.
4. CO4 : 8-10 bulblets / plant, Resistant to thrips.
5. MDU 1 : Resistant to thrips, tolerant to lodging due to thick erect leaves.
6. Agrifound Red : Light red in colour.

- Delay in irrigation may result in seedling mortality upto 80-90%
- Critical stage of irrigation is Bulb formation.
- Excess moisture induces incidence of purple blotch especially when vegetative growth is over.
- Green onions are harvested when they are just ready for earthing.
- Yield of Common onion 250-300q/ha.
  Pickling type 160-200q/ha.
  Multiplier onion 150-180q/ha.
- Onions are cured when neck is tight and dried scales rustle.
- Curing is done to remove excess moisture from the outer skin which helps to reduce the infection of disease.
- Delhi market prefers big sized bulbs.

- Jute bags are used for onion packaging for sending to distant market by Trucks.
- Onion should be packed in 14-15 kg capacity cane baskets for export purpose. (Bangalore Rose and Multiplier onion)
- Temperature is more important than day length in seed production.
- Day length is more important than temperature for bulb production.

Crop Improvement

(1) Introduced Varieties –

1. Early Grano from USA
2. Brown Spanish
3. Barmuda Yellow

(2) Selection Varieties - Most of varieties in India had been developed through mass selection which is common in cross pollinated crops.

I.	Pusa Ratnar	II.	Pusa Red
III.	Pusa White Round	IV.	Pusa White Flat
V.	N-53	VI.	Arka Kalyan
VII.	Arka Niketan	VIII.	Arka Bindu
IX.	Arka Pragati	X.	Nasik Red
XI.	Patna Red	XII.	Agrifound Dark Red
XIII.	Bangalore Rose		

(3) Selfing - Pusa Madhvi

Brown paper bag and three rings muslin cloth bags are used for selfing

(4) Hybrid – (1) Arka Pitamber – White × Red

(2) Arka Kirtiman

(3) Arka Lalima

- Male Sterible lines have been isolated in Pusa Red at IARI (Good storage and wide adaptability)

(5) Small onion varities – 1. Bangalore Rose
2. Arka Bindu
3. Agrifound Rose

- Largest cultivated onion varieties – Orange and Yellow coloured.
- Onion is used against sunstoke.
- Hybridization is done when we want to introduce character from other varieties.

  1. Intervarietal Hybridization – Very common
  2. Interspecific Hybridization – Very rare

- Utilized in *A cepa* x *A fristulosum*

(6) Long day type

(1) White Spanish (2) Cream Gold
(3) Early Lockier
(4) Brown Spanish–Best in areas 1000 MSL, 28 tons/ha yield.

- Seed sowing – September-November
- Transplanting – March
- Har.est – August.

  Long day varieties crop only be grown in hills.

- Rabi onions are harvested one week after 50-70% neck fall.
- Kharif crop is harvested when leaves colour changes to slightly yellow and tops starts drying.
- Storage : 0° C temperature and 65-70% R.H. for 6-8 months.

- Storage models had been created by National Horticultural Research and Development Foundation (NHRDF) and National Agricultural Cooperative Marketing Federation of India Ltd. (NAFED).
- For Kharif season onion spraying of M.H. 25 PPM 75-90 days after transplanting check the sprouting up to satisfactory level.
- MH should not be used in Rabi crop; it would not work at higher temperature.
- At BARC, Trombay irriated bulb with gamma rays at very slow dose level (4,000-9,000 Krods) offered an effective physical method to control sprouting of onions.
- Export is canalized through the NAFED and its associates shippers.
- From India large onion is being exported to Malaysia, Dubai, Qatar, Kuwait, Singapore, Srilanka and Mauritius
- Small onion- Bangladesh, Singapore and Malaysia.
- Big onion are exported from Maharashtra, Gujarat, and Tamil Nadu ports
- Non bolting cultivars are -Early Grano, Texas.
- Export to Bangladesh and Nepal is from U.P. and Bihar by road.
- 90% Export Varieties - N-53, N-2-4-1, Ballary Red, Patna Red, Agrifound Dark Red, Agrifound Large Red, Pusa Red, Pusa Madhvi, Arka Niketan.
- Export potential small onion variety – Agrifound Rose, Arka Bindu and Bangalore Rose.
- Export potential Multiplier Onion varieties – CO4, CO3, and Agrifound Red (10%)
- Bangladesh prefers 3-4 cm diameter of bulbs for import.
- Gulf countries and Far East countries prefers 4-6 cm of bulbs.

- European and Japan – 6-7 cm
- Malaysia and Singapore – 2.5-3.0 cm.
- Major diseases of onions: pink rot, *Fusarium* basal rot, *Botrytis* leaf blight, neck rot, white rot, smut, downy mildew, *Alternaria* purple blotch, black mold, *Stemphyllium* blight, smudge and bacterial soft rots.
- Insect pests of onions: onion thrips, onion maggot, aster leafhopper and vegetable leafminer.

■ Physiological disorder –

1. Sprouting of bulbs – In white cultivar more than purple and yellow

   Causes –

   1. Excess soil moisture at maturity
   2. Supply of excess "N"
   3. Bolting (1) Low Temperature
      (2) Poor "N" Supply
      (3) Late seedling transplanting

- Poor seed germination is the main constraints in onion because seeds loose their viability with in a year under ordinary storage condition.

"GARLIC"

- Next to common onion most widely used of the cultivated Allium is probably garlic.
- Frosty hardy crop
- China ranks Ist in area and production.
- India ranks IInd in area and IIIrd in production.

- Egypt rank I[st] in productivity (21.5 t/ha.) followed by USA (18.6 t/ha.)
- Indian Productivity – 4.1 t/ha.
- Per capita availability of garlic is highest in Korean Republic (11.4 kg/Year)
- In India – 0.340 kg/ capita/ year
- In India, Madhya Pradesh is the leading state accounting 22.2% area and 24.40% production with average yield of 3.91 t/ha.
- Gujrat has the highest percentage share of production i.e. 28.98%.
- In India yield/ha. is highest in Punjab (17.20 t/ha.) followed by Haryana (12.80 t/ha.) and Maharashtra (6.85 t/ha.)
- Garlic has higher nutritive value than other bulb crops.
- Alliin – A colourless, odourless, water soluble amino acid present in uninjured garlic.
- Alliicin - On injury of cell, an enzyme, allinase comes in contact with alliin and causes its breakdown into a sulphar containing product alliicin which is the antibacterial substance of garlic and has typical odour of fresh garlic.
- Diallyl disulphide – Responsible for pungency in garlic which is found in alliin in an allyl radical form.
- Garlic contains 0.1 – 0.4% essential oil.
- Chief constituents of oil are diallyl – disulphide (60%) diallyl trisulphide (20%) and allyl-propyl disulphide (6%).
- True Garlic odour – Diallyl di-sulphide.
- Garlic oil is an item of commerce.
- For recovery of oil, fresh garlic is the best
- Fresh garlic bulb yields 0.1-0.2% essential oil on steam distillation
- Whole plant 0.05-0.9%

- Diallyl, dimethyl and allyl methyl sulfide, disulphide and trisulfide and minor –n– propyl group- All formed by thermal decomposition of allicin and its homologues.
- Garlic is a perennial species.
- Garlic differs from *A. Cepa* as the leaf base dose not store food.
- Cloves – Well developed axillary bulbs within the thin foliage leaves.
- Garlic is a sexually sterile diploid.
- Garlic is propagated by vegetative method i.e. by cloves and in some cases inflorescence bulbils (Jones 1983).
- *Allium longicuspis* – Wild ancestor of garlic.
- Garlic is mainly used for flavouring and seasoning.
- 50% production of garlic comes from Madhya Pradesh and Gujarat.
- Garlic is considered to possess antibacterial, antibiotic, antitumour, antiviral, antifungal,anti-cordidal,antimycotic, antithrombotic, fibrinolytic, antirosclerosis crosis, platelet aggregation, inhibitory, hypoglycemic, cytotoxic and lipid lowering properties. Eg. Allicin, ajoen, diithins and essential oils are principle contributors in this direction.
- Indore, Mansore (MP), Angul (Orissa), Junnar Taluka (Maharashtra), Manipur, Etah and Etania (UP), Sikar, Jhunjhunu, Jhalawar, Ajmer and Udaipur (Rajasthan) are major garlic growing pockets.
- In India small cloved garlic varieties are being grown in plains, recently large garlic cultivation has been introduced in North Indian plains.
- Garlic is a frost hardy crop, requires cool and moist period during growth and relatively dry period during maturity of bulbs.
- Garlic can be grown well up to 1000-1300 MMSL (Mean Sea Level).

- In India mostly short day types are grown.
- Garlic should be planted early to promote vegetative growth under short day condition and cool temperature.
- Critical day length for bulbing is 12 hours.
- Expose of dormant cloves or young plants around 20° C or lower for 1-2 months hastens subsequent bulbs.
- Garlic requires medium black, fertile loamy soil with good content of potash.
- Sandy or loose soil results in Bulbs lighter in weight.
- Highly alkaline and saline soils are not suitable for its cultivation.

Recommended Varieties of Garlic

S. N.	Variety	Features
1	Yamuna Safed (G-1)	It is a selection from a local collection from Delhi. The Bulbs are compact with attractive white creamy flesh sickle shaped. 25-30 cloves/plant. Tolerant to insect pest and disease like thrips, purple blotch and stenophyllium blight, TSS 38%, dry matter 39.5%. Maturity 150-160 days. Yield 150-175 q/ha. Recommendation for cultivation in all India.
2	Yamuna Safed-2 (G-50)	It is a local collection from Karnal. Bulbs diameter 3.5-4.0 cm suitable for North India, tolerant to purple blotch and steno-phyllium disease. The bulbs are compact with attractive white creamy flesh. Number of cloves 35-40 / plant. TSS 38-40%. Maturity 150-160 days. Yield 150-200q/ha. with dry matter % of 40-41.

3	G-282	This is a local variety from Dindigul (Tamil Nadu). The leaves are wider than other varieties, bulbs creamy white and bigger sized 5-6 cm having 15-16 cloves/plant. Dry matter 39-43%, TSS 39-43%. Maturity 125-130 days. Yield 175-200 q/ha. It has a medium storability. Suitable for export.
4	Agrifound White	First garlic variety released by central variety release committee for growing in western, southern and central parts, TSS 41%, dry matter 43% bulb diameter 3.5-4.5 cm, Number of cloves/bulb is 20-25. Selection from a local cultivar of Bharsharif of Bihar. Bulbs are silvery white. Recommended for Rabi in areas free from purple blotch and stenophyllium blight.
5	Agrifound Parvati	It is a selection from a local variety from Hong Kong. Bulbs are white with pinkish tinge. Tolerant to common diseases, medium storability. Long day type, as such suitable for North Hilly areas, Bulbs bigger in size 5-6 cm in diameter, 10-16 cloves/bulbs, suitable for export.
6	Godavari	Selection from Jamnagar collection. Bulbs pink coloured medium size. Tolerant to mites. Duration 130-140 days.
7	Sweta	Selection from Gujarat collection. Bulbs white duration 120-130 days.
8	Ooty 1	Bulbs bigger in size, dull white, duration 120-130 days.

■ Other varieties –

- G-51 – Tolerant to purple blotch and stenophyllium disease.
- Godavari

- Sweta
- HG-1
- HG-6
- Pusa Sel. 10 : IARI New Delhi
- LCC 1 – PAU Ludhiana
- ARU 52- UPKAS, ALMORA
- Other local varieties – Madrasi, Tabiti, T56-4 and Jamnagar Fawari and Rajalle Gaddi grown in Bellary District.
- Varieties released from NHRDF.

  Agrifound White,

  Yamuna safed (G1)

  Yamuna Safed – 2 (G50)
- All these varieties are small bulb size and have number 20-30 of smaller cloves.
- G 282 and Agrifound Parvati – Bigger bulb and bigger cloves.
- Propagation – By cloves, some times bulbils are also used.
- Spacing – 15 × 10 cm – Bigger size cloves.

  10 × 7.5 cm – Small size cloves.
- Planting time – September – November – Main season

  March – April – Hills
- About 500 kg cloves of 8-10 mm diameter are required to plant in one hectare.
- Excessive 'N' results in thick neck and sprouting before harvest.
- Borax upto 10kg/ha. increases bulb size and yield.
- $MnSO_4$ 0.1%, boric acid 0.2%, $CuSO_4$ 0.02% and

$ZnSO_4$ 0.02% stimulate dry matter accumulation in cloves.

- Garlic is produced only in one season – Winter (Rabi).

  Harvesting – March-April in Plains.

  July-Sep. in South Hills and North Hills.
- Close spacing cause purple blotch disease.
- In India, harvesting is done manually by hand digger.
- Yield – 100-200q./ha.
- 25-50% losses take place in garlic if it is not stored properly.
- Curing is done to remove excess moisture and to allow the bulbs to become compact and go into dormancy.
- Storage – 0.5-2.2^0 C temperature and 60-70% relative humidity.
- Storage life is prolonged and loss in weight is reduced by spraying 3,000 ppm M.H on crop before harvest.
- For export quality 40-60 mm size diameter and 10-15 cloves in each bulb are required.
- Longer day length – Higher yield.
- Physiological disorder – 1. Sprouting

  2. Rubberization – Nilgiri hills of Tamil Nadu
- Storage life of garlic is prolonged and weight loss is reduced by spraying MH 3000 ppm, 3 weeks before harvest.
- Irradiation with 6 Krad of cobalt 60 gamma rays has been recommended for successful storage of garlic.
- Garlic is a condiment crop.

"LEEK"

- Non-bulb forming tetraploid plant of onion family.
- Grown only in kitchen garden in India.
- Used as salad and in flavouring the soups.

■ Varieties –

1. PPL–1 : Selection, Good replacement for green onion.
2. London Flag.
3. American Flag.

Seed rate – 5-6 Kg/ha. in nursery during September-October.

- Blanching is an important cultural operation in leek cultivation which improves the quality of crop.
- Yield – 250-300q./ha.

6

LEGUMES (PEAS AND BEANS)

Common	B.N.	Family	2x	Origin Name
Pea	*Pisum Sativum*	Leguminaceace	14	Central Asia, Abessina, Meditteranean
Cowpea	*Vigna unguiculata*	Leguminaceace	22	Africa
Cluster Bean	*Cyamopsis tetragonoloba*	Leguminaceace	14	India / Africa
French Bean	*Phaseolus Vulgaris*	Leguminaceace	22	S. Mexico and Cent. America
Lima Bean	*Phaseolus Lunatus*	Leguminaceace	—	Guate mala
Broad Bean	*Vicia faba*	Leguminaceace	12	Central Asia, Med. South America, Europe
Indian bean/ Lab-Lab Bean	*Dolichos lablab /Dolichos purpureus*	Leguminaceace	22, 24, 20	India
Winged Bean	*Psophocarpus tetragonolobus*	Leguminaceace	18, 22	Madagascar in Africa
Soyabean	*Glycine Max*	Leguminaceace	22	Eastern half of N. China
Jack Bean	*Canavalia ensiformis*	Leguminaceace	22	Mexico, C. America
Sword Bean	*C. gladiata*	Leguminaceace	22	Tropical Asia

"PEA"

- In India, pea is grown mainly as a winter vegetable in the plains of North India and as a summer vegetable in Hills.
- Rich source of digestible proteins (7%) along with carbohydrates and proteins and minerals like Ca and P.
- It is used as a fresh vegetable or in soup, canned, processed or dehydrated.
- Average yield in India – 6.12 t./ha.
- Average yield of world – 6.1 t./ha.
- *Pisum elatius* – Wild species or ancestor of P*isum sativum.*
- Pea is a diploid (n = 7, 2n = 14).
- Two sub species :

  Pisum arvense – Field pea, coloured flower.

  Pisum sativum – Garden pea, sweet pea, horticultural pea.
- *P. arvense* is used as pulse, seed round or little angular, hard and whitish, plants are very hardy, resistant to drough and frost.
- *P. sativum* – Used for vegetable (table purpose) and canning purpose, seeds wrinkle, sweet and green, tender pods are used.
- Grown as early as the stone age.
- Harvested organs : Mature or immature seeds and/or young, tender pods grown either for the fresh market or processed for canning or freezing.
- Edible-podded peas sometimes referred to as a separate variety (*P. sativum* var. *macrocarpum*) (e.g. snow and snap peas).
- The field pea, *P. sativum* var. *arvense*, is grown for drying, green manure and as a forage crop.

- The anti-nutrient factor in pea is cyanide containing compounds.
- Pea is a self-pollinated crop.
- Green pea – 7.2 g protein and dry pea – 19.7 g/100g.
- Pea is a herbacious winter annual.
- Pea protein – 25% amino acid and sugar 12%.
- *Pisum sativum var – hortense* contribute 67% of total production.
- Varieties differ for maturity (60 - 75 days), pod and seed colour (light and dark green), plant height tall (indeterminate) and short (determinate), wrinkled (sweet) or smooth (starchy) seed surface.
- Gregor Mendel conducted experiments with pea plants, including analysis of the inheritance of traits such as wrinkled and smooth seed, led to the science of genetics.
- Direct seeded sowing is done.

Recommended Varieties of Peas

S. N.	Variety	Features
1	Arkel	Introduced from UK, dwarf (50cm), pods dark green about 8 cm long, slightly curved and well filled, 100 seed weight 17g, in early sowing escapes the attack of powdery mildew but in late planting highly susceptible to powdery mildew, also susceptible to collar rot, Grains very sweet and wrinkled when dry. Suitable for sowing in mid October. Maturity 55-60 days. Yield 100-125 q/ha.
2	Early Badger	Early dwarf, wrinkle seeded variety, pods pale green, 7.5 cm long pods. Suitable for early October sowing. Maturity 60-65 days. Yield 50-60 q/ha.

3	Meteor	Plant dwarf 5-60 cm tall, round seeded variety suitable for early October sowing, maturity 55-60 days. Yield 90-100 q/ha.
4	Pusa Pragati	Early, powdery mildew resistant variety. Average pod length 10 cm with 9 seeds per pod. Maturity 60-65 days. Yield 67 q/ha.
5	Jawahar Matar-4	Plant height 70 cm green pods, stem and foliage, pods well filled mature seeds wrinkled, green. Maturity 65-70 days. Yield 60-70 q/ha.
6	Early December	Plants bushy 70 cm tall with white flowers. Pods round to oval shaped, 5 cm long, 1 cm broad and five seeds per pod. Maturity 65-70 days. Yield 60-65 q/ha.
7	PM-2	Plant height 50-60 cm. Fruits setting starts from 6 node. Pods green with 6 seeds/pod and seeds wrinkled when dry. Escapes powdery mildew. Maturity 55-60 days. Yield 100-120 q/ha.
8	Arka Ajit	A mid season variety. Pods 8-9 cm long; seeds bold green, sweet shelling percentage 55, resistant to both powdery mildew and rust, duration 90 days and yield 10 t/ha. It is recommended for UP, Rajasthan and Karnataka.
9	Arka Sampoorna	The plants are bushy erect. Pods are flat, perchmentless, crisp sweet, resistant to powdery mildew and rust. Duration 80 days and yield 8 t/ha.
10	Arka Karthik	Fruits bushy erect and resistant to both powdery mildew and rust. Pods are long (11-12 cm) with 8-10 green, sweet seeds. Crop duration 90 days. Pod yield 11 t/ha.
11	Kashi Nandini	This is the earliest variety in the seed chain of the country and grown in the northern

		plains (25 October – 25 November), Plants erect (60 cm), pods 8-9 cm long. Shelling quality 47-48%, duration 55 days. Yield 110-120 q/ha. Tolerant to leaf miner and pod borer. Suitable for Punjab, Uttrakhand, UP, Bihar, J&K, Karnataka, Tamilnadu and Kerala.
12	Kashi Udal	Early maturing, 60-65 days duration, yield 110 q/ha.
13	Kashi Mukti	Early maturing 60-65 days duration, yield 110-120 q/ha.
14	Kashi Shakti	High yielding medium maturing sown in November in plains shelling quality 48-49%, yield 100-120 q/ha.

■ Crop improvement –

- Arka Ajit was developed by IIHR by crossing Bonneville to powdery mildew and rust resistant lines and then again back crossing to Bonneville.
- Though 25% heterosis is reported for yield in pea, F1 hybrid seed production is almost restricted because of the high cost involved in manual emasculation; pollination and availability of less number of seeds per pod unlike the fruit of solanaceous crops.

Classification of varieties according to maturity –

1. Early Maturing

a. Arkel

b. Asauji – Smooth seeded variety.

c. Alaska – First early canning variety.

d. Meteor

e. Early Giant – Suitable for hilly areas, seeds sweet in taste.

f. Harbhajan – Extra early variety, first picking 50 days, 100 seed weight 12 g., highly susceptible to powdery mildew.

g. Jawahar Matar-3 : 100 seed weight is 18g, susceptible to powdery mildew but may escape due to earliness.

h. Jawahar Matar-4 : Susceptible to powdery mildew but may escape due to earliness.

i. Jawahar Pea-54 : (JP-54) Double cross of (Arkel × JM-5) × (46C × JP-501), resistant to powdery mildew.

j. Kanwari – Tall, double poded variety, a mucilage excretion on the pod is a characteristic feature of this variety, largely grown in hills and plains.

k. Little Marvel – Introduced from England (Chelsea x Sutlon Alaska) dwarf variety.

l. Matar Ageta – 6 Long pods, 12-15 in number, 50% pods can be harvested in first picking.

m. Pant Matar-2 – This variety suffer badly due to powdery mildew attack.

n. Pant Sabzi Matar-3 – Arkel × GC 141, susceptible to powdery mildew.

o. VL Ageta Matar- by VPKAS, Almora (Pant uphar × Arkel), free from the incidence of powdery mildew.

p. Early Superb -- From England.

q. Early Badger – Good for canning of green pea, resistant to Fusarium wilt and tolerant to heat and drought.

r. Yates Early crop – Australian variety.

s. Kelvendon Wonder

t. Lucknow Boniya – First variety to come to Delhi market from the plains in the season.

2. **Main Season Variety –**

1. Bonneville – Highly susceptible to powdery mildew disease.

2. Jawahar Matar-1 :– Bigger pod, highly susceptible to powdery mildew.

3. Jawahar Matar-2 – Greater progress × Russian 2, Highly susceptible to powdery mildew disease

4. Jawahar Pea 83 (JP-83) – (Arkel × JP-829) × (46C × JP–501), resistant to powdery mildew.
5. Jawahar Matar 71 - (Arkel × SP-827) × (JP 501 × JM1), resistant to powdery mildew.
6. Pant Uphar (IP-3) – Highly susceptible to powdery mildew disease.
7. P-88 – Highly susceptible to powdery mildew.
8. VL-3* – Susceptible to powdery mildew, resistant to wilt.
9. Lincon – Better suited for hilly regions of North India also suitable for late sowing.
10. JP-19- Whole pod are consumed, no need of removing the peelings. This quality of green pod is due to lack of sclerenchymatous tissues. This character of pod not only saves yield loss but also the burden of house wives.
11. JP4 (JM6) - Resistant to powdery mildew.
12. FC-1 - Resistant to powdery mildew.
13. PRS-4- Moderately resistant to powdery mildew.
14. Arka Ajit- Resistant to powdery mildew and rust.
15. UN 53 (6) – Snap pea (Whole pod edible).
16. Ooty 1- Pure line selection, resistant to white flies.
17. Sylvia – Whole pod edible.
18. Perfection New Line.
19. Thomas Laxtan – Most suitable for freezing.
20. Alderman – Most suitable for kitchen garden.
21. Palam Priya – Tolerant to powdery mildew.
22. Solan Nirog – Resistant to powdery mildew.
23. Duke of Albany – Also known as Telephone, need stacking, produce Ist flower on 12-14th node.
24. Khaper Kheda – Popular cultivar of M.P.
25. Delwiche Commando – Resistant to wilt.
26. N.P 29 - Suitable for dehydration purpose.

- Optimum temperature for germination – 22° C.
- Soaking seeds in GA_3 (10 ppm) for 12 hours gives highest yield.
- Hot weather interferes with pollination and seed setting, affect the number of pod/plant, pod weight and lowers the quality of pod produced due to the conversion of sugars into hemicellulose and starch.
- **Sowing time-**

  October - November – North Indian Plains.

  March – Hills for Summer crop.

  May – Hills for Autumn crop.
- **Seed rate –** 100-120kg. : Early.

  80-90kg. : Mid and late.
- Phosphatic fertilizer enhances yield and quality of pods.
- Prolonged application of DAP causes sulphur deficiency.
- Gypsum or pyrite – Complete the deficiency of sulphar.
- Excess moisture causes - Yellowing of crop.
- Pea sown earlier or later suffers from wilt or powdery mildew attack.
- Critical stage of irrigation is flowering, fruit setting and grain filling.
- Foliar application of 0.1% ammonium molybdate at the time of flowering showed favourable effect on yield and quality of peas.
- Spraying 15 ml of 10 M solution of growth regulators CCC at the five node stage of development has favourable effect on the growth and yield.

  Harvesting – Early (50 Days after sowing).

  Mid (60-65 Days after sowing).

  Late (70-75 Days after sowing).

- In processing industry the maturing of pea is tested with the help of tenderometer. If harvest is delayed, surface become coarse which brings down the market value of product due to the conversion of sugar into starch, the quality deteriorates especially when the temperature is very high.

  Yield – Early 2.5-4.0 t/ha.
Mid 6-7.5 t/ha.
Late 8-10 t/ha. } Green pod.

- The shelling percent ranges from 35-50%.
- Maturity – Measured by Tenderometer (tenderness).
- Storage 0° C with 90-95% relative humidity.
- Marsh spot – Due to Mn deficiency.
- *Rhizobium leguminosarum* – Application advisable to reduce the quantity of nitrogenous fertilizers.
- Major diseases of pea: *Fusarium* wilt and root rot.
- Major insect pests of pea are :
  - loopers (contaminant).
  - alfalfa caterpillars (contaminant).
  - stink bugs (contaminant).
  - ground beetles (contaminant).
  - pea aphids (virus vectors)

"COWPEA" (Southern pea, Black eye pea)

- Used for tender pods and dry seeds used as pulse for culinary purpose.
- Pods are rich in protein, vitamin B and minerals.
- It is used as fodder or green manure crop.
- Most probable progenitor of cowpea is *var menensii.*
- Cowpea belongs to sub-family fabaceae.

- Flowers are white coloured.
  1. Common Cowpea – 20-30 cm long pods and small seed size.
  2. Catjang bean (*Vigna catjang*) – 7-13 cm long pod, 5-6 mm kidney shaped seeds.
  3. Asparagus bean or yard long bean (*Vigna sesquipedalis*) – 30-90 cm fleshy pod / yard long bean with elongated kidney shaped 8-12 mm long seeds.
- It is grown as catch crop, drought tolerant and sensitive to water logging.

Recommended Varieties of Vegetable Cowpea

S. N.	Variety	Features
1	Pusa Phalguni	Bushy dwarf variety suitable for sowing in February-March in the northern plains. Pods dark green, erect and 10-20 cm long. Maturity 60 days (first picking). Yield 70-75 q/ha.
2	Pusa Barsati	Suitable for growing during rainy season. Pods light green, 26-28 cm long and pendant. Maturity 45 days (first picking). Yield 70-75 q/ha.
3	Pusa Dofasli	Bushy suitable for growing both in spring-summer and rainy seasons. Pods light green and about 18 cm long. Maturity 45-50 days (first picking). Yield 75-80 q/ha. It is a photo-insensitive variety, developed by IARI, New Delhi.
4	Pusa Komal	Bushy, suitable for growing during spring-summer and rainy seasons. Pods light green, thick, fleshy and 20-22 cm long.

		Resistant to bacterial blight disease. Yield 120-120 q/ha. It is a photo-insensitive variety developed by IARI, New Delhi.
5	Pusa Rituraj	Bushy, 40-50 days for first picking, light green pods, seeds brown, Photo-insensitive variety.
6	Cowpea-263	Bushy, dwarf, medium green pods, 45-50 days for first picking, developed by PAU, Ludhiana, tolerant to moasaic virus.
7	C-152	Resistant to root knot nematode, tall, erect plants, 90-100 days for maturity, oblong light brown seeds. Developed by IARI, New Delhi.
8	Narendra Lobia 1	Bushy, 45-48 days for first picking, green pods and bold seeds.
9	Kashi Shyamal	Dwarf bushy variety, suitable for both kharif and rabi season. Yield 48 days after sowing. Pods yield 70-80 q/ha. and the plants are tolerant to golden mosaic virus, recommended for Punjab, UP, Bihar and Jharkhand.
10	Kashi Gauri	Dwarf, bushy type, photoinsensitive, suitable for spring summer and rainy season. Resistant to golden mosaic virus and leaf spot diseases. Yield 10 -12 t/ha. Green pods. Duration 40-48 days.
11	Arka Green	Plant are tall, vigorous, bushy with small vines and photo insensitive. Leaf colour light green, flowers purple, pods light green, long thick, fleshy and stringless. Suitable for vegetable purpose. Tolerant to heat and moisture stress. Duration 90 days. Pods yield 18 t/ha. Recommended for HP, Maharastra, Karnataka, Kerala.

12	Arka Suman	Plants, erect bushy, photo-insensitive, pods medium long, fleshy, crisp without parchment. Duration 70-75 days, pod yield 18 t/ha. Recommended for Karnataka.
13	Arka Samrudhi	Plant erect, bushy, photo-insensitive, early with pods above the canopy. Pods green, medium thick, medium long, round fleshy without parchment. Good cooking qualities. Duration 70-75 days. Yield 19 t/ha. Recommended for UP, Bihar, MP and Maharastra.

■ Five distinct sps. of cowpea :

A. Dekinotiana } Wild
B. Mensenensis } Wild
C. Ungiculata } Cultivated in India
D. Cylindrica } Cultivated in India
E. Sesquipedalis } Cultivated in India

■ Crop Improvement

1. Introduction –

(1) Phillippines Early – Trailing growth habit.

(2) Yard Long Bean- Species *sesquipedalis*, late maturing variety.

2. Selection

a) Pusa Barsati – Early variety, Suitable for rainy season.

b) Pusa Phalguni – Dwarf variety, most suitable for February – March (Spring).

c) Arka Garima- Bushy, pole type and photo-insensitive, evolved by back cross (*unguiculata* x *sesquipedalis*), more suitable for rainy season.

d) Arka Samrudhi – Bushy, photo-insensitive.

e) Arka Suman – Bushy, photo-insensitive.

3. Hybrid

a) Pusa Dofasli – Pusa Phalguni x Phillipines, Bushy suitable for both summar and rainy season, photo insensitive.

b) Pusa Rituraj – Bushy and prolific bearer, being highly photo thermo insensitive, can be grown during rainy as well as summer season, suitable for dual purpose.

c) Pusa Komal – Pusa Rituraj × P-246, dwarf and bushy, Resistant to bacterial blight, Suitable for sowing summar as well as rainy season.

d) Pusa Sukomal

■ Other Varieties

a) Cowpea 263 – Suitable for rainy as well as summer season, comparatively resistant to mosaic and free from bean mosaic virus.

b) Narendra Lobia – Being photo insensitive suitable for rainy as well as summer season.

c) Sel. 1552 – Resistant to bacterial blight.

d) Bidhan Barbati 1 – Resistant to cowpea mosaic and cowpea golden mosaic virus.

e) Bidhan Barbati 2 – Semi-determinate.

f) IIHR 220 – Resistant to rust.

- Frost is harmful to this crop.
- Germination is better at 12-15° C.

Seed rate – Rainy Season – 12-15 kg/ha.
Summar Season – 20-25 kg/ha

Sowing time – Rainy Season – June-July
Summer season – February - March

Spacing – 45-60 RXR : Rainy Season
15 cm PXP : Rainy Season
30-30 cm RXR : Summer season
15 cm P × P : Summer season

- Cowpea has strong tap root system.
- The pole type varieties need support; these varieties are often restricted to kitchen garden.
- Spraying of MH (50-200 ppm) before flowering improves pod set and yield.
- Picking of pods 15 days after pod setting is ideal for vegetable purpose.
- Compared with other picking stage the pod at this stage are longer and less fibrous with more protein content and are more succulent.
- Yield – 50-80 q/ha. Green pod.
  12-15 q/ha. Seed yield.

"CLUSTER BEAN"

- Grown for green pod as vegetable, dry seed, forage and green manure. Some varieties are used for extraction of gum.
- Diploid 2n = 2x = 14.
- *C. senegalenis* – Ancestor of guar.
- Guar is one of the hardest among legume vegetables.
- Seed contain gum like mucilaginous substance called "guar gum" or "galactomannon" which is used in textiles, paper industry and cosmetic industry.
- Green pods are rich in Vitamin A and Iron.

Recommended Varieties of Cluster Bean

S. N.	Variety	Features
1	Pusa Mausami	Pods attractive smooth bright green and 10-12 cm long. First picking in 65-80 days after sowing. Suitable for rainy seaon/ kharif season.
2	Pusa Sadabahar	Selection from local variety "Jaipur" of Rajasthan, Non-branching type. Pods green and 12-13 cm long, tender and fiberless. First picking in 45-55 days after sowing. Suitable for both summer and rainy seasons.
3	Pusa Navbahar	Pusa Mausami x Pusa Sadabahar, Pods 15 cm long and of better quality. Non-branching type. Suitable for both summer and kharif seasons, susceptible to bacterial blight and lodging.
4	P-28-1-1	Smooth green and long pods. A photo insensitive variety. Flowers in 30-32 days in summer and 40 days in kharif. Suitable for both summer and kharif seasons.
5	P28 1	Developed by NBPGR, Pusa Navbahar × IC11521, suitable for both rainy and summer, photosensitive.
6	Sharad Bahar	Developed through S.P. selection, profused branching, single plant produce about 133 pods.
7	Giant type	Good quality seeds with high content of guar gum and grown in Gujarat.
8	Dwarf type	Good quality pod which can be used as vegetable, Grown in Punjab and Uttar Pradesh.

■ Varieties –

Seed rate : 25-30 kg/ha

Sowing Time : June-July – Main Season.
February-March – Early crop.

"FRENCH BEAN" (Rajmah, Farash bean)

- It is also known as kidney bean, haricot bean, snap bean, navy bean, string bean, garden bean.
- Three types of French bean are there.
  1. Bush type with short internodes.
  2. Semi pole type with longer internodes.
  3. Pole type – Longer internode than semipole type.
- Edible portion is 94% of the pods which contains 1.7% protein, 0.1% carbohydrates, 4.5% fat, 1.8% fibre and 0.5 mineral.
- Dried beans are rich in protein and closely compared with meat.
- In India grown for tenere vegetable while in USA grown for processing in large quantities.
- *Phaseolus oborigineus* – Progenitor of French bean.
- Diploid 2n = 2 X = 22.

  Four cultivated species.

  1. *P. vulgaris* - Common red kidney bean
  2. *P. coccineus* - Scarlet runner bean
  3. *P. lanatus* - Lima bean
  4. *P. acutifolius* - Latifolius-Moth bean
- All the species are self-pollinated except *P. coccineus* which is generally cross-pollinated.

Recommended Varieties of French Bean

S.N.	Variety	Features
1	Giant Stringless	A bushy variety, pods green, long, slightly curved, meaty and stringless.
2	Contender	Plants bushy bearing light green, fleshy oval and thick pods. High yielder than Giant stringless in the plains. Maturity 50-55 days (first picking). Yield 80-95 q/ha. Resistant to powdery mildew.
3	Pusa Parvati	A bushy variety bearing light green round and meaty pods. Maturity 45 days (first picking). Yield 80-85 q/ha. Resistant to powdery mildew.
4	Tender Green	Bushy type bears light green, round and oval pods.
5	Kentucky Wonder	A pole type with green, fleshy round and slightly curved pods, Maturity 60-65 days (first picking). Yield 100-120 q/ha.
6	Pusa Himlata	A Pole type pods medium long, round straight light green meaty and stringless. Seeds white, Maturity 60 days (first picking) .
7	Kashi Param	Determinate plants. Yield 120-140 q/ha. Suitable for J&K, HP, Uttrakhand, MP, Maharastra.
8	Arka Komal	Has a green, fleshy, tender long and straight pod. Produces 200-250q/ha. of green pods in 70-75 days. A very popular variety developed at IIHR, Bangalore.
9	Arka Anoop	A pedigree selection from Arka Bold × Arka Komal Plants bushy photonsensitive, resistant ot both rust and bacterial blight. Pods long (17-18 cm) flat and straight. Duration 70 days and yield 20 t/ha.

10	Jampa	A Mexican variety reported to be outstanding in performance in Maharashtra. The plant has a slight tillering habit. It flowers in about 47 days and the first picking is obtained after two months of sowing. In all, eight pickings within 40 days are obtained. The pod is flat, smooth, non stringy and pale green. The seeds are black, smooth small and flat. It is resistant to wilt.
11	Pant Anupam	Selection from exotic germplasm. Pods medium long, round, straight green. It is resistant to rust, common bean mosaic and agriculture spot. Developed by GBPUAT, Pantnagar.
12	Pant Bean 2	Pods flattish round, strinlgess, both for green pods and dry seeds, cross between Turkish Brown X Cotender. It is resistant to rust and common bean mosaic. Developed by GBPUAT, Pantnagar.
13	YCD 1	Selection from local type Shevroy hills. Pods slightly flat, green dual prupose developed by TNAU Coimbatore.
14	VL Boni 1	It is a rust resistant variety. Pods medium long, stringless, Pale green. Developed at Almora, UP.
15	Lakshmi	Hybridization between Contenders X Local Pole variety. Pods round, green stringless. It is resistant to angular leaf spot.

■ Varieties

(A) On the basis of fiber

(1) String type – Pod contain fiber, indigenous varieties are string.

(2) Stringless type – Pods are free from fibre. Most of improved varities are stringless type.

(B) On the basis of growth habit

(1) Pole type – Tall, Indeterminate in growth and requires support.

(2) Bush type – Plants dwarf and bushy, popular because of compactness, easy harvest and short duration.

1. Pole type –

1. Pusa Himlata
2. Kentucky wonder – Introduced from USA.
3. Lakshmi - Tolerant to angular leaf spot disease.
4. SVM 1 – Contender of (*P. vulgaris*) × (*P. multiflorus*), resistant to angular leaf spot.
5. VL Lata Bean 12
6. VL Lata Bean 17
7. Tweed wonder

2. Bush Type

1. Pusa Parvati – Developed through X-rays irridiation of an American variety, wax pod, resistant to mosaic and powdery mildew disease.
2. Pant Anupam – Early, resistant to angular leaf spot and moderately resistant to mosaic virus.
3. Arka Komal – Very good for transport and good keeping quality.
4. Arka Suvidha
5. IIHR 220- Resistant to rust, Pure line selection variety.
6. IIHR 909 – Contender x Blue crop.
7. Jampa (Maxican variety) – Highly resistant to wilt and withstand warmer conditions.
8. Contender – Early, tolerant to powdery mildew mosaic.
9. Premier – Introduced from Sweden, late sowing, susceptible to mosaic and wilt, suitable for hills.
10. Bountiful – USA introduced, pods borne on main stem in clusters.

11. Pant Bean – Moderately resistant to common virus.
12. VL Boni-1
13. TKD
14. KKL – For elevation of 1800-2400 m.
15. YED
16. Ooaty – Moderately resistant to leaf spot, anthracnose and pod borer.
17. Giant stringless.

- Most of French bean varieties are day neutrals except semi-pole varieties which are short day types.
- French bean is a warm season crop that can't tolerate frost/sensitive to frost. Optimum temperature range from – 15-25° C.
- Plants drop/shed their blossom or young pod in very hot or rainy season.
- Pole type – Generally grown in heavy rainfall areas of Chhotanagpur of Uttar Pradesh.
- It can't withstand water logging.

Seed rate –	50-75 kg – Bush type/ha
	25-30 kg - Pole type/ha
Sowing time –	July-September
	January-February – Plains
	March-April – Hills
	September-November – Southern India
	October – North-east and Central Zone.

Spacing –

30 × 5 cm	Bush type (2,50,000 plants / ha.)
90 × 7.5 cm	Pole type (80,000 plants / ha.)

- Foliar application of B, Cu, Mo, Zn, and Mg @ 0.1% – Enhance pod quality and yield.
- Seed should be inoculated with strains of *Rhizobium phaseoli* for quick nodulation and high yield.
- The seed should be treated with it's culture @ 30 g. sugar solution/kg. seed.

- It is sensitive to both water excess and water stress condition.
- Cultivation or working among plant, where foliage is wet should be avoided, because spores of anthracnose and angular leaf spot disease are easily dispersed under wet condition.
- In temperate climate, pole beans are grown in combination with corn or okra as the stalk is used for support of the vines.
- Application of GA_3 (50 ppm) on 30 days old plants is effective in increasing plant height, number of leaves and pod yield.
- 2-4 foliar spray of growth retardant CCC (400 ppm) reduces vegetative growth, causes early flowering and increases yield.
- Flooding in french bean cause anoxia leading to root rot.
- Maturity standard – AIS (Alcohol insoluble solids).
- Yield – 8-10 tonnes - Bush varieties/ha.
  12-15 tonnes - Pole varieties/ha.
- Storage 5-7° C for 20-25 days.
- In India bulk is consumed as fresh.
- In USA bulk is processed, more than 80% (White colour is preferred).

■ Physiological disorders –

1. Delayed Flower initiation and development– Occurs below 10° C where fertilization may not occur, producing small and misshaped pods.
2. Blossom drop and ovule abortion – High temperature (35° C) is the main cause.
3. Transverse cotylendon cracking (TVCC) – Occurs when dry seed of beans are sown in wet soils. Seed moisture content less than 12% enhance cracking of cotylendon.

White seeded cultivars are more prone to this malady.

Resistant cultivars – Show 60-80% germination.

Susceptible cultivars – Show 20% or less germination.

4. Necrosis of hypocotyl – Low "Ca" content in seed after germination and soil rich in Ca and Mg are affected by this problem.

"BROAD BEAN"

- Also known as faba bean, horse bean, bakla bean.
- Mainly cultivated in Latin America.
- It is the only bean which is grown as a winter crop and autumn crop.
- Pollinated by insects.
- The pollen grains and green pods cause allergy to some people which turn in serious illness known as "Favism" (Haemolytic anaemia).

Varieties –

[A] Based on seed colour –

a. White seeded varieties – Less soft.

b. Green seeded varieties – Soft and suitable for freezing.

[B] On the basis of pod size –

1. Long pod -

   a. Red Epicure.

   b. Imperial White long pod.

   c. Imperial Green long pod.

   d. Master piece Green long pod.

   e. Aquadule Claudin.

2. Small poded -
   a. Imperial White Windsore.
   b. Gaint Four Seeded Green Windsor.
   c. Imperial Green Windsor.

- It can withstand temperature as low as 4° C, mainly grown at higher elevation.
- Seed rate : 70-100 kg./ha.
- Yield – 7-10 tons/ha.
- Pusa Sumeet – Improved variety of Broad Bean, developed at IARI, New Delhi.

"INDIAN / DOLICHOS BEAN" (SEM)

- Also known as hyacinth bean, lablab bean or field bean.
- Crop is multipurpose, can be used as pulse vegetable and green manure.
- From India it was introduced to China, West Asia and Egypt.
- The term *Dolichos* is derived from Greek language means "Long pod" and lablab is an Arabic term means "Production of sound by dried seeds in the pods".
- Trypsin Inhibitor is present in mature and dark coloured seeds.
- On heating it is broken down into toxic cyanogenic glucoside which is soluble in boiling water. Therefore it is necessary to boil the mature seeds in water before consumption.
- Sem is of two types –

  (A) *Dolichos purpureus var- typicus* (Garden type)

  Soft edible, pod wall have less fibre.

(B) *Dolichos purpureus var – Lignosus* (Field bean). Grown for fresh dry seed as pulse, pod wall have high fibre content and have characteristics aroma.

Recommended Varieties of Dolichos Bean

S.N.	Variety	Features
1	Arka Jay	Developed by backcross and pedigree selection from the cross involving Hebble Avare and IIHR 93. Plants dwarf, bush type, erect and photo-insensitive. Flowers are purple and pods long, light green, slightly curved without parchment. It is a vegetable type with excellent cooking quality. Resistant to moisture strees and heat. Crop duration is 75 days and yield 12 t/ha. Recommended for Karnataka.
2	Arka Vijay	Plants dwarf, bushy erect, photo insensitive, dual purpose, pods with characteristic aroma, with parchment, tolerant to heat and drought with excellent cooking qualities, light aroma in pod. Developed by back cross by using herbal Avare -3 and Pusa Early Prolific, Duration 75 days and pod yield 12 t/ha. Recommended for Karnataka.
3	Pura Early Prolific	It is a pole type early variety. Light green pods. Suitable for autumn and spring. Duration 200-215 days. Yield 14 t/ha. Developed by IARI, New Delhi.
4	Pusa Sem-2	Pods dark green, first picking in 120-125 days. Highly tolerant to antracnose and viruses, insect pest, pod borer, June-July is the sowing time. Yield 15 t/ha.
5	Pusa Sem-3	Pods green, 15 cm long, first picking in 115-120 days. Highly tolerant to antracnose

		and viruses, insect pest, pod borer, June-July is the sowing time. Yield 17 t/ha.
6	Blue Lake	It is early variety, bushy, photo insensitive, dual type, spreading white flowers, 70-80 days for harvesting.
7	KonkanBhusan	Green and tubular pods, 100-180 days. Yield 6-8 t/ha.
8	Dasarawual	Green with purple tinge. Yield 8 t/ha.

■ Other Varieties –

1. Herbal Avare 3 – Determinate growth habit, photo insensitive pods are not edible but can be used in hybridization/ programme for incorporation of dwarf character.
2. Wal Konkon – Resistant to yellow mosaic virus.
3. Kalyanpur Type – 2
4. Jawahar Sem-37, 53, 79, 85 - For low rainfall areas.
5. Arka Jay and Arka Vijay – Can be grown throughout the year.
6. Climbing – Rajni, Deepali, KDB – 403, 405, 71
7. Pusa Early Rajni – An improved variety developed at IARI, New Delhi.
8. CO 10 – Mutant Variety

Seed rate – 10-12 kg – Climbing type.
20-30 kg – Bushy type.

Sowing time – June – July – North India.
July – August – South India.

- In North and Central India it is usually grown as mixed crop with ragi or sorghum.
- Yield – 3-5 tonnes/ha.

"LIMA BEAN"

- Also known as double bean or butter bean.
- It is also used to prepare wine.
- It is cross compatible with French bean.

■ Varieties –

1. **Pole Type –**

a. Florida Butter.	b. Challanger.
c. King of the Garden.	d. Karolina Butter.
e. KKI 1 – Butter bean.	

2. **Semi pole type –**

a. Hopi.	b. Wilbur.

3. **Bush Type –**

a. Handerson Bush.	b. Fordhook 242.
c. Baby Potato.	d. Baby Fordhook.

Sowing Time - June-July.

Seed Rate -	20-35kg/ha	Small Seeded.
	45-50kg/ha	Large Seeded.

Yield – 5-10 Tonnes/ha.

"WINGED BEAN"

(Asparagus Bean, Multiflorus Bean)

- Also known as four angled bean, Goa bean, Manila bean, Princess bean.
- It is a tropical legume of high value.

Pod	1.2-2.9% Protein
Leaves	6-15% Protein

- Pod show four sides wings protruding from angles hence the name winged bean.
- It is photosensitive and requires short day for normal flower induction.
- It is a diploid 2n = 18, 22.
- Mainly grown in Tamilnadu, Kerala, Karnataka, Goa, Orissa, Maharashtra, West Bengal, and North Eastern States.
- Multipurpose legume, every part of plant like green pods, seed, leaves, flower and tuberous roots are edible.
- Main breeding object – Development of varieties for photoinsensitivity.

■ **Varieties** – IIHR sel, 20, 60, 71

WBG2 – North East Parvatiya Anusandhan Kendra, Meghalaya

"SCARLET RUNNER BEAN" (*Phascolus coccineus)*

- It is L.D.P. (Long Day Plant).
- Consists of Protein (17%), carbohydrates (65%) and 2% edible oil.

"JACK BEAN" (*Canavalia ensiformis)*

- Seeds are utilized to produce urease enzyme.

"SWORD BEAN" (*Canavalia gladiata*)

- Seed contain some toxic substance like saponins therefore it is available to use the seeds after cooking.

7

LEAFY AND SALAD VEGETABLES

"PALAK/SPINACH BEET/ BEET LEAF"

- It is mostly grown in Uttar Pradesh, West Bengal, Maharashtra and Gujarat.
- It is primarily used as pothers.
- It is a rich source of vitamin A and vitamin C (70 mg/ 100 g.)
- It is rich in nutrients 3.4 g protein, 0.8 g fat, 380 mg Ca, 30 mg P and 16.2 mg Fe.)
- Spinach beet, swiss chored and garden beet has 9x = 2x = 18.
- Leaves of this crop might have been first used in Bengal and hence it is known as *B. valgaris var, Beg.*
- Palak leaves contain low oxalic acid.

C.N.	B.N.	Family	Ch. Number	Origin	Seed rate	Sowing Time
Palak/Beat leaf/ Spinach beet	*B. Vulgaris variety bengalensis*	Chemopodieae	18	India-China	10-15 kg winter 20-30 kg summer	August - November Plains March-May Hills
Spinach	*Spinacia Oleracea*	Chenopodiaceae	12	Central Asia (Iran)	37-45 kg/ha. winter	September-October August - October Hills
Amaranthus	*Amaranthus Sp.*	Amaranathus	32	India	1.5-2.0 kg/ha. summer 1 kg transplant	March July-Kharif
Lettuce (salad crop)	*Lettuca Sativa*	Compositae	18	Med. Region	400-500g (50-75m.area)	September-November
Basella (Poi) / Indian/ Malabar Spinach	*B. rubra var –alba green leaves* *B. rubra var-rubra red leaves*	Basellaceae	24	South Asia (India)	12-15kg/ha	March-May Plains Mar-April-Hills October-November Southern
Potulaca	*Portulaca oleracea*	Portulacacea	54			
Parsley	*Petroselinum hartease*	Umbelliferac	22	E. Europe	250-300g/ha.	February-April
Agathi	*Sesbania graniflora*	Leguminacea	24	India	7-8 kg/ha	Rainy season
Water leaf/ kalmisag	*Ipomoea acuatica*	Convolvulaceae	10	E&S Asia	60-100 kg/ha	April and May
New Zealand Spinach	*Tetragonia expanse*	Aizoaceae				
Chicory	*Chichorsum intyless*	Asteraceae	18	Europe, Australia	3-4 kg/ha	March
Sorrel	*Runnex vesicarius*	Polygonaceae	18	Sudan		Any time

Difference between spinach beet and spinach –

S.No.	Attributes	Spinach Beet	Spinach
1.	B.N.	*B. vulgaris var-beg.*	*Spinach oleracea*
2.	Ch.	2x = 18	2x = 12
3.	Sex forms	Bisexual	Monoecious, Dioecious Hermaphrodite
4.	Leaves	Have entire margin	Have lobed margin
5.	Seed	Monogerm	Multigerms

Crop Improvement - Mass selection and hybridization.

Recommended Varieties of spinach beet

S. N.	Variety	Features
1	Pusa Bharati	Late variety, It has higher vitamin C content and β carotene than Pusa Jyoti, Leaves 25 cm long and 14 cm in breadth, smooth, tender, puer green without red pigmentation. Can also be grown during kharif and summer seasons. Performs best in winter and early spring seasons. Has high nutritive value. Maturity 30-40 days. Yield 500 q/ha. A variety released by IARI, New Delhi
2	Pusa Jyoti	A giant leafed strain produces large, green, thick, tender, succulent leaves, good taste when eaten raw as salad. Plants very vigorous, quick growing, regenerate quickly after each cutting. Gives 6-8 cuttings. Rich in calcium,

		iron and ascorbic acid, can be grown throughout the year. Maturity 40 days. Yield 490 q/ha.
3	Pusa Harit	Suited to hills, plants upright, vigorous, uniformly thick, green, slightly crinckled big size leaves, heavy yielder. Late bolting habit and wide range of adaptability to varying climate, tolerant to alkaline soil. Maturity 35-40 days. Yield 425 q/ha.
4	All Green	Produces uniform green, medium sized, tender leaves, yields about 123.5 q/ha of green material. As a winter crop, gives about 6-7 cuttings at 15-20 days interval. Maturity 40 days. Yield 120 q/ha. Developed by IARI, New Delhi.
5	Jobner Green	Produces uniform green, large, thick, green, succulent tender leaves, strong flavour. Maturity 40-45 days. Yield 300 q/ha. Evolved by RAU, Jobner.
6	Pusa Palak	Hybridization between Swiss chord × local Palak. Leaves uniform green with red pigmentation late bolting. Developed by IARI, New Delhi.
7	Pant Composite	Composite variety, leaves green, high yielding, yield tolerant to cercospora leaf spot. Yield 27.5 t/ha. Developed GBPUAT, Pantnagar.
8	Punjab Selection	Selection from local type. Leaves light green, thin, long narrow smooth, stem with purple pigmentation. Yield 37.5 t/ha. Developed by PAU, Ludhiana.
9	Punjab Green	Selection from local type. Yield 37.5 t/ha. Through selection, leaves free from sourness and succulent, mild purple pigment on stem, Low oxalic acid content which is desirable traits in greens. Developed by PAU, Ludhiana.

10	Palak No 51-16	Selection from local type. Leaves green, several cuttings, late bolting.
11	HS-13	Selection from a local type. Leaves light green, medium sized, quick growing. Developing by HAU, Hissar.
12	Arka Anupama	Late bolting variety with medium large, thick, dark green leaves. It has got excellent cooking quality and it is rich in vitamins and minerals. Four cuttings can be taken over a period of 65-70 days. It is resistant to cercospora leaf spot under epiphytotic condition. Yield 41 t/ha.

■ Other Varieties –

1. All Green – Late bolting.
2. Jobner Green – Evolved through spontaneous mutation detected from a local collection S.I. No. 5.
3. Pusa Palak – Hybridization, (swiss chord x local palak).
4. Pusa Jyoti – Developed by polyploidization, result of induce mutation using All Green as a source, very big leaves with dark green colour.
5. Pusa Harit – Hybridization (Sugar beet x Local palak) heavy yielder, late bolting, has low chilling requirement.
6. Punjab selection – Slightly sour taste.
7. Arka Anupama – Resistant to cercospora leaf spot, regenerate faster.
8. Banerjee Giant – Development by S.P. Banerjee through hybridization (Local Palak x Sugar beet).
9. Pant Composite – 1 – Tolerant to cercospora leaf spot.
10. HS-23 - Mass selection.
11. Palak no. 51-61

- Spinach beet is a winter season crop.
- It can tolerant frost better than other vegetable.

- It can also tolerant high temperature but causes premature bolting.
- Palak is tolerant to slighly alkaline soil and also highly tolerant to salts.
- Spacing 20 x 5 cm (RXR and PXP).
- Palak is a leafy vegetable which requires more "N" for crown growth, apply "N" @ 20-25 kg/ha after every cutting as top dressing.
- During summer – Only one harvesting is recommended.
- Washing of leaves is not advocated because rotting may take place at centre of bundle.
- Storage – 0° C temperature, 90-95% relative humidity for 10-14 days.

"SPINACH" (Vilayati Palak)

- It is a cool season vegetable and an annual crop.
- It is extensively cutivated in USA, Canada and Europe.
- Leaves are rich in Vitamin - A (5580 IU), Vitamin – C (28 mg). Folic acid (123mg), Iron (17.4mg), Ca (190mg) and P (42 mg) / 100g.
- Spinach is related to Swiss chord, sugar beet, table beet, pig weed and saltbeet.
- In China it was introduce from neighbouring hilly kingdom Nepal in 647 AD.
- Spinach is a long day crop/plant (LDP).
- This plant tolerates frost better than most other vegetables.
- This plant is sensitive to acidity.

In spinach 4 sex types exist – Tetramorphic.

1. Male – Plants bloom early and hence are poor yielders.
2. Vegetative Male
3. Female – Give fairly high yield because they flower late and have longer vegetative phase.
4. Monoecious

Recommended varieties of Spinach

S. N.	Variety	Features
1	Virginia Savoy	This is a smooth seeded variety having blistered, large, dark green leaves with round up. Plants are upright and vigorous in growth. Maturity 30 days. Yield 80 q/ha.
2	Early Smooth Leaf	This is a smooth seeded variety and produces small light green leaves with a pointed apex. Maturity 25-30 days. Yield 60-70 q/ha.

- Virginia Savoy – Prickly seeded variety.
- Early smooth leaf – Smooth seeded.
- Khara Lucknow, Khara palak, Banerjee Giant and Banarasi (Katvi Palak) are popular desi varieties.
- Being leafy vegetable, spinach require higher dose of "N" for increasing leaf yield.
- The seed rate in spanich is higher because 50% of plants are male with poor growth and it is not possible to remove them before seed stalk formation.
- Prickly seeded cultivars are best suited for the autumn winter crop in the hills.

- Smooth seeded cultivars are suitable for spring summer in hills and autumn sowing in plains.
- Thinning is done to keep plant spacing within rows at 5.0-7.5 cm.
- Spinach is ready for harvesting in 40-50 days after sowing at 5-6 leaf stage or just before seed stalk formation.
- Spinach has a large leaf surface ratio to weight ratio and has very high respiration rate.
- Storage 0° C – 90-95% relative humidity.
- Yield – 8-10 tonns/Ha.

"AMARANTHUS" (Chaulai)

- It is one of the cheapest leafy vegetable in tropical and subtropical parts of the country.
- In India, grown during summer and rainy season.
- It fits well in crop rotations because of it is short duration and large yield per unit area.

  Among leaf vegetable types

  1. *Amaranthus tricolor* (Badi chaulai) – Main cultivated species in India.
  2. *Amaranthus blitum* (Choti chaulai).
  3. *Amaranthus tristis* – More popular in South India.

- Amaranthus is a C4 plant which can make efficient use of CO_2 and supress its photorespiratory loss.
- Amaranthus is susceptible to water logging.
- It can be grown in soil as high as 10 pH.

Recommended varieties of Amaranthus

S. N.	Variety	Features
1	Pusa Kirti	Leaves green with broad ovate lamina. 6-8 cm long with 4-5 cm wide. Petiole 3-4 cm long, stem : leaf (1:3.4). The stem is green and tender. Suitable for growing in spring-summer (March to June) season in the plains. Maturity 30-35 days. Yield 500 q/ha.
2	Pusa Kiran	Leaves glossy green with broad ovate lamina. The lamina 7-9 cm long and 6-7 cm wide. The petiole 5.0-6.5 cm long, stem glossy, green. Suitable for growing in kharif (July to October) season in the plains. Maturity 25-30 days. Yield 300 q/ha.
3	Pusa Lal Chaulai	The uppersurfaces of the leaves deep-red or magenta, lower surfaces purplish red. The lamina 8.5 cm long and 6.5 cm wide. The petiole 4.2 cm long. The stem deep-red in colour. Suitable for sowing in both spring-summer and kharif seasons in the plains. Maturity 25-30 days. Yield 500 q/ha. in spring-summer and 400 q./ha. in Kharif seasons.
4	Co-1	Dark green in colour, grows vigorously, high yielder, released by TNAU, Coimbatore.
5	Co-2	Dark green lanceolate leaves, 20-25 days duration, 10-11 t/ha. Developed by TNAU Coimbatore.
6	Co-3	Green Leaves, 20-25 days for harvest upto 3 moths multi cutting 10-12 t/ha.
7	Co-4	A dual purpose type for greens and grains. Plants dwarf 20-25 days for leaves, 80-90 days for grains. Yield 7-8 t/ha. (Green) and 2-2.5 t/ ha. (Grain)

8	Lal Sag	Grown popularly in many states, belongs to *A. mangostanus,* high yielding Indian variety. Produces seed early has small flowers.
9	Arka Suguna	It is a pureline selection from an exotic collection from Taiwan (IIHR-47), leaves are light green, broad and succulent stem. First harvest in 25-30 days after sowing and 5-6 cuts in 90 days. It is moderately resistant to white rust and yield 25-30 t/ha. It is recommended for Karnataka.
10	Arka Arunima	A new mutant which is resistant to white rust and rich in calcium and iron and low in ant nutrient factors like oxalate and nitrates. It is multicut, purple amaranth variety. Its leaves are broad and dark purple in colour. It becomes ready for first cutting in 30 days after sowing. Two subsequent cuttings can be taken at 10-12 days, interval with a total yield of 27.4 t/ha. It grows well in kharif and rabi summer seasons.
11	Badi Chaulai	By selection, suitable only for kitchen and container gardening. Plants tall 2-3 cuttings suitable for summer and kharif sowing. Developed by IARI.
12	Choti Chaulai	By selection, highly suitable for commercial cultivation. Plants dwarf suitable for spring-summer season. 6-7 cuttings. Developed by IARI.
13	Amt 105	Green leaves, stem green with reddish base. Suitable for central zone of Kerala. Yield 13 t/ha.
14	Amt 237	Green leaves and stem. Suitable for Kerala. Yield 15-16 t/ha. (Greens)

■ **Other Varieties** –

It is a cross- pollinated (anemophilous) crop.

1. Pusa Kiran – Suitable for rainy season, stem : leaf (1-4.6).
2. CO1 – Especially suitable for late harvesting.
3. CO2 - Especially suitable for early harvesting.
4. CO3 - Especially suitable for clipping havesting, local Name – Araikeerai.
5. CO5 – Tetraploid and very vigorous growth.

- Pusa Kiran – *Amaranthus tricolor* × *Amaranthus tristis.*
- Pusa Kirti – *Amaranthus triocolor.*
- Badi Chaulai – *Amaranthus tricolor.*
- Chhoti Chaulai – *Amaranthus blitum.*
- Pusa Lal Chaulai – *Amaranthus tricolor.*
- CO1 – *Amaranthus dubius.*
- CO2 – *Amaranthus tricolor.*
- CO3 – *Amaranthus tricolor.*
- CO4 – *Amaranthus hypochondriacus*, Dual purpose variety useful for grain and leaf.

  Dried seed with 6% moistures content are stored, yield 100-500 q/ha.
- It can be stored for 6 days at 24-28°C temperature.

"LETTUCE"

- The Latin name (*Latuca*) is derived from the Latin root word "lac" or milk. "Lettuce" is derived from the French "*Laitue*" meaning "milk" and "*Sativa*" means grown from seed.
- Lettuce is major salad crop in North America and is also important in Australia and most countries of Europe and South America.
- It occupies the maximum area among salad crop.

- Its popularity is increasing in Africa, Middle East & Japan.
- In India it is grown in the kitchen gardens and to meet the demand of continental hotels.
- Harvested organ : leaves, eaten raw, often in salads.
- It is rich in "Vitamin A" (900 IU), Vitamin C (10 mg), "Choline (178 mg)", "Ca" (50 mg) and "P" (28 mg).
- If cooked most of the vitamin C content of leaves is lost.
- There are four important botanical varieties of cultivated lettuce.

1. Head type (*Latuca sativa var capitata*) – This is divided into 2 groups.
   (a) Butter head type – Produces relatively small and loose head, soft leaves; smooth texture; varieties bred for both outdoor summer conditions and greenhouse winter conditions; summer butterheads larger and firmer than the winter types; winter butterheads smaller and less compact; popular in N. Europe.
   (b) Crisp head type – Popular in North America and European countries, produces brittle textured and light folded large head (up to 1 kg.) which have excellent shipping and handling abilities, brittle or crisp textured; prominently veined leaves; wrapper leaves green; inner leaves whitish-yellow; predominantly outdoor types.
2. Leafy or Bunching type – (*Latuca sativa var - crispa*) This group includes non heading cultivars cultivated in America, Australia, Europe and India.
3. Cos or Romaine type – (*Latuca sativa var longifolia*) Produce elongated leaves to form a loaf shaped head. Elongated leaves developing into large loaf-shaped heads; slower to bolt than other lettuces; useful as a warm-weather crop; popular in S. Europe and the US.
4. Asparagus or stem type – (*Latuca sativa var – asparaginia*) Also known as celery lettuce, consumed after peeling.

Recommeded varieties of Lettuce

S. N.	Variety	Features
1	Great Lakes (Crisphead type)	Head large, firm with green leaves, outer leaves blistered. Fairly resistant to tip burn, susceptible to powdery mildew.
2	Slow bolt (Leaf type)	Leaf Broad, frilled, slightly yellowish green, crisp and tender, very late in bolting.
3	Chinese Yellow (Leaf type)	Early, leaves with light green, crisp and tender.
4	Imperial 859 (Crisphead type)	Head medium large, solid and well covered by heavily blistered outer leaves. Resistant to tip burn but susceptible to powdery mildew.
5	White boston (butterhead type)	Has soft, solid heads, inner leaves oily, soft greasy texture.
6	Dark Green	Narrow leaves with elongated upright heads.

■ Other Varieties –

1. Punjab Lettuce No. 1 – Non heading type.
2. Slow bolt – Leaf type, late in bolting, suitable for gardens.
3. Chinese Yellow – Leafy type.
4. White Baston – Butter head type.
5. Dark Green – COS type
6. Celtuce – stem type non heading.

- The Portuguese or the British most likely introduced Lettuce into India during the 16th century.
- *Latuca serriola* - Probable progenitor of cutivated lettuce.

- Lettuce is a self-pollinated, cool season crop. (13-16° C temperature).
- Sensitive to both day length and high temperature.
- Late spring days and high temperatures prematurely bring on flower initiation, causing bolting.
- Lettuce seed may go into dormancy when subjected to high temperature in the dark and its exposure to chilling at 4-6° C for 3-5 days results in breaking of dormancy (Thermo-dormancy).
- High temperature promotes seed stalk and causes bitter taste of leaves and induces tip-burn injury.
- Seed do not germinate when soil temperature goes above 22° C.
- Increase CO_2 enrichment (1000-1500 ppm) under glass house condition results in high yield.
- Mulching is recommended for lettuce.
- Lettuce for shipped is often kept fresh by vacuum cooling.
- Breeding objectives: resistant to diseases and pests, increased yield and uniformity, improved quality and bolting resistance.
- It is highly sensitive to acidic soils. Optimum soil pH: 6.5 to 6.8.
- Harvesting – 50-60 days in summer season.
  120 days in winter season.
- Storage – 0° C & 90-95% relative humidity for 2-3 weeks.
- It is better not to harvest the lettuce immediately after a rain or while the dew is on it, because the leaves being crisp and brittle, break easily while handling.
- Major insect pest are aphids, loopers, leafhoppers, cutworms and leafminers.

- Pre and post harvest spray of BA 5-10 ppm help delay senescence in storage and improves the shelf life.

Physiological disorders

1. Tip burn – Burning or scorching of lateral margins of inner leaves of mature head. Caused by unfavourable climatic factor and "Ca" deficiency.
2. Russet spotting – Ethylene injury.

"BASELLA"

- It is commonly called Indian spinach or Malabar spinach or climbing vine.

  Basella alba – Dark green and round to oval leaves.

  Basella rubra – Red aired leaves and stems.

  Other spp. – *Basella connifolia, Basella nigra, Basella saponica, Basella cordifolia.*
- Inflorescence – Axillary, solitary.
- Flower – Sessile in the axil of the leaves, regular bisexual, rose coloured or yellowish white.
- Vitamin A – (3250 IU).
- It can be propagated by both seed (12-15 Kg./ha) and stem cutting.
- Succulent leaves with petioles and tender stem, which are used as vegetables.

"PORTULACA"

- Summer season vegetable.
- Fresh branches and leaves are used as vegetable.

"PARSLEY"

- It is a biennial herbaceous plant.
- Both leaves and stem are consumed (Salad, garnishing, flavouring)
- Rich source of Iron, Vitamin A and C.
- It is a cool season, hardy crop which flourishes well both in temperate as well as in tropical climate.

  Three types of varieties are there.

  1. Plain leaf.
  2. Double curled.
  3. Moss curled – plants are dwarf, compact and bushy.
- Humburg – Plants tall, coarse and tough stem, leaves resemble those of celery shoots.

"AGATHI"

- It is a tropical tree valued for its leaves and flowers.
- It is grown as a standard for black pepper and betel vines, as shade plant for the coconut seedlings and as wind break for Banana in Tamil Nadu.
- The bark contains tannin and yield good fibre and gum.
- The crop grows best in black cotton soil.
- It can tolerate seasonal water logging.

■ **Varieties –**

1. Red Flower.
2. White Flower – Suitable for kitchen garden.

8

PERENNIAL VEGETABLES

Common name	Botanical name	Family	2n	Edible part	Origin	Seed rate	Sowing time
Asparagus	*Asparagus officinalis*	Liliaceae	20	Spears	Europe, Asia & Africa	3-4 kg/ha	July – Nov. Plains March – May Hills
Globe artichoke	*Cynara* scolymus	Compositae/ Asteraceae	34	Flower bud	Med. Region	Suckers or off shoots	Feb. - March
Rhubarb	*Rheum shaponti- -cum*	Polygona- -ceae	44 (4X)	Leaf stalk/ petiole	Asia (Siberia)	Crowns	
Drum stick	*Moringa oleifera*	Moringaceae	28	Green pods	Africa tropics & N.W. India	500-600 g/ha	June-August
Curry leaf	*Murraya koenigi*	Rutaceae	18	Leaves	Tarai tract of U.P.	–	Winter
Chekura- -manis	*Sauropus androgynus*	Eupharbi- -aceae		Green leaves	India & Burma region	2.5-3 kg/ha	Sept-Oct.
Ceylone spinach/ water leaf	*Talinum triangulara*			Leaves & tender shoots	Brazil		

"ASPARAGUS" (Shatawar)

- Asparagus is derived from Greek word "aṣpharagos" ("sprout" or "shoot").
- Origin: Coastal areas of Asia, Africa and Europe.
- First cultivated by Romans 2,000 years ago.
- Brought to N. America by European settlers.
- Cool season, cold hardy, dioecious vegetable (male and female flowers on separate staminate and pistillate plants, respectively).
- It is an herbaceous perennial and is grown in large area in USA, Germany and France, while Indian contribution is meager.
- Tender shoots called "spears" stem (cladophyllus) are used as vegetables and in preparation of soup.
- First harvest after 2 or 3 years. Plantings may produce for 15 to 20 years.
- Spears are harvested when they become 13 to 25 cm.
- The tender shoots contain a white crystalline substance, asparagine, which is used in medicine as diuretic in cardiac, dropsy and chronic gout.
- It was an introduction in other countries and is an important commercial crop in America.
- Asparagus is a dioecious herb with erect branches and grows to a height of 1-3 m, with winter dormancy.
- The roots are tuberous, occur horizontally inside the soil and serve as storage organs.
- The crown is made up of roots and rhizomes.
- The aerial stem (spear) arises from a bud on the rhizome.
- Male flower – Yellowish green and conspicuous.
- Female flower – Less conspicuous.
- The male and female plants are separate and the female plants produce berries with 1-3 seeds/fruit.

- Insect pollinated.
- Female flowers produce red berries.
- Male plants more productive than female plants.
- The main objective of asparagus breeding is development of male hybrids, fusarium and rust resistance, improved spear quality and high yielding ability.
- Being a dioecious crop with male and female plants, cross-pollination is the rule and the cultivars should be grown in isolation to maintain genetic purity.
- Male plants yield more and live longer than female plants and are comparatively more disease resistant.

■ Varieties –

1. Green – More popular and produced mainly for fresh market.
2. Light green - Used in processing.
3. White – Used in processing.

- **Other varieties** – Martha Washington, Marry Washington, Reading Giant, Palmetto, Colossal, Argentenil, Berr's Mammoth, Mammoth White, Jersey Queen, Jersey Giant, New Jersey, Improved perfection, UC 7266 , SL 831, Limbrass 22,126, Lucullas 1813J.
- Perfection – Improved variety released at IARI, New Delhi.
- Jersey Centennial – Vigorous, rust resistant and more productive than Marry Washington.
- Blanching is done to produce white asparagus.
- Green Asparagus contains more Vitamin-A than White.
- Pre cooling in asparagus is done by hydrocooling.
- Asparagus has two types of varieties, one with spears overcast with purple and the other with spear tips almost free from the purple overcast.

- Spears with open (Feathered) tips decay more rapidly, toughen and are unattractive.
- Asparagus is propagated through seeds or crowns.
- Crown becomes dorment during winter and formation of spears start in the beginning of spring.
- Storage – 0^0 C and 90% relative humidity for 3-4 weeks (spears).
- Mounding the soil to a height of 25-30 cm over the row is practiced to blanch the young spears and get white asparagus.
- Takes 3 year to obtain real yield.
- Major diseases are fusarium crown and root rot and cercospora leaf spot rust.
- For fusarium crown and root rot control: use certified, disease-free crowns and avoid fields with a history of the disease, carefully manage soil pH, fertility and drainage.
- Fusarium crown and root rot are promoted by low pH, poor drainage, and infertile soil.
- Hybrids are more tolerant to fusarium than open-pollinated varieties.
- Major insect pests are asparagus beetles, cutworms, tarnished and alfalfa plant bugs and asparagus aphids.

"GLOBE ARTICHOKE"

- Common name: Globe artichoke or French artichoke.
- Latin name: *Cynara scolymus*.
- Family name: Compositae (Asteraceae), Diploid (2n = 34).
- Origin: Central and Western Mediterranean over 2,500 years ago.
- Harvested organ: Immature flower bud, surrounded by many leaf-like bracts; the thickened bracts (phyllaries), heart (receptacle) and choke (flower buds) are eaten.

- Immature flower heads or buds are used as vegetable.
- Small heads are eaten raw or cooked.
- Large heads are eaten only after cooking.
- The thick receptacle known as heart is used for canning.
- Flower heads are considered useful in the dietary of diabeties.
- Leaves are bitter and considered to be useful in dropsy and rheumatism.
- The above ground portion dies every year during winter and the crown of individual plant decays after a year growth.

■ Varieties –

Green Globe, Purple Globe. Both are popular variety.

Violet de Provience – Fresh Market.

Catanese - Fresh Market.

Spinosa - Fresh Market.

Bianco Tarantino – For processing.

Precoce di Jesi - Processing

Bull, Tudella, E 15, Brindisino, Romonesco are some of other popular varieties.

- It is a cool season crop that can be grown successfully at a temperature of 12-18° C. Temperature below 28^0 C causes tough bud and tendency to spread occurs.
- A light frost may not destroy the edible quality of bud but causes blistered appearances.
- Freezing temperature (-2° C to 1° C) prior to harvest, kills the buds and above ground growth. This will result in delayed harvesting upto 6-8 weeks.
- Propagation –

By seed – Less popular, inferior quality.

By suckers or off shoots – Commercially used.

- The suckers or off shoots are removed when they are 30-45 cm in height.
- Perennial - plantings produce for 3 to 7 years. Mature plant grows to about 1.5 m.
- Leaves blue-green, sometimes spiny. Flower stalks bearing a central bud and 2 or 3 secondary buds also arise from the central crown.
- Plants grown form seeds come to bearing in about 8 months.
- Plants grown form suckers come to bearing in about 6-7 months.
- Harvest season – September - October is best.
- A mature plant will produce 12 or more stems and 45-50 edible buds.
- The buds can be stored at 0° C and 90-95% relative humidity for a period of 3-4 weeks.
- Pre cooling of buds at 5° C is practiced to achieve less weight loss, discolouration and incidence of decay than buds that are not precooled.
- Yield – 10-12 tonnes / ha.

■ **Physiological disorders –**

1. Scuffing
2. Split bracts
3. Detached epidermis
4. Bract wilting – All are marketing disorders.

"RHUBARB"

- Plant herbaceous, perennial, with large fleshy under ground rhizome, having a fibrous root system.
- Economic parts are large thick leaf stalk or petiole. They are used for sauces and pies.
- It was first grown in Italy.
- During late 18th century it was introduce in USA.
- Small greenish white flower are borne in panicles.
- Flower are protandrous, self-pollinated within a flower does not take place.
- Lower flowers are fertilized by pollen from upper flower.
- Rhubarb is a cold resistant plant.

- **Propagation –**
  1. Seed – Not recommended as the seed do not give rise to true to type plants, only used in development of new varieties.
  2. By division of crowns – Each crown should contain good strong bud.
- Where the freezing temperature is common, the crowns are planted during spring.
- Where the temperature is non freezing, the crowns are planted in autumn.

Planting pattern –

Depth – 5 cm below soil surface

1.2 - 1.8 m Row × Row

0.6 -1.2 m Plant × Plant

Varieties – Victoria (Oldest variety)

Linnaeus (Oldest variety)

- Mcdonald, Ruby Valentine, Sunrise are new varieties, alternatives with red stalk.
- Strawberry and Cherry Red – Also a popular variety of Rhubarb.
- Major insect – Snout beetle boring into stalk.
- Major disease – Foot rot / crown rot.
- Production of rhubarb stalks during the winter is called forcing of rhubarb in regions where the climate is suitable for production of vigorous crowns.
- Rhubarb is harvested from the second year of planting.

"DRUMSTICK"

- The Moringaceae is a single genus family with 14 known species.
- *Moringa oleifera* (syn. *M. pterygosperma*) – Most wildly grown and utilized species.
- It is commonly called "horse radish" tree (arising from the taste of a condiment prepared from the roots) or drumstick (arising from the shape of a pod) or ben oil tree (arising from the ben oil extracted from seed) and has a host of other country specific vernacular names.
- Popularly known as Ganigana, Mullakkai, Murraugi, Shajan and Muringa.
- Its tender leaves and immature or halfmature fruits are eaten.
- It is most popular for its distinct appearing flavouring fruits.
- Seeds contain an oil called ben or behen oil which has been much used for illumination, soap industry and highly priced for lubricating watches, computer and other delicate machinery etc.

- Seed contain 38-40% of non-drying oil which is clear and odourless and never becomes rancid.
- The oil cake is water coagulant and used for purifying effluent water from rivers, ponds and shallow wells and it is used as organic substitute for water purifying chemicals as aluminium sulphate (alum).
- In India, it is commercially grown in Tamil Nadu.
- It is a deciduous tree of 8-10 m height.
- Flower are white or creamy white.
- Zygomorphy of flower may be due to particular pollinators used in this aspect and it is akin to sub family caesalpinoideac.
- Anthesis 4:30 – 6:30 AM, 5:30 AM peak period.
- Anther dehiscence- 4:30 – 6:30 AM
- At full maturity, anthers are greenish yellow and after dehiscence they turn to pale colour.
- Pollination occurs through honeybees.
- Pollen viability is 72% at anther dehiscence.
- The fruit set due to artificial pollination using fertile pollen may go even upto 100% while it is only 11-15% under natural pollination.
- It is a tropical plant, highly susceptible to frost and high temperature exceeding 40°C causes flower shedding.
- It can not grow in stiff clay soil.

■ Varieties –

1. Jaffna –Yazphanam type, introduced from Sri Lanka, highly suited for coastal tract of Tamil Nadu, Kerala, Karnataka, pods are long (60 - 90 cm length), 400 pods/tree.
2. Moolanoor Murangai
3. Chavakacheri Murungai – Ecotype of Jaffna type, 90-120 cm, damage occurs during transportation due to 90-120 cm length of pod.

4. Chemmurungai–Pod tip red colour, grown in home gardens.
5. Palmurungai–Ecotype of Jaffna.
6. Punamurungai–Ecotype of Jaffna.
7. Kodikanal Murungai – Annual type, pods 20-25 cm in length, predominantely cultivated in betelvine garden, propagated by seed.
8. Kattu Murungai – Released from Annafarm at Kudumiyamalai (Tamil Nadu), annual type, development through pureline selection, propagation by seed, plants are like shrub and harvested easily.
9. Palamedu Murungai – Released from Tamil Nadu as seed propagated moringa.
10. KM – Selection from an annual type, propagated by seed, Developed at Annafarm, Kudumianmali of Pudukkottai district in Tamil Nadu.
11. PKM 1 – Annual type developed at Periyakulam (Tamil Nadu).
12. PKM 2 – Released during January 2000 at Periyakulam (Tamil Nadu).

    Cross – (MP 31 × MP 28)

    - Propagation – Perennial type by limb cutting (1-1.5m), Annual type by seed.
    - Fruit weight – 230 g containing 10-20 seeds each.

"CURRY LEAF"

- Ground curry leaf with mature coconut kernel and spices form an excellent preserve.
- It's leaves are used mainly to improve the taste and flavour of foods, rich in "Ca".
- Its leaves, roots and bark are credited with tonic, stomachic and carminative properties.

- Leaves are reported to cure piles and allay heat of body.
- Green leaves are eaten as raw for treatment of dysentery.
- An Influsion of roasted leaves is used to stop vomiting.
- Decoction of leaves is drunk for dropsy.
- Volatile oil (curry leaf oil) is produced by steam distillation of fresh leaves, (2.6%) under high pressure (90 IB/Sq. inch).
- Oil is deep yellow in colour with a strong spicy colour and pungent clove like taste.
- A crystalline glucoside "Ksenigin" from the leaves and "murayin/muragin" from flower are extracted which are industrial products.
- It is a sefl-pollinated crop.
- *Murraya paniculate* – Indigenous to Burma.
- *Murraya exotica* - Indigenous to India.
- It is cultivated in Coimbatore, Salem and Trichirapalli district of Tamil Nadu on a commercial scale.
- Flower cymes, berries purplish black when ripe and two seeded tree bears flower from February - May.
- Grows well in warm climate.

Varieties –

1. DWD 1 (Suvasini) – 0.1629 mg of chlorophyll content per gram of fresh leaf, highly aromatic oil 5.22%, sensitive to winter season, evolved at University of Agri. Sci. (UAS), Dharwad, clone of single plant root suckers.
2. DWD 2 – Good aroma, oil 4.09% evolved at UAS, Dharwad, seeding progeny, winter insensitive, high yield then DWD – 1.
3. Senakanbu – Mainly grown in Tamil Nadu, red petioled.

- There are two types of curry leaf viz broad leaved and small leaved, Small leaved are more fragrant and hence used for extraction of essential oil.
- Propagation – By "seed", but also by "root suckers".
- Polyembryony has been reported and about 14% of the fruit contains two seeds per fruit.
- Normally fruits for seed are available during July-August. Each fruit contains 2-3 seeds.
- Spacing 4 × 4 M.
- 1 Year old seedlings are planted in main field.

"CHEKUMANIS"

- It is a perennial, small shruby leafy vegetable.
- Due to high nutritive value it is called "Multivitamin Greens".
- It is occupying prominent place in almost all household kitchen garden of Kerala.
- It is reported to be first introduced into Kerala from Malaysia in 1953 and is known in Malayalam as Madura Keera.
- The inflorescence is axiliary with small reddish flowers.
- Flowers are monoecious.
- The crop is highly cross-pollinated and entomophilous because of protogynous and monoecious nature of flower.
- The leaves are very rich in protein (6.8-7.4%) as compared to amaranthus (3.2%), and also rich in minerals and vitamins.
- Leaves and tender shoots – Mainly used as leafy vegetable either as salad or after frying.

- The leaves are also said to be used to give a light green colours to pastry and to fermented rice in the Dutch East Indies and Java for preparation of soup.
- When the plants are raised under shade, it produces broader leaves.
- Propagated by seed and stem cutting (Soft wood and semi hardwood) (500 ppm IBA/IAA treated before planting).
- For planting a hectare area about 1, 00,000 cuttings are required.
- It can be trained as hedge or even on bowers and on trellises.

"CEYLON SPINACH OR WATER LEAF"

- It is a soft mucilaginous leafy vegetable grown in tropics.
- Being a shade loving species it is suitable for cultivation under other crop canopy also.
- It was introduced to Java from Brazil.
- Short day condition (SDP) favours flowering and fruiting.
- The crop is highly self-pollinated.
- It grows upto an altitude of 1000 M, most humid conditions favour optimum growth and development of crop.
- Propagation – By seed or herbaceous stem cuttings.
- Flower – Pink colour.

9

CUCURBITS

■ Sex forms in cucurbits –

1. Monoecious – Where staminate (Male) and pistillate (female) flowers are separately produced on the same plants eg. Cucumber, Watermelon. Bottle gourd, Bitter gourd, Ash gourd, Ridge gourd, Sponge gourad, Snake guard, Pumpkin etc.
2. Androecious – Where only staminate flower are present on the plant.
3. Andromonoecious – Where staminate and hermaphrodite (perfect) flowers are produced separately on the plant. eg muskmelon (desert), Snapmelon Phoot Long or serpent melon, Kachri, Vellarikkai.
4. Gynoecious – When only pistillate (female) flowers are present on the plant, eg Cucumber.
5. Gynomonoecious – Where pistillate and perfect flower are produced separately on the same plant. eg. Cucumber.
6. Gynoandromonoecious – (Trimonoecious) where staminate, pistillate and perfect flowers are produced on the same plant. eg. Cucumber, Muskmelon, Ridge guard.
7. Hermaphrodite – Where only bisexual or perfect flowers are produced. eg. Cucumber (original forms),"Satputia" cultivars of Ridge gourd, Muskmelon.
8. Dioecious – Where staminate and pistillate flowers are produced separately on separate plants. Example 1. Parwal (Pointed gourd), 2. Ivy gourd, 3. Kakrol.

Common name	Botanical name	Ch. (2n)	Origin	Features
Cucumber	*Cucumis sativus*	14	India	• Ethral increase female flowers • 2nd Most cultivated
Water melon	*Citrullus lanatus*	22	Tropical Africa	• Bitterness due to Cucurbitacin E • Heavy dull – Maturity
Round melon	*Citrullus lanatus Praecitrullus fistulosus*	22	Tropical Africa	
Bottle gourd	*Lagenaria siceraria*	22	Tropical Africa	• Processed products: Kofta and petha – most popular
Bitter gourd	*Momordica charantia*	22	Indo-burma	• Bitter taste – Momorcidin • Rich in iron
Pointed gourd	*Trichosanthus dioca*	22	India (Bengal and Assam)	• Propogated by vine cuttings
Snake gourd	*Trhichosanthus anguina / Cucumerina*	22	India	
Musk melon	*Cucumis melo*	24	Tropical Africa (Sahara desert)	• Maturity indices Netting stage • except Hara Madhu
Long melon	*Cucumis melo Var. utillisimus*	24		
Snape melon (Phoot)	*Cucumis melo var. momordica*	24		
Gherkin	*Cucumis anguira*	24		
Ash (wax) gourd	*Benincasa hispida*	24	Java, Japan	

Ivy gourd (Scarlet gourd)	*Coccinia indica /groundis*	24	India	
Ridge gourd	*Luffa acutangula*	26	Asia, Africa	
Sponge gourd	*Luffa cylindrical*	26		
Chow-chow (Chayote)	*Sechium edule*	24	Southern Mexico & Central India	
Spine gourd / kakrol	*Momordica dioca*	28		
Pumpkin (Sitaphal)	*Cucurbita moschata*	40	Mexico, Peru	
Summer squash	*Cucurbita pepo* (Chappan kadu)	40	Mexico	
Winter squash	*Cucurbita maxima*	40	India,	Bangladesh
Sweet gourd	*Momordica cochinchensis*	26	S. Asia, S. China N.E. Australia	

Pollination mechanism –

- All cucurbits are highly cross pollinated crops (Entomophillus)
- Cucurbits are day neutral plant.

  Genetic Male Sterility is found in – Muskmelon, Water Melon

  Cytoplasmic Male Sterility is found in – Cucumber

New Cultivars –

1. Pusa Uday – Cucumber
2. Pusa Ujjwal, Pusa Shakti - Ash gourd
3. Pusa Sneha – Sponge gourd
4. Pusa Shandar – Snap melon
5. Pusa Sandesh – Bottle gourd

Seed rate, planting distance and depth of sowing in different cucurbits.

Crop	Seed Rate (kg/ha)	Planting distance		Depth of seed sown (cm)
		Row to Row (cm)	Hill to Hill (cm)	
Watermelon	3.7-5.0	250-350	90-120	2-2.5
Roundmelon	3.7-5.0	180-240	60-120	1.5-2.0
Muskmelon	3.0-7.0	180-240	60-120	1.0
Snapmelon	2.4-3.7	180-240	60-120	1.0
Longmelon	2.4-3.7	180-240	60-120	1.0
Cucumber	2.5-3.7	150	60-90	1.0
Bottlegourd	3.0-6.0	200-300	100-150	2.5

Bittergourd	4.5-6.0	150-250	60-120	2-2.5
Ridgegourd	3.7-5.0	180-250	60-120	1.5-2.0
Spongegourd	2.5-3.7	180-250	60-120	1.5-2.0
Snakegourd	5.0-6.0	180-250	60-120	2.0-2.5
Ashgourd	5.0-7.0	250-300	60-120	2.0
Pumpkin	6.7-9.0	250-300	100-180	2.0-2.5
Bush Squash	8.0-10.0	60-75	45-60	2.0-2.5

■ Public sector hybrids of cucurbits

Name of hybrid	Source
Muskmelon	
Punjab Hybrid-1, MHL-10	Ludhiana
MHY-3, MHY-5	Durgapura
Pusa Rasaraj, DMH-4	IARI
Watermelon	
Arka Jyoti	IIHR
RHRWH-2	Rahuri
Cucumber	
Pusa Sanyog	Katrain
RHRWH-2	Rahuri
AAUC-1, AAUC-2	Jorhat
Bottle gourd	
NDBGH-4, NDBH-7	Faizabad
Pusa Manjari, Pusa Hybird-3	IARI
Pusa Meghdoot	
PBOG-1, PBOG-2	Pantnagar
Summer squash	Katrain
Pusa Alankar	

Pumpkin

Pusa Hybrid-1	IARI

Bitter gourd

Pusa Hybrid-1, Pusa Hybrid-2	IARI
RHRBGH-1	Rahuri

- **Anthesis** –
  1. Muskmelon – Morning 5.30 – 6.30 am.
  2. Cucumber – Morning.
  3. Sponge Gourd – Morning.
  4. Bottle Gourd – Evening 5-8 pm.
  5. Snake gourd – Evening 6-9 pm.
  6. Ridge gourd – Evening 5-8 pm.
- First hybrid for cucumber was developed in Japan (1940).
- First commercial hybrid in vegetable was developed by IARI.
  1. Pusa Meghdoot (1971) -By Choudhary and Singh.
  2. Pusa Manjari (1971)- By Choudhary and Singh.
- Recalcitrant seed – Pointed gourd, Chow-Chow.
- Pumpkin – 9:1 (male : female Ratio).
- Highest Cu – Pointed gourd.
- Highest Cl – Cucumber.
- Dr. Kihara (Japanese Scientist) developed Pusa Bedana (Tetra 2 × Pusa Rasal) (2n = 33 triploid seedless) in 1972.
- Ash gourd – Longest storage life among cucurbits.
- Flower Colour
  (1) Pointed gourd – White
  (2) Snake gourd – White
  (3) Bottle gourd – White
- Except three cucurbits other bears yellow colour flowers.

- Punjab Hybrid – Cultivar of muskmelon, First commercial vegetable hybrid which was developed by using male sterility cross between MS1 × Hara Madhu.
- Amphidiploid – (*Cuurbita moschata* × *Cucurbita maxima*).
- Autotriploids – (In Watermelon and Sugarbeet), Triploids (3x) by crossing between Tetraploid (4x) × Diploid (2x), Developed by Dr. Kihara (1972).
- India is an important supplier of fresh vegetable to South East Asia and Gulf countries.

"WATER MELON"

- Annual, Dicotyledon, Trailing vine, Diploid (2n = 22) (but note that some cultivars are tetraploid, and seedless water-melons are triploid).
- Water melon crop is grown in garden lands and river beds in India.
- The fruit has cooling effect and is used as an expectorant, diuretic and stamachic and in allaying thirst.
- The fruit juice is consumed by many after adding a pinch of salt and black pepper.
- Fruit contains 95% water, 0.2% protein, 0.3% minerals and 7% carbohydrates /100g fresh weight.
- It is also a rich source of Iron.
- The seed is also eaten as a snack after shelling and roasting.
- *Citrullus colocynthis* occurs from India to tropical Africa.
- This is also known as Tumba in Bikaner and other arid tracts.
- They are cooling, purgative, anthelmintic, antipyretic and carminative in action.

- They cure a wide range of ailments such as ulcers, asthama, bronchitis, jaundice, enlargement of spleen, dyspepsia, constipation and anaemia.
- The ground roots are applied externally to control inflamation of breasts and pain in the joints and uterus.
- Water melon is indigenous to tropical Africa more specifically South Africa.
- It is truely wild or native only in sandy dry areas of South Africa chiefly Kalahari Desert.
- Cirtrullus genus belongs to the subtribe Benincasinae, tribe Bininace and sub family cucurbitaceae.
- *Cirtrullus vulgaris* is now renamed as *C. lanatus*.
- The species classification of citrullus and their relationship have been studied in relation to cucurbitacin content (bitter principle) by South African investigators.
- *(1) C. vulgaris, (2) C. colocynthi and (3) C. ecirrhosus – Cucurbitacine E.* is the main bitter substance (Tetracyclic Terpenes).
- *C. naudinianus* – Cucurbitacin-B together with E and their derivatives are the main bitter substance.
- From the results of morphological and cytogenitical studies it is observed that all these four species were cross compatible with each other and several factors like geographical isolation, difference in flowering habit, generic difference and structural changes in chromosome would have contributed to the maintenance of their identity.
- Pangalo (1930) and Goldhousen (1938) – Reported that bitter and nonbitter wild races of vulgaris should be distinguished and treated with *C. edulis* and *C. colocynthoides*.
- The citroides comprising the citron or preserving melon produce fruits with hard inedible flesh and greenish or tan seeds.

- The citron or preserving melon (var - *citroides* of Bailey) can be considered as intermediate stage between primitive bitter and cultivated sweet forms.

Recommended varieties of Water Melon

S. N.	Variety	Features
1	Sugar Baby	Fruit round, 7-8 inches in diameter, weight 3 to 5 kg. rind bluish green with faint black stripes, thin, hard, and tough. Flesh medium red, firm, crisp, sweet (TSS 11-13%) fine texture. Seeds relatively few, very small, dark, tan mottled with black colour. Maturity 85-90 days. Yield 250-300 q/ha.
2	Pusa Bedana	A seedless watermelon. Dark rind with faint strips, somewhat triangular shape, tough rind, red flesh with remanints of false seeds, TSS 12-13%. Fruit weight 5-6 kg, produces 3-6 fruits per vine, very good keeping quality. Maturity 115-120 days. Yield 200-250 q/ha.
3	Arka Jyoti	F1 hybrid between IIHR 20 (a Rajasthan selection X Crimson Sweet), An early season crop, Flesh colour deep pink, very sweet (12 to 14% TSS) Average yield 800 q/ha. Has excellent keeping and transport quality. Duration 90-100 days. Suitable for J&K, West Bengal, Punjab, Rajasthan, Maharash--tra, Karnataka, Tamilnadu. Developed in 1973. It performs well under south as well as north India.
4	Arka Manik	A selection from advanced generations of a cross between IIHR 21 × Crimson Sweet by IIHR Bangalore. Average fruit weights 6 kg excellence granular texture, high TSS

	(12-15%) deep crimson flesh, pleasant aroma. Average yield 600 q/ha. Good keeping and transport quality. Duration 90-100 days. Resistant to powdery mildew, downy mildew and anthrocnose and Bloosom end rot (multiple resistances), It stands well in transport and storage, Most suitable for South India.

■ Crop Improvement –

(1) Introduction:-

a. Asahi Yamato – Introduced from Japan, TSS 11-13%, fruit 6-8 Kg in weight.

b. Sugar Baby – Introduced from USA, TSS 11-13%, very sweet in taste, suitable for North India.

c. Improved Shipper – Introduced from USA, 8-9% TSS

d. New Hemisphire Midget – 1.5-2.0 kg weight, suitable for home gardens.

e. Dixie

f. Fucken

(2) Selection –

a. Special No. 1 – Early maturing, red flesh and red seed, selection made by P.A.U, Ludhiana.

b. Durgapura Meetha – Good keeping quality, TSS 11%, dark red colour, matures 125 days after sowing (Late cultivar).

c. Dugapura Kesar – Late cultivar, flesh is yellow in colour, moderate sweet, large seeds.

d. Pusa Rasal – Old cultivar.

(3) Hybridization –

a. Pusa Bedana (Triploid) – Tetra-2 × Pusa Rasal (Seedless variety), 3-4 fruit/vine, Developed by Dr. Kihara (1972).

b. Arka Manik – (IIHR 21 × Crimson Sweet) followed by modified back cross method.

c. Arka Jyoti – F1 hybrid (IIHR 20 × Crimson sweet).

(4) Polyploidy breeding –

Polyploids were produced by treating seedlings at the cotyledon stage with an emulsion of 0.2% colchicine, stearic acid, morphaline and lanoline.

- Fruits produced on triploid plants had seeds devoid of embryo.

(5) Resistance breeding –

- Arka Manik highly resistant to powdery mildew and moderate resistant to anthracnose.
- Resistance to red pumpkin beetle has been reported in lines Sl. – 120, Afghan.

Locally grown cultivars are :

- Farrukhabadi, Moradabadi – UP.
- Mateera - Rajasthan (grown in rainy season around Bikaner).
- Katagolan – Cultivar of Jamuna river bed whose flesh is not sweet but it keeps well for 2-3 months at ambient temperature during July-September.
- Private sector hybrid – Madhu, Milan, Mohini.
- MHW 4, 5, 6, 11, 12, 15, NS 246, 295, Suruchi, Samtrupti, Amruth and Century 2-Appealing appearance, good quality flesh, transportability, resistance and tolerance to wilt.
- Arid region of Rajasthan are best suited for production of quality fruits.
- High temperature is beneficial during ripening.

- Water melon requires hot dry climate and a long growing season preferably with warm days and cool nights.
- It can't withstand frost or very low temperature.
- Temperature for seed germination : 25-30° C.
- Temperature for fruiting : 24-27° C.
- Plant growth is optimum under 28-30° C.
- Sandy loam is suitable for early crop.
- Loam – High yield potential.
- Alluvial river beds are also good for water melon crop.
- Excess humidity cause powdery mildew, downy mildew, anthracnose and virus diseases and pest such as fruit fly.
- Cool nights and warmer days – Ideal for accumulation of sugar in fruit.
- Seed rate – 3.0 - 3.5 kg for small seeded types.
  5.0 kg for large seeded type.
- Spacing – 2.5 – 3.0 m. (Row × Row).
  90 – 120 cm. (Hill × Hill).
- Sugar baby – 2.0 m (Row × Row).
  1.0 m (Hill × Hill).
- Plant to plant spacing could further be reduced to 0.6 m to accommodate around 16,600 plants/ha, if single plant is retained in place of 2.
- Spacing – 30 × 60 cm for sugar baby for Zone IV at NDUAT at Faizabad.

Sowing time – Mid Feb. for Northern plains in garden land.
Nov. – January for river beds.
Dec. – January for Central & South India.

Sugar Baby – (December 20 - January 20) under Maharashtra conditions gives highest yield.

Asahi Yamato performs best during April – May in Northern Hills.

Mateera – August - September in North-West Rajasthan, a rainy season crop of a local watermelon because of absence of heavy rainfall.

In some areas where early crop is desired, seeds can be sown in alkathene bags (10 × 15 cm) of 100-200 gauge.

- 4-5 seed in a hill is sown and finally two vines are retained / hill.
- Training and prunning – The excess vine growth can be pruned manually to restrict vegetative growth and promote higher female : male flower ratio.
- Yield and fruit quality are improved with prunning.
- Fruit thinning is useful and retaining of 2 fruits /vine improves fruit size as well as fruit yield.
- 3-4 main shoots per plant are retained on Sugar Baby for optimum production.
- Yield loss upto 30% have been observed due to weeds.
- In watermelon to increase the flowering and fruit yield ,exogenous application of chemicals such as TIBA 25-250 ppm, boron (3-4 ppm), Mo (3-4 ppm), Ca 20-25 ppm is recommended at 2 true leaf stage, Repeat the spray at 4 true leaf stage.

Maturity –

- Colour change of the fruit portion which rests on the ground from white to creamy in case of light skin colour or to yellow in case of dark skin colour.
- Metallic sound when fruit is tapped denotes immaturity.
- Heavy dull sound indicates ripening.
- Tendril dries and the ground spot has turnes from white to yellow.

- Watermelons reach maturity approximately 40 – 45 days after blooming; 90-120 (Days after sowing), highly dependent upon cultivar.
- Immature watermelons have turgid tendrils; a completely wilted tendril can be indicative of over-maturity.
- Fruit is rich in sugar (soluble solids); a sugar content of 17% is considered excellent quality.
- Fruit should be cut cleanly from the vine to avoid stem damage and prevent stem-end.

  Yield – 200-250 q/ha.

  300- 400 q/ha. in hybrid/high yielding varieties.
- Cylindrical or oblong shape like that of Charleston gray is suitable for transportation in trucks.
- Most of cultivars have been deep-pink or pink or pale-pink flesh colour with slightly redish-tinge containing largely lycopene and anthocyanin pigments.
- Uniform development of pink colour from centre to rind is necessary.
- More often in Indian cultivars "White heart" seen at the central portion indicating poor quality.
- Hollow heart and fibrous flesh signify over maturity.
- Average TSS for watermelon : 8-12%.
- Seed maturing and edible maturity coincide with each other.
- Water melon from Uttar Pradesh and Andhra Pradesh are transported to Kolkata market.
- Fruit is chilling sensitive and should not be stored below 7°C; optimum storage temperatures are 7°C to 10°C.
- Major diseases of watermelons – bacterial fruit blotch, damping off, anthracnose, downy mildew, *Fusarium* wilt, gummy stem blight, various viruses (e.g. watermelon mosaic

virus) (transmitted by aphids and cucumber beetles; effective control of these diseases involves control of the insect populations).

- As in tomato, calcium deficiency can result in a disorder called "blossom-end rot".
- Major insect pests: cucumber beetle, aphids, mites and sap beetles.
- Cucumber beetles are of lesser concern to watermelon growers than cucumber and muskmelon growers because watermelon is not susceptible to bacterial wilt.
- Protection of watermelons from cucumber beetles is necessary when plants are small and high beetle populations are feeding on stems, or when beetles are feeding on the fruit.
- Pesticides application toxic to honey bees ,during flowering can decrease pollination and fruit set.

"MUSK MELON"

(Cantaloupe, melon casaba, winter melon)

- Mature fruits (*Cucumis melon* 2x = 24) are round in shape and 8-16 cm in diameter.
- Nevertheless some oblong, flat and oval fruit shaped varieties are also available.
- In Ayurveda and Unani medicines muskmelon have cooling effect on the body system.
- The roots have emetic and purgative properties.
- The fruits are sweet and musky in flavour. Fruit is rich in Vitamin A, B and C, Ca, P and Fe.
- Immature melons are used fresh in salads, cooked (soup, stew, curry, stir-fry) or pickled.

- Mature fruit eaten fresh as a desert fruit, canned or used for syrup or jam; dehydrated slices (lightly processed) for short-term or moderate storage can be reconstituted; and the pressed juice can be canned.
- Melon seeds are a dietary source of unsaturated vegetable oil and protein, and may be lightly roasted and eaten like nuts.
- Muskmelon provides a wholesome food.
- For export the fruits should have good keeping quality.
- Jaffrey (1980) classified the entire genus cucumis in two subgenera.

1. Subgenus *cucumis* x = 7, 3 or 4, Sino – Himalayan species including *C. sativur* (*C. hystric x C. muriculatus*).
2. Subgenus *melo* x = 12 species mostly in Tropical and South Africa.

a. Anguria group – Dioecious, monoecious and andromoenocious perennial or annual with yellowish or brown fruit, closely related and partially fertile interspecific hybrids. *C. anguria, C. dispaceus, C. prophetarum. C. myriocarpus, C. sacleuxii.*

b. Metuliferus group – Monoecious, annual with red spiny fruits, *C. metuliferus*.

c. Melo group – Monoecious, andro monoecious perennial or annual with smooth fruit. *C. Melo, C. sagittatu, (syn. C. anglolensis), C. dinteri, C. humifructus.*

   1. *C. anguria* – Resistant to nematode.
   2. *C. ficifolius* - Resistant to nematode.
   3. *C. metuliferus* – Resistant to nematode.

- Pickling melon – *C. melo* var *conomon.*
- Mango or Lemon Melon – *C. melo* var *chito.*
- Melons may be andromonoecious (perfect and staminate flowers), gynoecious (pistillate flowers), gynomonoecious

(perfect and pistillate flowers), hermaphrodite (perfect flowers), or monoecious (pistillate and staminate flowers).

- Monoecious and andromonoecious are the most common.
- Pollinated by honey bees; 1 to 2 bee colonies per acre recommended.
- Hybrid gynoecious melons (with multiple disease resistances earliness, uniformity of harvest and fruit quality) are becoming increasingly popular.
- Two major genes control sex expression in melon (*A* and *G*):

  A-G- = monoecious

  A-gg = gynomonoecious

  aaG- = andromonoecious

  aagg = hermaphroditic

  gy interacts with *A* and *G* to produce a stable gynoecious phenotype.

Recommended Varieties of Musk Melon

S. N.	Variety	Features
1	Pusa Sharbati	Fruits round, straw coloured, netted and striped skin, flesh salmon orange, firm, thick with small seed cavity. TSS 11-12 %. Fruits have good keeping quality. Maturity 80-85 days. Yield 150-170 q/ha.
2	Pusa Madhuars	Fruits flattish round, sparsely netted, yellowish green, smooth skin with salmon orange flesh. Very sweet, TSS 12-14%. Maturity 95-100 days. Yield 180-200 q/ha.
3	Durgapur Madhu	Fruits oblong, pale green smooth rind, light green flesh with big seed cavity, TSS 12-13%. Maturity 85-90 days, Yield 160 q./ha.

4	DMDR 2	Fruits oval-round with green strips, yellowish smooth skin, orange flesh, Average fruit weight 950-1000g. TSS 11-12% Resistant to CGMMV and downy mildew diseases. Maturity 85-90 days. Yield 160 q/ha.
5	Hara Madhu	Fruits round, tapering towards the stalk end with prominent green sutures, rind thin, smooth, turn pale yellow at maturity. Flesh green, thick, very juicy, melting and sweet. TSS 12-15% Maturity 110-120 days. Yield 125-150 q/ha.
6	Pusa Rasraj	The first monoecious hybrid, Fruits slightly oval, skin smooth, non striped, dull white with green and sweet flesh. TSS 11-12%, Maturity 70-75 days. Yield 200-250 q/ha.
7	Punjab Hybrid	Fruits round with distinct sutures. Rind thick, netted, creamish yellow. TSS 12-13%. High keeping and transport quality, moderately resistant to powdery mildew and fruit fly. Maturity 75-80 days. Yield 160-170 q/ha.
8	Kashi Madhu	Fruits round. Harvesting starts from 90 DAS - 105 days. Tolerant to powdery and downy mildew and muskmelon mosaic virus. Medium maturity, yield 250 q/ha. in 105 days. Recommended for UP, Punjab, Bihar, Jharkhand.
9	Arka Rajhans	An early-medium variety, medium-large oval fruit, weight 1 to 1.5 kg dirty white, fine nets, excellent keeping and transport quality. Flesh thick, white and sweeter. TSS 20 to 14%. Highly resistant to the powdery mildew. Yield 250-300 q/ha. 85-90 days duration. Recommended for Southern regions of the country.

10	Arka Jeet	Fruits small, attractive orange yellow, round weight 400 to 800g. Very sweet, TSS 15 to 17% with high vitamin C (41.6mg/100g fresh weight), an improvement over the local Lucknow Bati strain of Uttar Pradesh. Yield 140-150 q/ha. Duration 90 days recommended for southern regions of the country.

■ Crop Improvement –

1. Selection –

a. Durgapura Madhu – Very early cultivar, Oblong, 500-600 g. /fruit weight, very sweet, 13-14% TSS, seed cavity big.

b. Hara Madhu – Late maturing variety, round, 1 kg fruit weight, TSS 13%, keeping quality poor, never reaches at full slip stage.

c. Pusa Madhuras – Mid season variety, 12-14% TSS, poor keeping quality, suitable for growing in North Indian plains.

d. Arka Rajhans – Mid season variety, large oval fruits (1kg), fruit transportable quality, keeping quality is excellent, tolerant to powdery mildew.

e. Arka Jeet – Early cultivar similar to Lucknow safeda of Uttar Pradesh, dwarf habit 300-500 g, flat and small, orange brown skin, white flesh, big seed cavity, TSS 12-14%, Excellent flavour and high vitamin C content.

2. Hybridization

a. Pusa Rasraj – F1 hybrid (Monoecious 3 × Durgapura Madhu), developed by using monoecious sex forms as female parent, 11-12% TSS, 1-1.2 kg fruit weight, yield 25 ton/ha.

b. Pusa Sharbati – Kutana × Resistant (contaloups) No. 6 of USA, 11-12% TSS, early cultivar.

c. Punjab Rasila – WMR 29 × Hara Madhu, Resistant to downy mildew and powdery mildew, TSS 11%, fruit wt. 600 g.

d. Punjab Sunehri – (Hara Madhu × Edisto), 11-12% TSS, early maturing.

e. Hisar Madhur – Pusa sharbati × 75.

f. Punjab Hybrid – F1 hybrid between a male sterile line (MS1) and Hara madhu, vine long 2-2.5 m.

Private sector hybrid – Shweta and Swarna

Heterosis Breeding –

- Expression of quality character (TSS %) is highly influenced by environmental factors.
- The andromonoecious nature of muskmelon involves the tedium of emasculation of perfect flower in the female parents for hybrid seed production.
- Further more monoecious line reduces the extent of self pollination in the female parent when seed production is done under open pollinated condition with adjacent row of male and female parents.
- Three stable gynoecious lines 86-104, 105 and 118 lines have been developed at IARI.
- Till now four true breeding homozygous monoecious lines (M1, M2, M3 and M4) have been developed.
- Andromonecious varieties are used to study the extent of heterosis in muskmelon.

Two commercial hybrids have been developed.

1. Pusa Rasraj – Using monoecious sex forms as female parent at IARI.
2. Punjab Hybrid – Using male sterile stock as female at PAU.

- **Polyploidy breeding** – Attempt was made to induce polyploidy in muskmelon as early as in 1953 by using colchicine.
- Quality of tetraploid strains was superior to that of diploid but other character like productivity, size, fertility and number of seed, the tetraploid strains were inferior to the diploid so polyploidy has no scope for muskmelon improvement.

Resistant breeding –

- Cantaloupe, PMR-5, 6, FM 5 (Cornell 83-275-61), Gulf Stream, Jacumba, Perlita - Resistant to powdery mildew under natural epiphytotic condition.
- PMR6 and campo have same locus for resistance.
- Gulf Stream differs from campo by two factors. Budama type 1 3 and 3, Phoontee Goomuk Nakkados. Kaya, Ex 1, Annamalai are resistant to downy mildew.
- In cucumber a gynoecious sex form WI 998 have been isolated in a stable condition in muskmelon in USA.
- Harela, Durgapura Madhu and Punjab Sunehri – Resistant to both *fusarium oxysparum* sp. *melonis* and *F. solani* (fusarium wilt).

  1. *Cucumis figareic* (GBNR 1804).
  2. *C. zeyheri* 1 (73252 Hg).
  3. *C. zeheri* 2 (GBNR 1053).
  4. Smiths perfect.
  5. VAC 1.
  6. *C. callosus*.

  All there six species are resistant to *Fusariun oxysporum species melonis*.

  Florida 67, PMR 6, Jacumba, PMR, Honey Dew, Bulgariams-10, Bacharacha, Durgapura Madhu, Gulf stream,

Kabul Melon and Ludhiana Col, 2 – Resistant to Cucumber Mosaic Virus.

1. *C. figarei*
2. *C. ficifolius*
3. *C. meeusii*
4. *C. argicanus - 1*
5. *C. zeyheri 2* - Immune to CGMMV based on confirmation made by back inoculation technique and transmission electron microscopic studio.

- By using phoot (*C. melo var. memordica*) as resistant source, five GMMV resistant lines namely VRM 5-10, 29-1, 31-1, 42-4, 43-6 have been developed.
- Cassba and PI 70683 – Resistant against red pumpkins beetle
- *C. melo var callosus* – Resistant against fruitfly.
- Long tap root system is adapted to the growth of cucurbits in riverbeds.

Local Cultivar –

1. Lucknow Safeda.
2. Baghpat Melon of Meerut.
3. Jaunpuri Netted.
4. Mau Melon of Azamgarh.
5. Tonk Melon – from Rajasthan.
6. Kharri Melon – Hoshangabad (Madhya Pradesh).
7. Goose Melon and Kavit Jam or Neel (Maharashtra).
8. Sharbat – e - Anar (Andhra Pradesh).
9. Bathesa – A.P.

- Non desert type –
  1. Phoot (Snap melon)
  2. Kachri or Kachauri
  3. Dosakaya.
  4. Budamkaya – Called cucumber and vallerakai (a salad cucumber type) of Tamil Nadu.

- Muskmelon is susceptible to frost.
- Excess humidity will promote disease like powdery mildew, downy mildew, anthracnose, virus disease and fruit fly.
- For good fruit quality and sweetness in muskmelon dry weather during fruit development is necessary.
- Cool night and warm days are ideal for accumulation of sugar in the fruits.
- Maturity hasten if nights are warm.
- Optimum temperature for seed germination : 24 - 29° C.
- Seedlings become ready for transplanting at 2-4 true leaf stage.

  Seed rate - 1 kg by dibbling method.
  2.5-3.75 kg by direct sowing in kera and pora method.
- 2-3 seeds/hills are sown.
- High temperature at the time of fruit maturity and ripening increases the sweetness.
- Desert muskmelons are grown only in those seasons when the weather is dry and warm during fruit development.
- During rainy season non-desert forms of muskmelon are grown primarily in most parts of the country.

  Sowing time –

  North India : November-January (riverbeds).

  South and Central India : October-November.
- November transplanted crop take significantly more days to fruit maturity than March transplanted.
- In muskmelon insitu sowing is practiced.
- The recommended seed rate for muskmelon is 3.0 - 7.0 kg/ha.
- To obtain highest yield of good quality fruits in cultivar Hara Madhu, treat the seed for 8 hours with a solution containing nitrophenolate sodium + nitroguaiacol sodium at 0.5%.

Spacing – 1.5 - 2.5 m. (Row × Row)
60 - 120 cm (Hill × Hill)

- 350 cm Row × Row 60 cm Plant × Plant (for cultivar Durgapura Madhu, Rajasthan) for zone VI by AICVIP.
- A close spacing of 30 cm is recommended for Hara Madhu under Punjab condition.
- Soaking of seed in ethaphon at 480 mg/L. for 24 hours improves germination in muskmelon at low temperature.

Training and Pruning –

- In muskmelon the highest yield and good fruit quality are obtained with prunning (as the first hermaphrodite flower is borne on secondary branch arising from the eighth node the secondary branches are pinched off upto the seventh node) plus stacking.
- Removing the secondary shoots upto the seventh node on the main stem is found optimal to improve plant growth and fruit set and induce early flowering.
- Maximum number of fruits with highest TSS is produced between 9th and 12th nodes on the main stem.
- Application of 128 – 180 kg Nitrogen/ha significantly reduce the perfect : staminate flower ratio.
- N and K as foliar spray @ 1.5% increases the number of hermaphrodite flower in cultivar Hara Madhu. Potassium treatment also increases TSS, ascorbic acid, flavour, taste and juiciness of fruits.
- Irrigation at 0.6 – 0.8 PEC (Pan Evaporation Coefficient) results in good vegetative growth and higher yield of musk melon cultivar "Hara Madhu".
- Muskmelon is a climactric fruit.
- Among cucurbits muskmelon is most suitable for daria cultivation.

Effect of Plant Growth Regulations (PGR) –

- In muskmelon application of Ethrel (250 ppm) increases the fruiting and the yield.
- Exogenous application of silver thiosulphate (300-400 ppm) induces the male flower.
- These PGR should be applied twice at 2nd true leaf stage and secondly at 4th true leaf stage.

Harvesting

- In muskmelon where fruits have to be picked at full maturity stage ready to be consumed as desert fruit becomes ready in 30-35 days.
- *Cucumis melon – var momoridica* (phoot) takes about 27-28 days from anthesis to harvesting.
- Yield – Pusa Rasraj (F_1) : 25 t/ha.
  Pusa Madhuras : 12-16 t/ha,
  Pusa Sharbati : 10-13 t/ha.
- Muskmelon is harvested at full slip stage except Hara Madhu cultivar.
- Full slip stage – Lucknow Safeda, Durgapura madhu, Cantaloupes of USA.
- In netted cantaloupe green colour of vein tract between the netting changes into yellow and netting become pale-or-dirty white. These cantaloupes can be picked at 3/4 slip stage for long distance transportation.
- Honey Dew and Casaba – Picked on the basis of external colour because netting do not separate at maturity.
- Long or serpent melon – Carpel separation in cross section.
- More juicer the flesh, the less will be the keeping quality of the whole fruit.

- The sweetness is entirely due to reducing and non reducing sugar in flesh chiefly glucose, fructose, sucrose, mannose etc.
- Salmon- Orange colour flesh contains carotenoid pigment.
- Green colour contains chlorophyll, white flesh have none.
- Half slip stage – Harvest for distant market to avoid losses from over ripening and decay.
- Flesh of muskmelon is used in ice-cream in western countries.
- Candy can be prepared from muskmelon.
- Pickling melon is used in making pickles.
- Jagger and Scott (1937) developed powdery mildew resistant contaloupe 45, PMR-45 of muskmelon.
- Muskmelon should have 10% TSS for Indian palate.
- TSS range – 11 -17% in most of cultivars.

"CUCUMBER" (Gherkin)

- It is the second most widely cultivated cucurbit after watermelon.
- Origin : India.
- Latin name: *Cucumis sativus* L. , Family: Cucurbitaceae.
- Fruits are used for salad and pickling, leaves are used as vegetable.
- In temperate countries, it is grown in glasshouse.
- Fruit and seed have cooling properties.
- Seed oil is used as antipyretic.
- Fruits are good for people suffering from constipation, jaundice and indigestion.

- Fruit contains – 0.4 % Proteins/100 g fresh wt.
  2.54% Carbohydrate/100g fresh wt.
  1.5 mg Fe/100 g fresh wt.
  2 mg Vitamin – C/100g fresh wt.
- India is considered to be the home of cucumber.
- *Cucumis hardwickii* – Small bitter cucumber with sparse and stiff spines found wild in the foothills of Himalayas which is the progenitor of cucumber.
- Free hybridization with cultivated sativus with no reduction of fertility in F2 generation indicates that *C. hardwickii* might be a feral or more likely proginator of the cultivated cucumber.
- Used for fresh consumption (slicing cucumber), or for preservation, marinated with vinegar, salt, or spices (picking cucumber).
- Warm season, frost-susceptible vine crop having separate staminate and pistillate flowers.
- Optimum temperature: 20° to 30° C.

Recommended Varieties of Cucumber

S.	Variety	Features
1	Poinsette	Fruits dark green, 20-25 cm long, resistant to mildew diseases, heavy yielder, suitable for summer season. Maturity 40-50 days. Yield 180 q/ha.
2	Pusa Uday	Early, 13-15 cm long, light green, non-wrinkled, suitable for both spring summer and rainy season. Maturity 48-52 days. Yield 140-155 q/ha.
3	Pusa Sanyog (F_1 hybrid)	Very early, fruits 28-30 cm long, cylindrical, attractive very heavy yielder. Maturity 50 days. Yield 200 q/ha.

■ Crop Improvement –

1. Introduction –

a. Japanese long green – Suitable for temperate region, extra early, by IARI, Katrain, suitable for hill region.

b. Straight Eight – Early cultivar, suited to hills.

c. Poinsett – From USA, Multiplied by (NSC) National Seed Corporation, developed originally at Charleston (S. Carolina) USA, resistant to powdery mildew, downy mildew, anthracnose and angular leaf spot (multiple res.)

d. China Long

2. Selection.

a. Sheetal – Recommended for Kankan region, Developed by KKV, Dapoli (Maharashtra), suitable for high rainfall areas.

b. Khira 75 – Developed at Dr. VSPUHF, Solan, HP.

c. Khira 90 – Developed at Dr. VSPUHF, Solan, HP.

3. Hybridization –

a. Himangi – Poinsett × Kalyanpur Ageti, developed by MPKV, Rahuri (Maharashtra), resistant to bronzing, fruit white in colour.

b. Phule Shubhangi – (Poinsett × Kalyanpur Ageti), green skin, tolerant to powdery mildew, developed by MPKV, Rahuri (Maharashtra).

c. Pusa Sanyog – It has gynoecious female parents (Japanese Gynoecious line × Green long (Napel).

- All of them are mostly restricted to the temperate region of the country, developed at IARI, Katarain.
- Private Sector Hybrid – Priya.
- The open pollinated cultivars of cucumber are monocecious in sex-form.

- The first F_1 hybrid, Pusa Sanyog developed in India in 1971 has a gynoecious female parent.
- It is suited to both temperate and subtropical regions of the country.
- This hybrid has shown heterosis for early yield, total yield and number of fruits per plants.
- Presence of gynoecious sex forms has made possible phenomenal progress in exploitation of F_1 hybrid under glasshouse condition.
- Temperate gynoecious lines are unstable for gynocey under high temperature and long photoperiodic condition hence three tropical gynoecious lines had been developed namely.
    - Gyc1 (87-304)
    - Gyc2 (87-3/9)
    - Gyc3 (87-338)

    All three developed at IARI, New Delhi
- Gyc4 (PKG -1) Developed at MPKV, Rahuri (Maharashtra) having parthenocarpic gene in it.
- DCH 1, 2 PH-9, 18 having involved tropical gynoecious lines as female parents are being tested in multilocational trials of PDVR.

4. Resistant Breeding –

The Indian variety Banglora was the starting point for developing downy mildew resistance in cultivar in the United States, Korea, Dutch and GY 598 – All these are less susceptible to red pumpkin beetle.

- *Meloidogyne incognita* resistant variety – GY 5937-587.
- Three different flower types namely staminate (male), pistillate (female) and hermaphrodite (both male and female) occur in cucumber.

- Main sex types in cucumber:
  - Monoecious plants; separate staminate and pistillate flowers on the same plant.
  - Androecious plants; staminate flowers only.
  - Gynoecious plants; pistillate flowers only.
  - Hermaphroditic plants; hermaphrodite flowers only.
  - Andromonoecious; separate staminate and hermaphrodite flowers on the same plant.
  - Dioecious; male flowers only on one plant and female flowers only on another.
- Three major genes determine sex in cucumber (*M*, *Acr* and *a*):

Gynoecious	=	*MM, AcrAcr*
hermaphrodite	=	*mm, AcrAcr*
monoecious	=	*MM, acracr*
andromonoecious	=	*mm, acracr*
androecious	=	*acracr, aa*

- *a* is responsible for androecy and has an effect in *acracr* plants only.
- The symbol *F* (*female*) is used synonymously with *Acr*.
- *F* locus has recently been shown to be very closely linked to a gene encoding ACC synthase, a key enzyme of ethylene biosynthesis.
- Identification of a 1-aminocyclopropane-1-carboxylic acid synthase gene linked to the *female* (*F*) locus that enhances female sex expression in cucumber.
- *Acr (F)* may be modified by an intensifier gene *In-Fe (In-F)*.
- *Tr* determines formation of trimonoecious plants (with staminate, pistillate and hermaphoditic flowers on the same plant).

- If a purely female variety is grown, there is need to provide an appropriate pollinator.
- Parthenocarpy is the ability to develop fruits without pollination.
- Parthenocarpy is controlled by an incompletely dominant gene *Pc*.
- Generally slicing cucumber have white spine colour and pickling cucumber have black spine colour.
- Balam Khira of Saharanpur (UP) regions is akin to pickling type of west.
- Pickling type – Fruits small, seed cavity develops more slowly, an important character affecting suitability for storage in brine.
- In pickling type thin skin with a lighter colour is preferred.
- K 90, K 75, Solan hybrid – Developed from Dr. YSPUHF, Solan.
- Poona Khira - Produce small size pale green fruits.
- Balam Khira – Similar to pickling type of USA.
- Cucumber is a thermophilic and frost susceptible crop.
- It is essentially a warm season crop.
- Excess humidity promotes disease like powdery mildew, downy mildew, anthracnose and pest such as fruit fly.
- Cucumber prefers slighty low temperature compared to melon and hence it is grown extensively in subtropical reigons.
- Optimum temperature : 26.4° C for growth.
- Optimum temperature : 13-18° C for anthesis and dehiscence of flower.
- If the temperature goes above 30° C the female flower production is reduced considerably.

Sowing Season –

November – January in North India (River beds)

June – July for rainy season

Mid February (Garden crops)

April – May (Hills)

- In North India seeds can be sown in alkathene bag (15 × 10 cm) of 100-200 gauge to avoid low temperature.
- Transplanting is done at 2 leaf stage.

Seed rate – 2.5-3.5 kg/ha.

Spacing 1.5-2.5m (Raw × Raw)

60-90cm (Hill × Hill)

Cultivar Shinko Ago – Spacing of 100 × 60 cm and pruning of plants to a single stem gives the highest yield.

- Cultivar Poinsett (Monoecious) spacing 60 × 60 cm under greenhouse.

Effect of Plant Growth Regulators –

- Application of Ethrel (150-200 ppm) increase the number of female flower, fruit set and in turn increase the fruit yield.
- NAA and ethephon promote pistillate flowers.
- GA (1500-2000 ppm), $AgNo_3$ (200-300 ppm) and aminoethoxyvinylglycine induce the male flowers on gynoecious cucumber.
- This enable maintenance of the gynoecious line in pure form and by using this gynoecious line, phenomenal heterosis has been achieved in cucumber.
- These PGR are applied at 2 true leaf and spray is repeated at 4 true leaf stage.
- Fruits of cucumber attain edible maturity with in a week from anthesis of female flower.

- Thermosensitivity – Sensitive to low temperature of January - February and high temperature of April-May.
- In salad or slicing cucumber dark green skin colour should not turn brownish-yellow or russeting, white spine colour is also a useful indication for their edible maturity.
- For commercial purpose, cucumber is harvested at immature condition 5-7 days after pollination.
- Crook-neck-shaped fruits are produced because of imperfect pollination and fertilization and also due to fly attack.

  Storage : 10° C. with 85% relative humidity.

■ **Physiological disorders –**

1. Chilling injury – When fruits are exposed to temperature below 10° C for prolonged period.
2. Pillow – Due to "Ca" deficiency.

- White spine colour indicates maturity in slicing cucumber.
- Black spine colour indicates maturity in pickling cucumber.

1. Japanese Long Green.
2. Straight Eight.
3. Pusa Sanyog.

All three varieties are suitable for temperate region.

- Light intensity and day length influence anthesis more than temperature. Temperature influence another anther dehiscence and pollen fertility.
- GA at higher concentration induces maleness but at the low concentration (10-25 ppm) increase the number of female flower.
- Pollinated by bees : 2 to 3 bee colonies per acre are recommended.
- Bacterial wilt is transmitted by striped cucumber beetles.
- Open pollinated cultivars of cucumber are monoecious.

- Pusa Uday – Suitable for both spring and Kharif season, developed by selection.
- Sex expression in cucumbers is modified by low fertility, higher temperature; longer light period. All these induce maleness.
- MH @ 100 ppm, NAA @100 ppm, Ethrel @ 150-200 ppm and "B" and Mo @ 3 ppm once at 2 leaf and second at 4 leaf stage increase the number of female flower, fruit set and in turn increase the fruits yield.

Varieties	
• Palmetto • PR - 27 • Santee • Palmer	Resistant to downy mildew
• Tokyo Long Green • Chinese Long • Wisconsin • Table Green	All these are resistant to Cucumber Mosaic Virus (CMV)

- Cucumber is characterized by monoecious sex expression but andromonoecious, gynoecious and trimonoceious are also found.
- Gynoecious lines are used in hybrid seed production because these lines do not require roguing of monoecious plants.
- Bitter principle in cucumber is cucurbitacin (tetra – cyclic-tri-terpenes).
- Bitterness in cucumber is due to metaxenic effect, when bitter pollen fertilizes non-bitter ovules, the resulting fruit become bitter, this phenomenon is called metaxenia.
- Preservation is done by brining cucumber in solution containing 5-8% Nacl during fermentation.
- During brine storage bloating, softening, bleaching and development of off-flavours and odours are the major defacts which usually occur.

- Bloating occurs due to production of excess CO_2 in cucumber.
- Cucumbers undergo lactic acid fermentation during storage which offers the advantages of acid fermentation.
- *C. foetidissima* – A xerophytic perennial sps; origin from Northern Mexico, also called buffalo gourd. Seeds rich in oil and protein.

"BOTTLE GOURD"

- Schwein furth has identified remains of *Lagineria siceraria* from Egyptian tombs dated about 3000-3500 B.C.
- A theory of trans-oceanic diffusion or drift of bottle gourd has been suggested to explain pre-columbion distribution in Tropical America.
- The wild species occurs only in East Africa and Mada gascar.
- *L. siceraria* (Syn *L. vulgaris*, *L. leucantha* (Bottle gourd) calabash, white flowered Zucca melon.
- *L. sphaerica* (Perennial wild Sp.) – Used for hybridization. The hybrid is sterile and possibility of transfer of desirable traits to cultivated species requires investigation.
- It is commonly grown in India, Ethiopia, Africa, Central America and other warmer regions of the world.
- The fruits of bottle gourd are used for making sweets halva, kheer, petha, burfi and pickles.
- Kofta is a most popular preparation of bottle gourd.
- A decoction made from the leaf is a very good medicine for curing jaundice.
- As a vegetable it is easily digestible therefore used even by patients.

- Fruit has cooling effect, cardiatonic and diureatic.
- Pulp is used for overcoming constipation, cough, nightblindness, and as an antidote against certain poisons.
- Seed are used in dropsy.
- The dry hard shell of the fruits have been used for making a wide range of articles of common use, including bowls, bottles, ladles, containers, floats for fishing nets, pipes and musical instruments.
- It is good for people suffering from biliousness and indigestion.
- Fruit pulp is very good source of fibre-free carbohydrate and fruit pericarp for crude fibre.

Recommended Varieties of Bottlegourd

S.	Variety	Features
1	Pusa Summer Prolific Long	Fruits long club shaped, curved thin 'neck', 40 to 45 cm long 20 to 25 cm thick at edible stage. Green in colour. Maturity 60 days. Yield 250q/ha. in summer and 275 q/ha. in Kharif.
2	Pusa Naveen	Fruits short, cylindrical and straight without curve or neck green colour, 25-30 cm long, 15-20 cm thick. Average fruit weight 800g. Suitable for sowing both in summer (February-March) and Kharif (June-July). Maturity 55 days. Yield 275 q/ha. in summer and 300 q/ha. in kharif.
3	Pusa Sandesh	Fruits round globular, medium size, green colour, Average fruit weight 500 g. suitable for sowing in both summer (February – March) and Kharif (June-July) Maturity 55 days. Yield 270 q/ha. in summer and 300 q/ha. in kharif.

4	Pusa Hybrid-3	Fruits long, slightly club shaped, straight without neck green colour, 35-40 cm long and 15-30 cm thick. Average fruit weight 1.2 kg . Maturity 55 days. Yield 425 q/ha. in summer and 470 q/ha. in Kharif.
5	Pusa Meghdut	A hybrid from cross between Pusa S.P. Long and S1-2. Fruit long. Light green, Yield about 258 q/ha.
6	Pusa Manjari	A hybrid from cross between Pusa S.P. Round and S1-11. Fruits round, light green. Yield 253 q/ha.
7	Arka Bahar	Economy-sized, medium long fruits (43cm), 26 cm girth, each weight about one kg. Skin light green, white tender flesh, yield 400-450 q/ha. Maturity 115-120 days in both the seasons.

■ Crop Improvement –

1. Selection

(i) Pusa Summer Prolific Long.

(ii) Pusa Summer Prolific Round- Prolific bearer.

(iii) Pusa Naveen-Fruit free from crookneck, weight- 850 g / fruit, cylindrical shape.

(iv) Punjab Long.

(v) Punjab Round.

(vi) Arka Bahar – Devoid (without) of crookneck, weight 1 kg/ fruit, pleasant aroma.

(vii) Samrat – Released in 1992 by MPKV, Rahuri (Mh.) good for box packaging, highest yield 40 t/ha.

2. Hybridization –

(i) Pusa Meghdoot – 1st F1 hybrid by public sector (Pusa Summar Prolific Long × Sel 2) released in 1971, fruits long, suitable for both spring and summer.

(ii) Pusa Manjari – (Pusa Summer Prolific Round × Selection 11), released in 1971, It has given 48% higher early yield over Pusa Summer Prolific Round.

■ Other varieties –

1. Pusa Sandesh – New variety suitable for spring, summer and rainy season.
2. Punjab Komal – Early variety, oblong fruit, 10-12 fruits/vine, tolerant to cucumber mosaic virus.
3. Pusa Hybrid – 3 – Recently released 1kg /fruit weight, fruits long, suitable for packaging in card board boxes for distant marketing, suitable for both spring summer and Kharif, very early in maturity, 40-45% higher yield than Pusa Naveen.

Yield	42.5 t/ha.	Spring / Summar Season.
	47 t/ha,	Kharif Season

4. Pusa Sneha – Improved variety over hitherto, very popular.
5. Pusa Supria – Suitable for spring and summer, kharif, high yielding variety.
6. Gutka
7. Harit
8. Varad
9. NDGB-2 : Highly preferred for diara land, long type convenient for packaging and transportation due to less bending, recommended for spring summar.
10. PBOG : Suited for Northern plains.
11. NDGB-4
12. Kalyanpur Hari Lambi – Suitable for summar and rainy season.
13. Phule BTG-1 – Cultivated in Maharashtra. Fruit long and Bottle shaped.
14. Rajendra Chamatkar-bottle shaped.
15. Kalyanpur Long Green – Developed by CSAUAT.
16. CO1 – Fruits typical bottle shaped.

- Sex form – Sex expression (Wide ranging) favours out breeding.
- Cucurbits do not suffer much from inbreeding depression.
- Breeding goal for bottle gourd are high yield, greater fruit number, great fruit weight, earliness, high female: male flower ratio, round long cyclindrical fruits, sparse hairs present on skin, non fibrous flesh at edible stage, non bitter fruit, attractive green fruit, resistant to disease and pest.
- The bottle gourd is a typical tropical plant which requires a hot and humid climate for best growth.
- It is highly sensitive to photoperiod (24-27° C).
- Higher temperature induces emergence of male flower and wide sex ratio.

  Temperature requirement –

  Night Temperature : 18-22° C.

  Day Temperature : 30-35° C. (For proper growth and development)

  Sowing Time : Dec-Jan in summar
  June-July in rainy season.

  Seed rate – 3-5 kg/ha. (450-500 seeds/100 g)

  Spacing – 2-3 m (Row × Row)
  1.0 m – 1.5 m (Hill × Hill)

- Brownish or whitish brown seed germinates well.
- Nitrogen deficiency – Yellowing of veins and foliage occur.
- Higher "N" produces more number of male flowers which is undesirable.
- Potassium deficiency – Reduce plant height, flower drop and checks fruits completely.
- In Maharastra bottle gourd is trained on bower system which increases yield.

- The Samrat cultivar of bottle gourd gives highest yield potential of 80 tonnes/ha on bower training system.
- Bower system prevents fruits of bottle gourd from rotting because fruits remain away from soil.
- High productivity on bower system is due to better use of sunlight by maximum number of leaves of side branches resulting in better assimilation of chlorophyll.
- PGRS – Ethrel (100-150 ppm), MH (400 ppm), Tri iodobenzic acid (50 ppm), "B" (3-4 ppm) increase fruit set at 2 leaf and 4 leaf stage. MH. (400 ppm) along with 100kg N/ha promotes the female flower production and increase fruit set.
- Tenderness and edible maturity are judged by pressing the skin and little pubescence persisting on skins.
- For hybrid seed production male flower buds of female parent are pinched before flower opening.
- The rows of female and male parents are kept in the ratio of 2:1 or 3:1.
- Processed product – Tutty fruity.
- Rainy season is very good for bottle gourd.
- Nail Test – Done for maturity.
- Summer season is suitable for seed production.

"BITTER GOURD"

(Balsam pear or bitter cucumber)

- Indo-Burma centre of origin has been reported by Garrison 1977.
- Bitter taste of fruit is due to momordicidin.
- Fruits turn to orange – yellow colour when ripe, rich in Iron, vitamin A, B, C and are inexpensive source of protein and minerals.

- The fruit is reported to have herbicidal effect and is laxative and easily digestible.
- Good for curing blood disease, rheumatism, diabetes and asthma.
- Cucurbitacin (bitter glucoside) – Helps in prevention of spoilage of cooked vegetable of bitter gourd.
- *Momordica dioica* - Spine gourd (Kakrol) 2x = 28.
- *M. cochinchinesis* – Sweet bitter gourd of Assam.
- *M. charantia* is incompatible with *M.diocea and M. balsamina*.
- White coloured varieties are less bitter in taste and preferred in South India.
- Fruits are exported to Gulf and European Countries.

Recommended varieties of Bittergourd

S.	Variety	Features
1	Pusa Do Mausami	Fruits medium long (15-18 cm), green, club shaped, unbroken ridges on the surface. Average fruit weight 80-100 g. suitable for sowing in February-March for summer crop and June-July for Kharif crop. Maturity 55-60 days. Yield 110-110 q./ha.
2	Pusa Vishesh	Fruits medium long, medium thick, glossy green, smooth broken ridges on the surface. Average fruit weight 100-150 g. suitable for sowing only in February-March for summer season crop. Maturity 60 days. Yield 125-130 q./ha.suitable for pickle and dehydration.
3	Pusa hybrid-1	Fruits medium long, small to medium thick, glossy green, smooth broken ridges on the

		surface. Average fruit weight 100 g. Suitable for growing in both summer and kharif seasons. Maturity 55-60 days. Yield 150-170 q/ha.
4	Arka Harit	Fruits short to medium long, thick, grayish green, smooth broken ridges on the surface. Suitable for growing in summer season. Maturity 60-75 days. Yield 75-90 q/ha.
5	Coimbatore	White medium long thin, broken ridges on the surface. Average fruit weight 60-75 g. Suitable for growing in kharif season. Maturity 60-65 days. Yield 70-75 q/ha. Released by TNAU. Coimbatore.
6	VK-1 Priya	Fruits are white and 35-40 cm long, heavy bearing. First picking at 60 days and yield 50 fruits/plant. Developed by Kerala Agricultural University.
7	Pride of Gujarat	Fruits are small, roundish, whitish, green in colour, weights 8-10 g.
8	Pant Karela-1	Highly resistant to red pumpkin beetle. Developed by GBPUAT, Pantnagar.

■ Crop Improvement –

A. Selection

1. Pusa-Do-Mausami – Fruits dark green, grow well in summer and rainy season, 12-15 t/ha yield.
2. Pusa Vishesh – For summer season.
3. Konkan Tara – Has good keeping quality (7-8 days under ambient temperature), suitable for export market.
4. Arka Harit – Spindle shaped, good for stuffing purpose (kharif and summer), rich in Vitamin C and Iron.
5. Arka Sumeet – Less seed per fruit.

6. Arka Sujat – Tolerant to downy mildew.
7. Priyanka.
8. Harkani.
9. Preethi.
10. VK 1 /Priya-Fruit 40 cm long.
11. Coimbatore Long white.
12. Co1.

B. Hybridization

1. Phule Green Gold – (Green Long × Delhi Local) Fruit dark green, tolerant to downy mildew.
2. Pusa Hybrid 1 – Early maturing, suitable for spring summer, for dehydration.
3. RHR BGH 1 – Tolerant to downy mildew, suitable for kharif and summer season.

C. Mutation

MDU 1 – Induced mutant with greenish white fruits, 30-40 cm fruit length, developed by irradiation of MC – 103 by gamma rays.

- Other varieties –
  Kalyanpur Baramasi
  Solan Green
  Solan Safed
- If temperature goes above 40° C. more male flowers are produced and plant become more prone to mosaic disease.
- Short day treatment – Reduces production of staminate flowers and increase number of pistil flowers.
- Low temperature (20° C.) – Increase production of female flower under short day condition.

Seed rate – 4-5 Kg/ha.

Sowing time –

April – July	In hills
June – July	Rainy season

	Jan – Feb	North India
	December-Jan	South India
Spacing –	1.5-2.0 m	(R × R)
	1.0 m.	(Hill × Hill)

- Application of azospirillium and phosphobacteria is also recommended to improve yield and quality of fruit.
- MH 50-150 ppm and CCC at 50-100 ppm increase female: male ratio but at higher concentration of 200 ppm CCC it is reduced.
- Ethrel at 25 ppm increase female flowers.
- GA at 60 ppm reduces the ratio of male: female flowers.
- MH (150-200 ppm) when applied at 2 leaf stage in variety Pusa Do Mousami induced the formation of pistillate flowers at the lowest node.
- Seed treatment with B9 at 3-4 ppm for 20 hours gave the highest number of female flower/plant.
- Soaking seeds in ethrel 20 ppm or GA_3, MH, $AgNo_3$ at 3-4 ppm gave the highest number of female flowers and fruits/plant.
- Immature tender fruits are harvested (60-70 Days after sowing).
- Fruits can be packed in poly propylene bags for extended shelf-life.

"PUMPKIN"

- It has high carotene content and good keeping quality. It is considered as a vegetable of immature value.
- Yerusseri – Prepared from immature fruits, is very popular in Kerala.
- The flowers of pumpkin are more nutritive than fruits.
- This crop is especially known for its low cost of production and long keeping quality.

- Medicinal use to reduce tape worm infection and its use as a diuretic are yet to be explored on large scale.
- The genomic structure of *Cucurbita moschata* is AABB which indicates that it is an amphidiploid.
- Both wild and cultivated species have same chromosome number 2n = 40.
- Five cultivated species
    1. *Cucurbita moschata* (Pumpkin).
    2. *Cucurbita pepo* (Summar squash).
    3. *Cucurbita maxima* (Winter squash)
    4. *Cucurbita mixta pang* (*Cucurbita angyrosperma*).
    5. *Cucurbita ficifolia* (Big leaf gourd or Malabar gourd)

(a) Xerophytic forms – *Cucurbita digitata, Cucurbita palmata, Cucurbita cylindiata, Cucurbita foetidissima.*

(b) Mesophytic forms – *Cucurbita ficifolia, Cucurbita sororia, Cucurbita lundelliana.*

- Pumpkin (*Cucurbita moschata)* appears to be more closely related to xerophytic forms and is considered as the axis through which the species under xerophytic forms are related to each other.
- The mesophytic forms are mostly cross compatible and *Cucurbita lundelliana* is seen as the key species in this group.
- *Cucurbita lundelliana* has wide spectrum of compatibility with the cultivated species particularly *Cucurbita moschata.*
- Among five cultivated species *Cucurbita moschata* (pumpkin), *Cucurbita pepo* (Summer squash) and *C. maxima* (Winter squash) are of great economic importance.
- Plant is monoecious.
- Flowers are unisexual, solitary or fasciculate and are lemon yellow to deep orange in colour.

- Isozyme studies in *Cucurbita moschata* revealed that there are 28 loci conditions in the isozymic phenotypes and this results together with the high chromosome number in the genus (2n=40) are taken as proofs for the allotretraploid origin of cucurbita.
- 85% area and 90% production of pumpkin is contributed by the state Orissa alone.

Recommended varieties of Pumpkin

S.	Variety	Features
1	Pusa Vishwas	Fruits spherical and medium. Average fruit weight 5.0 kg, fruit flesh thick and golden yellow in colour. Recommended for growing in northern plains of India, Western Himalayan region Sub humid Sutlej-Ganga alluvial plains, sub-humid to humid eastern and eastern unplands and humid to semi arid western ghats, Karnataka. plateau of the country. Maturity 120 days. Yield 400 q/ha.
2	Pusa Vikas	Fruits small and round flat, average fruit weight 2.0 kg. Fruit flesh deep yellow. Plant type dwarf (average length 2.0 m), Maturity 110 days, yield 250-300 q/ha.
3	Pusa Hybrid-1	Fruits round –flat and medium. Fruits flesh thick, golden yellow in colour and rich in vitamin A. Average fruit weight 5.0 kg, maturity 110 days. Yield 500 q/ha.
4	Arka Suryamukhi (*C. maxima*)	Fruit small, economy sized (1 kg), round with flat ends, ring colour deep orange yellow, with discontinuous thin cream lining along with shallow furrows. Linings originate from the blossom end, whereas

		the furrows originate from the stem end. Has exellent flavour. Firm texture, bright orange flesh colour and rich in vitamin C highly resistant to fruitly, bear's 8-10 fruits per plant. Yield about 335 q/ha. and the crop lasts for about 115 days.
5	Co-1	Fruits long, prominently ridged, light greenish yellow coloured flesh yield 141.3 q/ha. Duration of crop 125 days.
6	Arka Chandan	Fruit medium sized (2.5 to 3.5kg) round with slightly pressed blossom end, rind colour light brown with creamy patches at maturity. Have excellent flavor, firm texture and bright orange flesh, rich in carotene content. Bears 2-3 fruits per plant and yield about 325 q/ha in 115-130 days.
7	Co-1 (*C. moschata* Duch)	A late variety matures in 175 days, fruits, attractive, globular, medium large (7 to 8 kg) each vine produces 7-9 fruits, yield 288 q/ha. Released by the Agricultural Research Institute, Coimbatore.
8	Co-2 (*C. mosachata* Duch)	Fruit small sized, weight 2 kg with orange coloured flesh. Yield 226 q./ha. also developed at Coimbatore.
9	Kashi Harit	Short vines, fruits are spherical, green, weight 2.5-3 kg at green stage, 5 pickings can be taken in 70-75 days, 36-40 t/ha. yield for UP, Bihar, MP.

■ Crop Improvement

1. Selection

a. Arka Suryamukhi (1-1.5kg wt)– Variety of *Cucurbita maxima* (Winter squash) , resistant to fruit fly (D*aucus cucurbita*), Keeping and transport quality are good.

b. Arka Chandan (2-3 kg) – Rich in carotene (3331 IU/100g.) TSS 8-10%, pleasant aroma, keeping quality good, suitable for Western ghat and Karnataka.

c. Pusa Biswas/Vishwas.

d. Pusa Vikas – Suitable for spring summer in North India.

e. CO_2 – Fruit colour green.

f. Ambili – Pure line selection.

- The most popular cultivar In *Cucurbita moschata* is the butter nut squash with fruits relatively small, cyclindrical with pronounced bulb surrounding the seed cavity and bright orange flesh.
- No breeding depression has been observed due to the fact that the original form of cucurbits is hermaphrodite.

For F_1 hybrid seed production –

- Male and female parents are planted in the ratio of 1: 10 in isolation.
- Insect pollination is reported to produce hybrid seed. Two MS gene viz MS1 and MS2 have been reported in cucurbits.
- Genetic male sterility has been reported in pumpkin.
- Good flesh quality character can be transferred from *Cucurbita maxima* to *Cucurbita moschata* and resistance to squash bug and squash vine borers from *Cucurbita moschata* to *Cucurbita maxima.*
- Resistance breeding is mainly done by back cross method.
    1. *Cucurbita martinezzi.*
    2. *Cucurbita foetidissima* -Both are used to transfer resistance to CMV in *C. moschata.*
- A single gene when homozygous in *Cucurbita moschata* confers a high level of resistance to Yellow vein mosaic virus (YVMV).

- Short days low night temperature and high relative humidity is best for pumpkin production.

Sowing time	April – May	In Hills
	June – July	In Plains
	January – February	In Plains

- Channel and hill system of cultivation is most useful for higher yield (Scientific method).
- Seed rate – 4-5 kg/ha.
- In general high "N" under high temperature condition promotes maleness in flowering resulting in low fruit set and high yield.
- Excessive vine growth can be pruned manually to promote higher female to male ratio.
- PGRS – Ethrel 250 ppm (2.5 ml of ethrel in 10 litres of water) can be applied to increase the female flower production.
- Honeybees are chief insect pollinators for pumpkin.

Seed Production –

- The cross compatible species like *Cucurbita pepo* and *Cucurbita mixta* should be planted in isolation from *Cucurbita moschata.*
- The pollen grains of non crossable species sometimes stimulate the ovary of pumpkins that leads to development of partheocarpic fruits.
- Isolation distance – 1000 M.
- Male and female parents can be grown in a 1 : 3 ratio.
- Matures 75-180 days after seed sowing.
- Full mature fruits have a long storage life.
- Yield 20-30 tonns/ha.
- Storage 2-3 weeks at 15-20°C.
- Well mature fruits can be stored for 4-6 months is well ventillated chamber.

"SUMMER SQUASH"

(Vegetable narrow or field pumpkin)

- It is also called Vilayati Kaddu or Chappan Kaddu as fruits resembles a cover of earthern pitcher.
- In Hindi it is called Kumra, Safed kaddu.
- Immatures fruits are used as vegetables.
- It is used as bathing sponge.

Recommended varieties of Summer Squash

S.	Variety	Source	Remarks
1.	Punjab	Punjab Agrl. University, Ludhiana	It is inbred selection from the segregating local variety of Punjab. Plants are bush type, foliage thick and erect, leaves non lobed and green without white specks, mildy ribbed with flat stem end and attractive average fruit weight 80g and bears per plant about 10 fruits, early maturing and is ready for first harvest in about 60 days from sowing. It has as predominant female tendency, field resistance to downy mildew and red pumpkin beetle. It is high yielding and gives about 20-25 tons/ha.
2	Patty Pan	IIHR Bangalore	An Introduction from USA and recommended by IIHR for cultivation during 1972. A bushy-type and fruits are disc-shaped, chalky white, tender and very attractive at edible stage. A short-duration variety (80-90 days), yield 54 tons/ha.
3	Early Yellow	IARI, Regional Station	An early bush-type variety, Fruits medium sized, warted and tapering

	Prolific	Katrain	are towards stem end light yellow skin turns to orange-yellow on maturity. Flesh is tender at the stage it is consumed as vegetable.
4	Australian Green	IARI, Regional Station Katrain	An introduction, Very early bush-type variety. Green fruits are dark-green with longitudinal white strips all over, 25-30 cm long. 15-20 fruits/plant and very tender at edible stage. Yield 15-16 tons/ha.
5	Pusa Alankar	IARI, Regional Station Katrain	An F1 hybrid between Ec 207050 and Sel 1 Pt 8 (a derivative from cross between chappan and early yellow prolific),early maturing having uniform dark green fruits with light coloured stripes slightly tapering towards the stem. The flesh is tender delicious and fruits mature in 45-50 days. High yield 20-30 tons/ha.

■ Crop Improvement

1. **Introduction** – Patty Pan, Australian Green.
2. **Selection** - Punjab Chappan Kaddu 1 - Inbred selection has a predominant female tendency, resistant to downy mildew and red pumpkin beetle.
3. **F1 hybrid** –

   Pusa Alankar – EC 207050 and Selection 1.

   (A. derivative from cross between Chappan Kaddu × Early Yellow Prolific)

- F1 hybrid can be obtained from most inter specific mating between *Cucurbita pepo, Cucurbita mixta, Cucurbita moschata* and *Cucurbita maxima* but such hybrid are normally sterile.
- *Cucurbita moschata* was used as a bridge, to transfer disease resistance gene from *Cucurbita martinezii* to *Cucurbita pepo*.

- The three ways cross between *Cucurbita pepo* and (*Cucurbita moschata* × *C. martinezii*) is being used to transfer resistant to powdery mildew and CMV from *Cucurbita martinezii* to *Cucurbita pepo.*
- Washek (1982) achieved a cross of *Cucurbita pepo* with the multiple disease resistant *Cucurbita ecuadorensis* through embryo culture.

Sowing Time	February – March	
	June – July	
	April – May	: Hills
	October – January	

- Fruits are harvested at 1/3 maturity.
- Summer squash is also known as bushy cucurbit or ornamental cucurbits.

"Winter squash"

- Monoecious.
- Tolerant to frost.
- Commonly grown in hills variety.
  - (i) Arka Suryamukhi – Resistant to fruit fly.
  - (ii) Pink Banana.

"RIDGE GOURD" (Ribbed gourd)

- Both the species (*Luffa acutangula and L. cylindrica)* contain a gelatinous compound called "luffein".
- The genus derives its name from the product "*Loofah*" which is used in bathing sponges, scrubber pods, doormats, pillows, mattresses and also for cleaning utensils.

- Ridge gourd – Kali tori.
- The sponge gourd fruits contain higher protein and carotene than ridge gourd.
- Among the gourd vegetables only bitter gourd, bottle gourd and pointed gourd have export potential.
- Monoecious, young fruits are used as vegetables.
- Anthesis – Start in evening and remain open throughout the night and are ready for selfing and pollination in the next morning.
- Ridge gourd flower – Pale Yellow.
- Sponge gourd flower – Deep yellow.

Recommended Varieties of Ridge gourd–

S.N.	Variety	Features
1	Pusa Nasdar	Early maturing, suitable for summar and rainy season, produce club-shaped fruit, monoecious. Maturity 55-650 days. Yield 100-155 q/ha.
2	Arka Sujat	Lush green and tender medium longer (35-45 cm) fruits much prominent ridges and delicate aroma, good transport and cooking qualities. Duration 100 days and yield 53 t/ha. Recommended for Karnataka.
3	Arka Sumeet	Lush Green and tender long fruits (50-65cm) with prominent ridges and delicate aroma, good transport and cooking qualities Duration 110 days and yield 50 t/ha. Recommended for Karnataka.

■ Other Varieties :

1. Punjab Sadabahar – Rich in protein.
2. Satputia–Hermaphrodite in sex form (Bisexual flower).

3. Konkan Harita.
4. PKM-1 (Kharif and summer).
5. Col, 2.
6. Hybrid – Surekha, Arka Sujat.

- Sowing time as mentioned in other crops.
- Seed rate – 3.5-5 kg/ha.
- Commercial crop are trained on a kniffen system.
- The sex ratio can be regulated by exogenous application of growth regulators.
- NAA 200 ppm increases the female flower production.
- Harvesting is done when fruit is tender and still immature.
- Fruit 14-15 cm long, club shaped & surface is distinctly ribed.

"SPONGE GOURD"

- Fruits of sponge gourd are smooth and is called ghiatori.
- *L. cylindrica* (Syn. to *L. aegyptiiaca*) is also known as sponge gourd, towel gourd, smooth luffa, veg sponge (dish cloth gourd).
- It is monoecious climber.
- Anthesis occurs at 4.0-8.0 am, suitable for selfing almost throught the day.
- Fruits are 20-50 cm long, almost cylindrical in shape.

Recommended Varieties of Sponge gourd–

S.N.	Variety	Features
1	Pusa Chikni	Fruits smooth, dark green, suitable for both spring summer and rainy, early maturing high yielding, poor yielder. Maturity 50-55 days. Yield 110-120 q/ha.

2	Pusa Supriya	Short vine growth, fruits light green, smooth, tender, 12-14 fruits per vine. Gives a 2% more yield than Pusa Chikni, suitable for growing in both seasons. Maturity 40-45 days. Yield 100-120 q/ha.
3	Pusa Sneha	Fruits dark green, 20-25cm long, straight tolerant to high temperature. Suitable for cultivation in spring-summer and rainy seasons. Maturity 45-50 days. Yield 110-130 q/ha.

■ Other Varieties :

1. Phule Prajakta – Suitable for summer and kharif season.
2. Harita – F1 hybrid

- Seed Rate : 2.5 - 3.5 kg./ha.
- Application of Ethrel 250 ppm has been found to be beneficial.

"ASH GOURD"

- Ash gourd or wax gourd and white gourd is also known as petha kaddu.
- It is used in the preparation of Petha sweet.
- Agra Petha is famous all over India.
- Ash gourd is considered good for people suffering from nervousness.
- Ash gourd is an annual hispid (rough white bristle like hairs) climbing, herbaceous and monoecious plant.
- It is also known as Chinese preserving melon.
- Also used to make sugar candies.
- The surface of the fruit is covered with a bluish white "Waxy" ash.
- Ethephon (250 ppm) spray increase production of female flowers.

Recommended Varieties of Ashgourd

S.N.	Variety	Features
1	Pusa Shakti	Fruits oblong-ellipsoid with greenish-white flesh. Average fruit weight 7.0 kg. Ideal for easy packaging and long distance transportation. Long storability, more than 6 months at room temperature. High TSS, Vitamin C and minerals. Suitable for cooking as vegetable and also for preparing Petha (sweet candy). Maturity 140 days, Yield 250 q/ha.
2	Co1	Fruits globular, light green with ash coating. Suitable for growing in South India. Maturity 150 days. Yield 250 q/ha.
3	Co2	Fruits light green, small sized, average weight 2.5 to 3.0 kg also suitable for growing in South India. Maturity 120 days. Yield 340 q/ha.
4	Kashi Dhawal	Fruits oblong, harvesting in 100-105 DAS. Most preferred variety for sweet preparation. 55-60 t/ha. For UP, Punjab and Bihar.

■ Other Varieties :

1. Pusa Ujjwal.
2. APAU-Fruit Long.
3. Mudliar - By TNAU.
4. Pusa Shakti – New variety.

- Short day, low night temperature and humid climate are good for production of female flower.
- Planting time – June-July.
- Seed rate – 5-7 kg/ha.

- Fruit at maturity have white waxy surface.
- The fruit is mainly purchased by the confectioners who use them to prepare "Petha".
- Ash gourd can be stored longer than any other cucurbits.
- Disappearance of ash wool on the fruit surface is an indication of maturity.

"POINTED GOURD"

- Pointed gourd (parwal) is a perennial cucurbit cultivated mostly in Bihar, West Bengal and Madhya Pradesh.
- Immature fruit of this plant is a popular vegetable containing 2% protein, 0.03 % fat, 2.2% carbohydrate and 153 IU Vitamin-A.
- Fruits are used in making curry and also fried.
- Parwal is a dioecious climber with perennial rootstock.
- Fruit is particularly recommended during convalescence (regain health after illness).
- Parwal remains dormant during winter.

■ Varieties :

1. Swarna Rekha – High yielding.
2. Swarna Alaukik – High Yielding.
3. Shankolia
4. Dandali
5. Chhota Hill
6. Hilli
7. CHES Elite line – Resistant to fruit fly.
8. CHES hybrid – First parwal hybrid developed in country.
9. Faizabad Parwal 1.

10. Faizabad Parwal 3 - Excellent for culinary purpose.
11. Faizabad Parwal 4 - Recommendation for bower system.
12. Damodar.
13. Kazil.
14. Bihar Sharif

- Require humid and hot climate since it is susceptible to frost and cold. It can withstand water but not water logging.
- It remains dormant in winter and grows vigorously during spring.
- Rainfall is considered very good for this crop.
- Pointed gourd is vegetatively propagated through vine cuttings and root suckers.
- About 2000-2500 (1m. length) cuttings are required.
- Seed propagation is avoided in pointed gourd because of poor seed germination and since it is dioecious in nature produces male and female plants in equal proportion if they are planted from seed.
- In other words, 50% plants are non fruiting (i.e.male plants). In order to ensure maximum fruit set and yield only 10-12% male plants are maintained in the garden to ensure the source of pollinizer and remaining plants must be female.
- In bower system 8000 cuttings are required to raise one hectare crop with in a spacing of 1.0 × 1.5 m.
- In flat system – 4500-5000 cuttings, 1.0 × 2.0 m spacing.
- In Riverbeed system – 3500-4000 cuttings.
- February - March – Ideal time in upland for sowing.
- Mid. November – Ideal time in diara land for sowing.
- Lunda or lachhi method – Cuttings folded into a figure of "8" commonly known as lunda or lachhi.

- Moist lump method
- Straight vine method
- Ring Method
- Rooted Cutting
- Root suckers

Other methods

- Minimum of 10% male plants in the field are essential to ensure the availability of enough pollen grains required for fertilization and fruit setting.
- For every 15 female plants one male plant is must to get highest yield.
- Fruit availibility – February-October

"SNAKE GOURD"

- Fruits are rich in carbohydrates, proteins, minerals and vitamins A, B and C.
- It is monocecious, annual and herbaceous climber.
- Immature fruits are eaten.
- The fruit improves the apetite, acts as a tonic and stomachic and cures bilousness.
- The roots and seeds are used in the treatment of diarrhoea.
- Flowers opening occur in afternoon.
- India is native home of snake gourd.
- It is widely cultivated in South and Soutn East Asia.

Recommended Varieties of Snakegourd

S.	Variety	Features
1	Co 1	160-180 cm long fruits with white stripes on dark green skin, flesh greenish. Each vine gives 10-12 fruits, average weight 4-5 kg. Maturity 135 days. Yield 180 q/ha.

2	Co 2	Suitable for HDP, 30-35 cm long stout fruits, greenish, stripeless skin, ashy bloom. Vine length short, can accommodate more number of plants per hectare at shorter planting distance. Maturity 125-130 days. Yield 250-300 q/ha.
3	MDU – 1	A hybrid, fruit skin green with white stripes and medium length around 60 cm. Maturity 125 days. Yield 200 q/ha.
4	APAU Swetha	Long fruits, white skin, green fingures (stripes) maturity 125-130 days. Yield 280-300 q/ha.

■ **Other Varieties –**

1. CO4 – Early maturing.
2. TAL – 19 – Developed by KAU, Kerala.
3. Konkan Sweta 1 – Fruits have good flesh if harvested timely otherwise is becomes hollow.
4. MDUI – F1 hybrid.
5. PKM-1-extra long (200 cm).

Seed rate – 5-6 kg/ha, spacing 1.5-2.5 m. R × R, 60-120 cm P × P.

- Training is essential to this crop. The bower system of training is best for this crop.

- **Sowing time –**
  - Summer Season – December – January.
  - Rainy Season – June-July.
- Fruits are hand picked when they are still tender and about 1/4 –1/3 of their full size.
- Fully mature fruits are lighter, fibrous and hard.
- Just after harvesting fruits are sent to market as they do not have shelf life more than two days.

"IVY GOURD" (Little Gourd)

- It is most commonly used as vegetable in Southern and Central India.
- The fruits are rich source of carbohydrate, protein, vitamin - A and C.
- IVY gourd is dioecious in nature.
- IVY gourd is also known as Coccinia or kundru or Tondli" or little gourd or scarlet gourd.
- It is perennial in nature.
- The immature fruits and tender shoots are used as vegetable.

  Propagation – By stem cutting (25 cm long).

  Planting – June-July and February – March.
- To get good yield 5-10% of the male type vines should be planted in the field.
- For a commercial crop, plants are trained on a bower system constructed to suit the growth of vine.
- Harvesting is determined by change in colour from dark green to bright or light green.
- Semi hard wood cutting (5-20 cm long) 1.5 cm diameter collected from female parent are planted @ 2-3 cuttings per basin, 5-6 cm deep.
- *Coccinia grandis* is a dioecious; about 10% of the total population should be male for commercial crop.
- There are certain clones which do not bear male flowers but set fruit parthenocarpically.
- Parthenocarpy – Vegetative type is found.
- In parthenocarpic clones the fruit set is without pollination hence there is no need of separate male and female clones.

- During winter plants go in dormancy under North India, however under South India plants grow throught the year.
- Yield 200-300 fruits, weight 3-4kg/vine /year.

"ROUND MELON"

- It is also known as Round gourd, Indian squash, Squash melon or Tinda (*Citrullus vulgaris var fistulosus*).
- It has cooling effect to the body.
- The fruits are also considered good for curing dry cough and for improving blood circulation in body (Vitamin A).
- 27-30° C temperature is optimum for seed germination.

■ **Varieties –**

S.N.	Variety	Features
1	Arka Tinda	Recommended for Karnataka and grown in northern states. An early summer variety. Fruits round, light green with soft hairs. Tender flesh at marketable stage. Maturity 90-100 days. Yield 95-100 q/ha.
2	Punjab Tinda	Fruits medium in size, flat round and light green. Fruit surface shining and pubescent. Flesh white less seedy tender, good cooking quality. Maturity 60 days. Yield 45 q/ha.
3	Bikanery Green	Fruits round, tender and green in colour. Suitable as spring summer crop in Rajasthan.
4	Hisar Selection -1	Fruits oval, green in colour with hairs, very attractive.
5	Tinda Ludhiana	Flesh White, less seed, Good keeping quality.
6	Tinda Tonk	Grown both as solo and as an inter crop.

- Seed rate : 3.5 - 7.0 kg/ha.
- Picking should be done when these are still immature and small in size.
- M.H. (50 PPM) aqueous solution sprayed at 2 leaf and 4th leaf stage stimulates vine growth, giving more femaleness and enhance female flowering at lower nodes.
- Due to high water content fruits are spoiled early.

"CHOW-CHOW" (Chayote, Choco, Askas)

- It (*Sechium edule*) is an herbaceous, perennial climbing, monoecious vine (13m. length) having large tuberous roots.
- Seeds are delicacy.
- Chow-Chow is a single seeded cucurbit.
- Single seed is located near the broad end of the pear shaped fruit.
- High "Ca" content.
- Viviparous germination, highly sensitive to frost.
- Propagation - By sprouted fruit.
- Chow Chow is grown for its large tuberous roots.
- Principally grown for its pear shaped fruits.
- It is rich source among gourds in nutritive value particulary in carbohydrate (6%) and calorific value 29 Cal/100 g.

"PERENNIAL CUCURBITS"

- Kakrol or spine gourd (*Monordica diocia*) is dioecious.
- Remains dormant during winter, fruit ovate in shape (10-12 cm long).
- Rich in protein and Vitamin C.
- Propagation – Tuberous roots (5 lakh/ha).

- Seed : 3 - 4 kg.
- Being dioecious in nature about 10% of male plants should be planted for good fruit set.
- Seed oil – Used as an illuminant.

"KARTOLI" (Sweet gourd)
(*Momordica cochinehensis*)

- Dioecious in nature.
- Propagation – Tuberous roots.

"SNAP MELON" (Phoot)
(*Cucumis melo var momordica*)

- Still grown during summer and rainy season in North India.
- The fruits brust when snap melon is ripe therefore it is popularly called phoot.
- Immature fruits having light green to dark green colour are harvested for vegetable and salad purpose.

"LONG MELON" (*Cucumis melo var utillissimus*)

- Commonly grown in North India.
- Two types of long melon are very popular in Lucknow.
  1. Lalia ki ungalian (Fingers of Lalian) – Fruits small and thin.
  2. Manju ki Pasalian (Ribs of Manju) - Long thin fruit
- Variety – Arka Sheetal an improved variety.

S.N.	Variety	Features
1	Arka Sheetal	Fruit medium long, light green skin, shallow furrows, fruit weight 80-90 g. fresh crisp excellent flavour. Yield 350 q/ha. Duration 100-110 days. A selection from the local strains of Lucknow developed at IIHR. Recommended for Karnataka.

"SNAP MELON"

(*Cucumis melo var momordica*)

- Still grown during summer and rainy season in North India.
- The fruits brust when snap melon is ripe therefore it is popularly called phut.
- Immature fruits having light green to dark green colour are harvested for vegetables and salad purpose.

"LONG MELON"

(*Cucumis melo var utillissimus*)

- Commonly grown in North India.
- Two types of long melon are very popular in Lucknow.
  1. Lalia ki ungalian (Fingers of Lalian) – fruits small and thin.
  2. Manju ki Pasalian (Ribs of Manju) - Long thin fruit

■ **Variety –**

Arka Sheetal an improved variety.

10

POTATO & TUBER CROPS

C.N.	B.N.	Family	Ch. No	Edible part	Origin
Potato	*Solanum tuberosum*	Solanecae	48	Stem Tuber	S. America (Peru-Bolivia)
Sweet Potato	*Ipomoea Batatas*	Convolvulaceae	30,40, 90	Tuber	South America
Yam	*DIoscorea sps*	Dioscoreaceae	20,30,40, 50,60,70, 80,81	Tuber	South East Asia
Cassava	*Manihot Esculenta*	Euphorbiceae	36	Tuber	Brazil
Colocasia	*Colocasia Esculenta*	Araceae	24, 28	Corms	South Central Asia
Elephant Foot Yam	*Amorphophyllus Compinulatus*	Araceae	26	Corms	India
Yam Bean	*Pachyrhizus* spp.	Fabaceae	22	Corms	South America
Tannia	*Xanthosoma sagittifolium*	Araceae	26	Corms, leaves and petioles	South America
Queensland /Purple Arrowroot	*Canna edulis*	Cannaceae	18	Rhizome	South America
Jerusalem Artichoke	*Helianthus tuberosus*	Compositae (Asteraceae)	49-53	Tuber	Eastern United State
Coleus	*Solenostemon ratundifolius* (syn. *Coleus parviflorus*)	Lamiaceae	48	Tuber	Tropical Africa, Asia & Australia
Arrowroot/ West Indian Arrowroot	*Maranta arundinacea*	Marantaceae	18	Rhizomes	West Indies and Florida
Giant Taro	*Alocasia macrorrhiza*	Araceae	28	Rhizomes & shoots	Asia
Chinese Water Chestnut	*Eleocharis dulcis*	Cyperaceae	10, 16 18	Tuber	China

- Central Tuber Crop Research Institute was established at Thiruvananthapuram (Kerala).
- Project Directorate of Vegetable Research is at Varanasi (UP).
- Potato contains dimethyl pyrazine (aroma containing compound).
- Cassava is monoecious while Yams are dioecious.
- Respiration rate of potato is very low.
- Yam Bean – Also known as Misrikand, Potato Bean.
- Potato and Sweet Potato are C_3 plants.
- Fuyu – Product made from yam.
- Sago – Important food product from Cassava starch.
- Sweet potato and yams- Fruit type is capsule.
- Inflorescence of Cassava – Penicle.
- Highest Carbohydrates – Tapioca.
- Sweet potato is cheapest source of calorie.
- High fiber- Potato.
- Diosgenin contaning yam is useful in manufacture of cortisone and contraceptive drugs.
- Cercospara leaf spot of sweet potato was 1st reported in Africa.
- Shree Shaya – Multiple hybrid of Cassava.
- Central Potato Research Institute (CPRI) – (Under ICAR) Established in 1949 with headquarters at Patna which was later shifted to Shimla in 1956.
- AICPIP (All India Coordinated Potato Improvement Project) was started in 1971 with headquarter at CPRI Shimla (HP). It has 22 centres.

- International Potato Centre (CIP) – Autonomous Scientific institution in Lima (Peru) established in 1972 by agreement with Govt. of Peru.
- It is one of the 16 International research and training center supported by consultive Group on International Agriculture Research (CGIAR).
- The CGIAR is sponsored by FAO of the United Nations.
- UNDP – United Nations Development Programme.
- UNEP – United Nations Environment Programme.
- TPS – True Potato Seed.

"POTATO"

- About 50% of potato produced in world is utilized as human food.
- Potato is used as staple diet in many of the countries especially in the west.
- The potato was introduced in India from Europe in early 17th century, only fourty years after its introduction in Europe by Portuguese.
- Potato is grown in India in almost all the states except "Kerala".
- Potato is grown during short days of winter (October-March) in about 22% area in the plains.
- Potato is grown during long days of summer (April-September) in about 10% area in hills.
- Potato is a dicot plant.
- Genus Solanum contains nealy 2000 species.
    1. Pachystemonum – Divided into five sections of which potato contain most of tuber bearing species.
    2. Lepto stemonum

- Section Petota – Two Sub section (a) Estolinifera (b) Potatoe.
- All the cultivated sp. has been grouped under the series Tuberosa of sub section potatoe.
- About 72% of the species are diploid (2x = 24) and nearly 12% tetraploid (2x = 48), rest triploid (2x = 36), pentaploid (2x = 60) and hexaploid (2x = 72).
- The widely cultivated potato belongs to tetraploid sp. (2x = 48) *Solanum tuberosum.*
- First mention of potato in India began on the account of a bouquet given by Asaf Khan to Sir Thomas Roe in Ajmer (Rajasthan) 1615.
- Probable ancestor of all cultivated potatoes is *Solanus stenotonum.*
- The annual per capita consumption in the country is 15 kg as against the world average of 24 kg.
- Potato breeding station, Kufri (Simla) was established in 1935 under IARI.
- Central Potato Research Institute, Shimla was established in 1949.

■ Crop Improvement –

1. **Introduction** – Upto date from N. Ireland, most of popular varieties in India.
2. **Clonal Selection** – Kufri Red from Darjeeling.
   Red Round } Both are selection
   Kufri Safed } from var Phulwa
3. **Hybridization** –

1. Kufri Sindhuri – K.Red × K. Kundan.
2. K. Badshah – K. Jyoti × K. Alankar, resistant to late blight, early blight and Potato virus Z.

3. K. Jyoti – 3069 d (4) × 2814 a (1) resistant to late blight, immune to wart disease.
4. Kufri Bahar – K. Red × Gineke.
5. K. Lalima – K.Red × AG 14.
6. K. Swarna – K. Jyoti × VI n2, don't show cracking, highly resistant to late blight and cyst nematode.
7. Kufri Jawahar – K. Neelami × K. Jyoti, moderately resistant to late blight.
8. K. Sutlej – K. Bahar × K. Alankar, moderately resistant to late blight.
9. K. Pukhraj – Craigs Defiance × JEX / B-687, resistant to early blight and moderately resistant to late blight.
10. K. Chipsona 1 – MS/78-79 × CP 2416, resistant to late blight, suitable for processing / producing light – colour chips and finger fries.
11. K. Chipsona 2 – Novazo × QB/B- 92-4, resistant to late blight and tolerant to frost, suitable for processing giving light colour chips and frying.
12. K. Alankar, K. Jevan, K. Moti, K. Deva, K. Kundan, K. Chamtkar, K. Naveen, K. Khasigaro, K. Luvkar.
13. K. Sheetman -Phulwa × Craig's defience, resistant to frost.
14. K. Chandramukhi - Sd – 4485 × K. Kuber, early maturing.
15. K. Griraj – SLB/J132 × EX / A 680-16, resistant to late blight.

■ Other varieties –

Early Maturing –

- K. Chandramukhi (75 days)
- K. Lavkar suitable both for Rabi and Kharif season.
- K. Ashoka (75 days) – Replaced K. Chandramukhi.
- K. Jawahar (80 days).

Late Maturing

- K. Sindhuri (120-140 days) – Popular in Northern India.
- K. Swarna (120 days).
- K. Deva.
- K. Megha – for Meghalaya.

Mid. Maturing –

a. K. Jyoti.
b. K. Lalima.
c. K. Badshah.
d. K. Sutlaj.
e. K. Bahra.
f. K. Pukhraj.
g. K. Chipsona 1 (100 days).
h. K. Griraj (100 days).
i. K. Muthu – For Southern Hills.
j. K. Lalima.
k. K. Sutlej Replace K. Badhsah and K. Bahar.

- Varieties resistant to Late blight –K. Badshah -Also resistant to early blight, and Potato virus x.

Late blight resistant varieties

K. Jyoti – Also immune to wart disease.

K. Swarna – Also resistant to cyst nematode.

K. Jawahar.

K. Pukhraj.

K. Chipsona 1 and 2.

K. Megha, K. Jeevan, K. Kuber, K. Moti, K. Naveen, K. Kundan.

K. Badshah – also resistant to early blight, Potato virus (Multiple Disease Resistant).

Varieties for plains –

K. Ashoka, K. Pukhraj, K. Anand, K. Alankar.

K. Kuber - Resistant to black scurf.

K. Badshah – Multiple resistance: Late blight, early blight and potato cyst nematode and potato virus.

K. Jyoti – Have tendency to cracking and immunce to wart disease.

K. Swarna – Resistant to cyst nematode and late blight.

K. Sherpa – Resistant to cyst nematode and late blight.

K. Pukhraj- Resistant to early blight.

K. Sindhuri – Resistant to early blight.

K. Chipsona I and II- Suitable for processing (Chips), Also tolerant to frost and late blight.

(4) Mutation breeding – R-8 through x rays.

Triploid potato (Solanum × chaucha) cultivated at Peru and Bolivia.

Pomato – Potato × Tomato, Developed with the help of protoplast fusion by Melcher (1978).

- Late blight (*Phytophthora infestans*) is the most devastating disease occurring regularly in epiphytatic form every year in hills and very frequently in plains.
- Potato virus – Aphid are vectors, important constraint in potato productivity.
- Bacterial Wilt – Common in Mid hills and Assam, Meghalaya, Maharashtra and Orissa.
- Wart Disease – (*Synchytrium endobioticum)* common in hilly regions of Darjeeling (W.B.).
- Cyst forming nematode – Common in Southern Hills.
- Potato Tuber Moth – More attack in warm plateau areas of Maharashtra, Karnataka and Madhya Pradesh.

- Varieties containing low sugar and high dry matter are suitable for preparation of special value added products like chips, french fries, cubes and other dehydrated products.
- Potato is a self pollinated crop but it is vegetatively propagated.
- The cultivated tetraploid varieties are highly heterozygous.
- Selfing or inbreeding in potato leads to loss of vigour of the progeny and non flowering.
- Hybridization –Intervarietal or interspecific.
- The parents for hybridization are generally selected on the basis of their combining ability.
- Resistant genes are found scattered in wild and semi cultivated species.
- Available in center of origin and diversity in S. America.
- Potato flowers are hermaphrodite (bisexual), emasculation is done in selected female parents in evening.
- Pollination is done in morning in emasculated flower.
- The bacterial "Plasmid" which is free floating rings of DNA is used in carrying and transferring the derived gene.
- Two commonly used bacteria in this technique are *Agrobacterium tumefaciens* – Causing crown gall disease, *Agrabacterium rhizogenus* causing hairy roots in plant.
- "Restriction endonucleases" and "Ligase" which are special enzyme heip in cutting and sealing the DNA molecules at specific sites.
- In Potato the genetic transformation technique have been successfully used in developing varieties resistant /tolerent against viruses, insect herbicide, frost and salinity.
- Potato naturally flowers under cool climate and long day condition of more than 15 hours light (Hills).
- Hills are therefore ideal for hybridization work.

- Carbohydrates - Major constituents of potato.
- Potato is a short duration crop.
- Maximum area under Potato is in alluvial soils.
- Black Soils – Prone to cracking on drying and exposure of tuber to sun and tuber moth infestation.
- Excellent potato crop is grown under river bed system of cultivation in Deesa, Banaskantha in Gujarat.
- Ph 5.5-7.5 (Acidic to neutral) is suitable for potato.
- In Potato crop temperate region crops response to thermoperiods.
- Favourable temperature for tuber farmation – Below 21° C.
- Temperature above 21° C causes sharp fall in tuberization.
- Summer crop is taken in Hills.
- Winter crop – Tropical and Subtropical areas.
- For best yield potato need long day condition during growth and short day condition during tuberization.
- Good crop growth occurs when days are sunny and nights are cool.
- Most potato genotypes do not tuberize when the night temperature is more than 23° C (areas not suitable for cultivation).
- Maximum day temperature must be below 35° C.

Propagation – Traditionally through seed tuber.

- Each tuber has several eyes which (3-5 buds) developes into sprout.
- Potato tuber has a dormancy of 8-10 weeks.
- The multiple sprouting stage is the best stage for planting because it generate emergence of several strong stem occurs.

- Storage under diffused light – To obtain seed tuber with numerous vigorous green sprouts.
- Tuber are taken out 10-15 days before planting from cold storage and kept for 24 hours in pre cooling chambers.
- A stem number of 30 main stem have been found to be optimum.
- It can be obtained by using 40 g tuber at 60 cm distance between row and 20 cm between tubers.
- Tuber cutting is not recommended specially when producing a seed crop because it increases the incidence of damages of transmiting viruses and bacteria.
- Seed tubers accounts for nearly 40-50% cost of potato cultivation and being vegetatively propagated its quality are most important in production.
- In India prior to 1960, potato seed used to be imported to replace the local seed stocks of various regions.
- N. Hills used to be the major source of healthy seed production for planting potato crops in plains.
- Seed plot technique – For producing healthy seed in Northern India plains Punjab, Haryana, and Uttar Pradesh, important potato seed producing state in plains.
- The high hills were the traditional source of healthy seed as population of aphid remains low due to low temperature therefore incidence of virurses was quit low in hill grown seed.
- Hills account for only 5% of the area under potato.
- Seed Plot technique – Consist of raising the healthy seed crop during low aphid period, available in Northern plains during October - January and cutting of hulms and harvesting the crop before crossing damage level. (20 aphids / 100 compound leaves).

- Tuber dip treatment with 3% boric acid for 30 minutes after harvest and before storage of produce in seed production technique (SPT).
- CPRI Produces breeder seed (nearly 2600 tonnes) at four regional stations.

  1. Jalandhar (Punjab). 2. Modipuram (UP).
  3. Patna (Bihar). 4. Gwalior (MP).

- Seeds are produced in North-Western plains and kept in cold storage during April-September.
- Post cold storage is necessary to get rid of the sweating problem and brings about respiratory equilibrium of tubers.
- N. Hills crop is harvested in September–October.
- For breaking tuber dormancy tubers are soaked in 1% thiourea for 1 hour.
- Tuber size – 25-30 mm or 25-75g - Better size.

  35-40 mm or 45-50 g. - Ideal.

Plant population

35-45 mm 60cm × 20cm-23,000 plants/ha.

45-55 mm 60cm × 25cm-67,000 plants/ha.

25-35 mm 60cm × 15cm-1,11,000 plants/ha.

Seed rate – 20-25 q/ha. (15g.)

25-30 q/ha. (30g.)

30-35 q./ha. (45g.)

- Dehaulming – Removal of the top portion of potato in the seed crop to avoid the infestation of virus carrying insect vectors is called as dehaulming. It should be done either by cutting the haulms or killed by spraying of total killer herbicides eg. Peraquat.
- When aphid population reaches to critical number (20 aphids/100 leaves) chances for viral infection becomes very high.

Planting method and time

- Ridge and furrow method is the most popular method.
- In the plains the best time of planting is when the maximum temperature is between 30-32° C and 10-20 ° C.
- In hills the best time of planting is when maximum temperature is about 20-22 ° C and minimum temperature is about 12-15 ° C.

  Planting time

  Autumn / Winter / Spring – In plains

  Summer / Autumn – In hills
- Main crop in gangatic plains is planted in October-November.
- Dhaincha and Sanai – For green manuring, reduces N, P and K requirement of crop by 20-30%, improves yield of potato 3 tonnes / ha.
- "N" is the most important nutrient for potato (120-150kg). N application increases the plant growth, leaf area, tuber number and tuber size.
- Max "N" response is recorded in alluvial soils.
- "Phosphorus" deficiency is more pronounced in acidic hill soil.
- "Phosphorus" deficiency leads to dull dark green leaves without lusture.
- "K" deficiency causes shortening of inter nodes, leaves become dark-bluish green. In mild "K" deficiency, terminal leaves show-bronzing accompanied with necrotic spots.
- Response to "K" is highest in alluvial soil.
- Earthing up is done – 40 days after sowing (DAS).
- Both Calcium ammmonium nitrate (CAN) and ammonium sulphate are good source of "N" in potato crop.

- Irrigation is withdrawn 10 days before harvesting.
- Critical stage – germination, tuber formation, earthing up and bulking.
- Late blight attack on stem, leaves and tubers, appears in crop in the hills every year but in the plains it occurs occasionally.
- Brown rot cause wilting of plants hence it is also known as bacterial wilt.
- **Aphid** – Transmit viral disease.
- **Leaf hoppers** – Transmit mycoplasmal diseases.
- Cyst nematode is present in the Southern hills only.
- About 100-120 g true potato seed is enough to raise a seedling crop or if the commercial crop is to be produced using seedling tubers the produces of 40-45 g True Potato Seed (TPS) is enough to plant one hectare crop.
- Next year TPS provides better disease resistance due to higher heterogenity in population.
- Water requirement – 350-550 MM.
- Water forms nearly 80% of the potato tubers.
- The late sown crop in plains should be harvested latest by April end to avoid high temperature and charcol rot infection in tuber.
- Harvesting is done before the temperature rise above 30° C.
- Harvesting is done by potato digger.
- Nearly one fifth of the total potato production in the country is used as planting material.
- Drying and curing of skin – 10-15 days in heaps in cool place.
- After grading potatoes meant for use as seeds next year are treated with 3% boric acid solution for 30 minutes for protecting against soil borne pathogens like black scurf, common scab etc before storing in bags.

Storage – 2-4° C with high relative humidity.

- The low temperature checks sprouting and rottage, whereas high R.H. reduces weight loss in tubers.
- The total cold storage capacity in the country is adequate only to store nearly 35-40% of total potato production.
- Value Added products – Chips, flakes, french fries, finger chips, granules, disc, cubes and flour etc.
- For best yield, potato needs long day condition during growth and short day condition during tuberization.
- Solanin – 5 mg/100g.
- If solanin is greater than 20 mg/100g, it is unfit for consumption.
- Protein – 1.6%
- Vitamin C– 17 mg/100g.
- Potassium – 568 mg/100g.

} Nutrient content in Potato

■ Physiological disorders –

1. Internal Brown Spot – Appears particularly in light sandy soils which are not irrigated regularly (moisture deficiency). Variety Craig's defiance is prone to this.
2. Black Heart – When tubers are stored at 35-40° C inner tissue break down and becomes black.
   a. Lack of O_2 is the probable cause (Poor ventilation).
   b. At high storage temperature accumulation of CO_2 at the center of tuber might contribute to damage by decreasing respiration.
3. Hollow Heart – (Excess N_2 leades this disorder) –Irregular cavity in the center of tuber might contribute to damage, decrease respiration is probable cause.

   It appears in variety which produces over size tubers e.g. – K. Jyoti and K. Sindhuri.
4. Chilling injury –Caused by low temperature of about 0° C.

5. Freezing injury – Caused by low temperature (Ice formation in potato tuber when exposed to temperature between 1° C to 2° C).When cut in half a blue black discontinuous ring is observed in the vascular region.
6. Greening-Cause due to excessive exposure to sun rays.

"SWEET POTATO"

- Latin name: *Ipomoea batatas* L. (syn. *Batatas edulis*).
- Family: Convolvulaceae.
- Cultivars differ for ploidy level: 2n = 30 (diploid); 2n = 40 (tetraploid); 2n = 90, (hexaploid)).
- Dicotyledon, perennial with swollen, tuber-like roots.
- Sweet Potato occupies a prime place in terms of calories production / Unit area and time.
- It is grown as a starchy crop thoughout the tropical, sub-tropical and frost free temperate climate zone in World.
- In world scenario more than 80% of the area under sweet potato cultivation and production lies in Asia.
- India 6th rank in total area of cultivation.
- Among Asian countries China ranks 1st in area and production.
- India with 1, 30,000 ha (0.11% of the gross cropped area in India) is the largest sweet potato producer in South Asia and 6th largest producer in the World.
- Productivity of India is very low (8.2 tonnes/ha.) and is little more than half of the world average productivity.
- Orissa has the largest area.
- Orissa, UP and Bihar – Account for 89% area and 88% of production of sweet potato.
- It is used as vegetable and also in making chips.
- In USA about 60-70% of the sweet potato is utilized for human food.

- It is a source of starch and used in textile, paper, cosmetics and glucose.
- *Ipomoea trifida* – Probable ancestor of cultivated sweet potato.
- Sweet potato was introduced in Europe, probably in Spain at least 60 years before Irish Potato.
- A tuber similar to sweet potato is cited in Vedic literature, generally it is considered to have been introduced in India by Portugues in early 16th century.
- It contains 10% starch and 3-6% sugar.
- Tubers are good source of vitamin A, B and C.
- Yellow and orange fleshed variety contains more carotene.
- Sweet potato reached Japan from Okinanea.
- Sweet potato is an annual crop. The plant is perennial vine.
- Flowers are solitary or cymose (White to purple).
- Fruits are capsule (5-8 mm diameter), having brownish black seed with a very hard testa.
- Anthesis occurs at 4-5 AM.
- Rich in vitamin A. Varieties are distinguished by leaf shapes, skin colour and flesh colour.
- Most sweet potatoes are propagated either by planting sprouts that rise from the roots or by rooting vine cuttings.
- The sprouts, slips or draws, are obtained by planting stored tuberous roots in a hotbed and are cut for transplanting to the production field when they have six to eight well-developed leaves and a good root system.
- Warm season: optimum growth temperature: 21° to 29° C; chilling injury occurs when soil temperatures drop below 10° C.
- Harvested 130 to 150 days after transplanting.
- Mechanically harvested like Irish potatoes, but are more suceptible to bruising.

- Harvested tubers are often 'cured' for several days at 29° C and 85 to 90% relative humidity to promote suberization and healing of bruises and/or cuts.
- The cured tubers are stored at 13° to 15° C and 85 to 90% relative humidity.
- If properly cured and stored, disease-free roots can remain edible for 6 to 7 months.

■ Crop Improvement –

Clonal Selection - Which are self incompitable.

1. Pusa Saffaid
2. Pusa Lal
3. Pusa Sundhuri – Rich in carotene.
4. Jawahar Sakarkand 114 - Extra early variety.
5. Shree Nandini – Superior cooking quality.
6. Shree Vardhini – Rich in carotene (1200 IU/100g.).
7. Kalmegh – Assam, very early variety.
8. Samrat
9. Kiran – AP, spreading vine.
10. Gouri – Orissa
11. Sankar – Orissa
12. Varsha
13. Bhuban – Orissa
14. Shree Bhadra – Excellent trap crop for root knot nematode.
15. Rajendra Sakarkand 35
16. Rajendra Sakarkand 43 - Modertely resistant to cercospora leaf spot and tolerant to weevil.
17. Rajendra Sakarkand 47- Tolerent to sweet potato weevil and cercospora leaf blight.
18. Co1, 2, 3 - From Tamil Nadu.
19. KonKon Ashwini

"CASSAVA or TAPIOCA"

- Major starch crop of the tropics.
- Perennial shrub.
- Monoecious and cross pollinated crop.
- Female flowers open about 10 days before the male phase.
- Cassava grown from seed, develops a tap root system typical of dicots.
- Main constituent of tuber – Starch.
- Bitter principle – Cyanogenic glucoside (HCN).
- Yellow colour of flesh is due to Carotene.
- Cassava is a photoinsensitive crop.
- Introduced into India by Portuguese during 17th century.
- Shri Visakam Thirunal encouraged cultivation of popular variety to overcome food shortage.
- Nigeria – Major cassava growing country (contributes 20% of total area and 20% of total production of world).
- In India – Kerala 50%, Tamil Nadu 32%, Andhra Pradesh 9% area is under Cassava cultivation.
- In India cassava was first introduced in Kerala.
- Area under cassava in Tamil Nadu is less then in Kerala, but production in highest due to very high productivity i.e. 38 tonnes/ha.
- Cassava is grown as rainfed crop in Kerala and Andhra Pradesh, but in Tamil Nadu it is grown both as irrigated and rainfed crop.
- Cassava is highly heterozygous in nature.
- Vascular blue lining or "Vascular streaking" is present in Cassava.

- Important food product from Cassava is Sago.
- Gari, Farsirla, Fulgar and Macroni – Proceed product of Cassava.
- About 4.16% of total area and 7.79% of total production under vegetable is shared by Cassava.
- Improvement research work on Cassava was started since the establishment of Central Tuber Crop Research Institute (CTCRI), Thiruvananthapuram (1963).

■ Varieties –

1. Sree Harsha (1996) – Triploid clone (Diploid × Tetraploid Shree Shaya) maximum starch content of 39.05%, ideal for industrial purpose.
2. Sree Prakash (1987) – Resistant to corcospora leaf spot, yield 35-40 t/ha, drought resistant, starch content : 30%.
3. Sree Jaya (1998) – Good cooking quality, suitable for low land cultivation.
4. Sree Vijaya (1998) - Short duration (6 months), suitable for low land cultivation.
5. Sree Sahya (H2304) (1977) – Multiple Cross Hybrid nonbranching, involving five parent, yield : 35-40 t/ha.
6. NIdhi – Suitable for sandy loam soils.
7. Sree Vishakam – Hybrid variety.
8. H-165 (1971) – (Chadayamangalam Vella x Kalikalan), non branching type.
9. H-97 (1971) – (Manjavella x Brazilian Seedling).
10. H-226 – Highly susceptible to CMD, became dominant variety in cassava growing areas.
11. Col (1977) - Starch 35%.
12. Co2 – Starch 36.4%
13. Co3 – Yield 42.6 tonnes/ha., starch 35.5%.

- Cassava is the drought tolerant crop but cannot withstand frost and growth ceases at temperature below 10-12°C.

Propagation – By stakes (top to middle portion) 25-30 cm length, thickness 2-3 cm.

- Planting the stakes deeper results in swelling of the mother stem and reduction in yield of root and size.
- Vertical planting is preferred because more uniform callus tissue formation takes place.
- Spacing – 75 × 75 cm – Non branching type.
  90 × 90 cm – Branching type.
- Harvesting is done after 10-11 months after planting.
- Harvesting is done after 6-7 months for short duration variety.
- Delay harvest results in cooking quality deterioration.
- Yield – 35-40 t/ha. - high yielding variety.
  28-30 t/ha. - short duration variety.
- About 70% of total cassava production is used as food either directly or in processed form.
- Parboiled chips are a method of processing cassava tuber for storage (these chips are hard due to the partial gelatization).
- *Rhizophus orysal*, *Bacillus* sp., *Penicillium* sp. - These organism are responsible for rotting of cassava.
- Drying is the main mode of processing of cassava tuber to enhance its post harvest storage.
- Curing in cassava is done at 25-40°C and 80-85% R H.
- Cassava tubers are rich in carbohydrates (80-90% on a dry weight basis) and 20-30% starch on fresh weight basis.
- Cassava contains 70% ethanol soluble sugar and 1-6% free sugar.
- Cassava cultivars are divided into three categories based on cynide content in the edible portion of the root.
  1. Sweet or Non toxic (< 50 mg/kg. fresh weight)
  2. Intermediate (50-100 mg/kg fresh weight)
  3. Bitter or toxic (> 100 mg/kg fresh weight)

- Cassava plant accumulates cynogenic glucosides – liamarin and losaustralin which are synthesized from amino acid valine and iso-leucine, respectively.
- These glucosides release HCN on hydrolysis.
- Enzymes responsible for hydrolysis are lina marinase, b-glucoside and hydroxynitrile lyose.
- The cynogenic potential of cassava leaves is 5-20 times higher than that of edible parts.

"COLOCASIA or TARO"

- Two type of colocasia are there -
  1. Eddoe–*Colocasia esculenta var – antiquorum* – Arvi.
  2. Dasheen–*Colocasia esculenta var- esculenta* – Bunda.
- The eddoe type is commenly called arvi and dasheen as bunda.
- Eddoe type is most prevalent as vegetable.
- Africa rank 1st in both area and production followed by Asia.
- Colocasia tubers are rich in starch.
- Leaves and petioles are also used as vegetable.
- Colocasia tubers contain more protein, minerals, phosphorus and iron compared to other tuber crops.
- Eddoe type have bigger cormels that the mother corms where as Dasheen types have smaller cormels than the mother corms.
- Taro is recommended for gastric patient.
- Colocassia responds to K fertilizer better than nitrogen and phosphorus.

■ Varieties –

1. Sree Rashmi (C-149) is suitable for southern zone, improved variety.
2. Sree Pallavi (C-266) is suitable for southern zone, improved variety.
3. C - 16 is suitable for southern zone, improved variety.
4. Satamukhi is also suitable for southern zone, improved variety.
5. Satamukhi is suitable for western zone.
6. Sahar Shamukhi is suitable for western zone.
7. White Gauriya is suitable for eastern zone.
8. Kakakachu is suitable for eastern zone.
9. Panchamukhi is suitable for eastern zone.
10. NDC1, NDC2, NDC3 is suitable for eastern zone.
11. Saharshamukhi is suitable for eastern zone.
12. Kadma – Eastern zone.
13. Muktakeshi – Eastern zone.
14. Nadia local – Eastern zone.
15. Ahina Local telia – Eastern zone.
16. Jhankhri – Eastern zone.

- Propagation - By small cormels (20-25g weight), 1 tonne planting material is enough for 1 ha., Spacing 60 × 45 cm.
- Ideal time for planting – Rainy season.
- Taro leaf hopper (*Tarophagus proserpina*) causes great crop losses in its cultivation.
- The irritant properties of taro are due to the calcium oxalate crystals in corms and leaves of *C. escultenta.*
- Calcium oxalate content is 0.008%.
- Major crystal types are raphides and druses.

- Tuber becomes very hard to cook due to water staganation in field in both eddode and dasheen types.
- Harvesting 120-150 days after planting.
- Yield – 30-40 tonnes/ha.

"YAMS"

- Three species of yams are important and cultivated widely.

1. Greater yam/water yam – *Dioscorea alata*
2. Lesser yam / Chinese Yam / Asiatic
   (Both are major food yams of India)
   Yam – *Dioscorea esculenta.*
3. White yam - *Dioscorea rotundata* is extensively cultivated in Africa.

- Africa alone produce 90% tuber and cover 95% of the area.
- *D. composita, D. floribunda, D. spiculiflora, D. friedrichsthalli, D. deltoida* – Exploited for pharmaceutical purpose and are non-edible.
- Yams contain 18-20% starch (contain mainly anylopectin).
- Fugu is an important product made from yams.
- Yam flour is also used for human consumption as "Kokonte" in northern zones of W. Africa.
- *D. alata is* used for making wafers, cripies and biscuits.
- Diosgenin present in many wild species is used in drugs for oral contraceptives.
- Dioscorea genus is dioecious in nature.
- Fruit type – Trilocular capsule.
- In *D. alata* no seed dormancy exists.
- $2x = 20$ in most primitive section but in *D. alata* majority of male clones are reported to be tetraploid ($2x = 40$).

- Lesser yam (*D. esculenta*) is sweet and matures early than other yams.
- In *D. alata* bulb is a minor storage organ while in *D. bulbifera* the bulbiles is the major storage organ.
- Amylose content ranges from 10-25%.

■ Varieties –

A. Greater yam / Water yam / Ratalu –

1. Sree shilpa – First hybrid variety released in *D. alata*, released from CTCRI.
2. Sree Roopa – Cooking quality is excellent, for Kerala.
3. Sree Keerthi – Excellent cooking quality and taste, for Kerala.

B. Lesser Yam –

1. Sree Latha – Suitable for Andhra Pradesh, Assam, Kerala, Maharastra, West Bengal and Bihar.
2. Sree Kala – Excellent culinary quality.
3. Konkan Kanchan – Suitable for Maharashtra.

C. White Yam –

1. Sree Subhra – Excellent cooking quality, yield 35-40 t/ha.
2. Sree Priya – Yield 35-40 tonnes/ha., for Kerala.
3. Sree Dhanya – Dwarf bush variety, do not require stacking.
    1. Greater yam – Popularly known as ratalu, it possesses four or more wings on the stem which twines to the right, tuber skin is black or brown.
    2. Lesser yam – Popularly known as Suthani or Kangar, is usually thin, often pubescent and spiny which twines to the left. Tuber skin yellow-brownish with white flesh. After cooking sweeter than other yam.
    3. White yam – It has wingless stem which is twining to the right.

Propagation – By tuber pieces/ small whole tuber.

- Lesser yam – Tuber weighing 125-150 g.
- Greater yam and white yam – Tuber weighing 200-250 g.
- Propagation by vine cuttings is not advised because of slow tuber production.
- Yam tuber have long dormancy period (2-3 months) depending on species which is major problem in yam cultivation.
- Dormancy is broken by dipping tubers quickly in 4-8% solution of ethylene chlorohydrin followed by dry storage.
- To delay sprouting in storage, soaking of tuber for 10 hrs in 0.1% solution of malic hydrazide is done.
- Spacing – 75 × 75 cm For lesser yam.
  90 × 90 cm Greater yam & white yam.
- Anthracnose of Yam (*Colletotrichum gloeospoioides*) was first found in Rajasthan, responsible for 70-80% loss in yield, severe form in *D. alata*.
- Harvesting – 7-9 months after planting (Dec. – Jan.).
- **Yield** 25-30 tonnes/ha - Lesser yam.
  33-40 tonnes/ha - Greater yam.
  18-25 tonnes/ha - White yam.
- Tuber top portion is known as proximal.
- Tuber middle and bottom portion is known as distal.

"ELEPHANT FOOT YAM"

- It is also known as Amorphophallus or Suran or Jimmikand.
- Inflorescence – Spadix.
- In Japan it is used as food crop.

- Recommended as blood purifier.
- Paste of tuber is used to reduce pain in arthritis.
- Acridity causes irritation in mouth and throat which is due to the presence of calcium oxalate.

■ Varieties –

1. Gajendra – Cooking quality is very good, free from acridity, free from daughter corms, Yield 400-600 q./ha, suitable for East and North India.
2. Santragachi – Having several rough daughter corms (East India), slight acridity, suitable for East India.
3. Sree Padma – Improved variety.
4. Kovvur

Propagation – Cut or small tuber (high yielding and non acrid varieties)

- 50-100 g weight of tuber.
- Whole tuber should be preferred over cut tuber to minimize rotting.
- For seed production – 100-500 g tuber should be used which produces 500g. weight of tuber.
- Tuber dormancy can be broken by treating them with thiourea (0.1%).
- GA_3 and Ethrel are also effective in breaking dormancy.
- Best planting time – Feb-March.
- Harvesting – 7-8 month after planting (MAP) (November).
- Yield – 50-100 tonnes/ha.
- Water stagnation during crop growth results in tuber that is very hard to cook.

"YAM BEAN"

- B. N. – *Pachyrhizus* spp.
- Family – Fabaceae.
- Origin - South America
- Edible part – Root.
- Populartly known as Potato bean and Mishrikand.
- High sugar content in tubers imparts sweet taste when eaten as raw.
- Its starchy conical (as turnip) shaped fleshy tubers are eaten.

■ Varieties –

Two types of cultivars are available in market.

1. Mexican Type – Tuber is large 500-700g., less sweet compared to local, they have tendency to develop cracking on their tubers, less sweet preferred in Indian market.
2. Local Type – Tuber is small 200-300g., sweeter, white flesh, less fiber, no cracking.

- Rajedra Mishrikand 1– Improved variety, popular in North Bihar.
- Yield – 40-55 t/ha, matures in 110-140 days.
- Propagation – By seed (20-60 kg./ha).
- Yam beans bear pods 7-15 cm long with 8-10 brownish yellow tored seed.
- Mature pods containing seeds are source of a toxic substance "rotenone" which is harmful to grazing cattle.
- Spray of 2, 4-D (50 ppm) at flower initation stage induces flower thinning and increases tuber yield.
- Harvesting – Traditionally the trend is to harvest (110-140 DAS) the crop on the occasion of "Saraswati Puja" with the start of spring season because of market demand.

Yield – 18-20 tonnes/ha.- Local cultivars.
36-40 tonnes/ha.- Rajendra Mishri 1.

- Seed should be harvested after 240 days during March-April.
- Tuber cracking is the main problem due to delay harvest.
- Application of "K" reduces cracking of tuber.

"TANNIA"

- B. N. – *Xanthosoma sagittifolius.*
- Family – Araceae.
- Origin – South America.
- Edible part – Corms, leaves and petioles.
- Tannia or Xanthosoma is an herbaceous edible species.
- Suitable for intercropping in Banana.
- Popular in Maharashtra and Gujarat for making roll of leaves.
- Propagation – Corms and cormels (20-25 cm long).
- In Maharashtra and Gujarat where only leaves are harvested the main corms and carmels are left in the field.
- They act as new planting material.

"QUEENSLAND ARROWROOT / PURPLE ARROWROOT"

- B. N. – *Canna edulis.*
- Family – Cannaceae.
- Origin - Tropical America .

- Edible part – Tuber.
- Perennial herbaceous monocotylendon.
- Propagation – By rhizomes.
- Yield – 25-30 tonnes/ha.

"JERUSALEM ARTICHOKE"

- B. N. – *Helianthus tuberosus.*
- Family – Compositae (Asteraceae).
- Origin – Eastern United State.
- Edible part – Tuber.
- Cold-hardy, frost tolerant.
- Autumn-blooming sunflower.
- Harvested organ: Tuberous, potato-like roots (rhizomes).
- Native of N. America.
- Cultivated by the American Indians before the 16th century.
- Introduced to Europe in the 17th century.
- Two important sp. *H. annus* and *H. tuberosus* are economically important.
- Carbohydrate content – 8-19% (Inulin-Polymer of fructose).
- Jerusalem artichoke is commercial source of levulose used as sweetening agent by diabetic patients.
- Glucofructosans – From sucrose, are principle carbohydrate in tubers.
- Tuber also used for alcohal and beer beverage.
- Propagation – By small sets or tubers (50g. weight) (2-3 eye buds)
- Mammoth white French – Widely grown variety.
- Sutton's white – Widely grown variety.

"COLEUS"

- B. N. – *Solenostemon ratundifolius* (syn. *Coleus parviflorus*).
- Family – Lamiaceae.
- Origin – Tropical Africa, Asia and Australia.
- Edible part - Tuber
- Popularly known as Chinese potato or Koorka.
- Annual herbacoleus bushy plant.
- Sree Dhara – Improved variety, released by CTCRI.
- Propagation – By vine cuttings (10-15cm length).
- Yield 20-25 tonnes/ha.

"ARROWROOT/WEST INDIAN ARROWROOT"

- B. N. – *Maranta arundinacea.*
- Family – Marantaceae.
- Origin - Tropical America.
- Edible part – Rhizome.
- High quality starch (27.8%) is used as food for infants.
- Biscuits are known in every corner of India made by it.
- Its starch is also used as special glue and paste as a base for face powder and ice-cream stabilizer.
- Recently it is also used in production of carbonless paper for computer printout.
- Propagation – By small pieces of rhizomes (known as bits) (4-7 cm long, 2-4 nodes).
- 3 tonnes/ha planting material is enough.
- Inflorescence - Teminal bearing.
- Harvesting – 10-11 Months after planting.
- Yield – 4-7 tonnes/ha.

"GIANT TARO"

- B. N. – *Alocasia macrorrhiza* (Syn. *A. indica*).
- Family – Araceae.
- Origin – Asia (Srilanka).
- Edible part – Rhizomes and shoots (stem tuber).
- The fleshy aerial stem of giant taro is the primary product.
- The edible corm of tropical taro has higher starch content than either potatoes or sweet potatoes.
- The species cultivated in Assam and West Bengal as food crops are – *A. indica, A. macrorhiza, A. cucullata.*
- Propagation – By suckers, shoot tip and cormels.

"CHINESE WATER CHESTNUT"

- B. N. – *Eleocharis dulcis.*
- Family – Cyperaceae.
- Origin – China.
- Edible part – Tuber.
- Perennial leafless sedge commonly found in marshes and swams through the world.
- Edible tuber rich in starch (7.3%) and sucrose (6.4%).
- Starch grains are very large, upto 27 μ in size.
- In India it is sold in Kolkata under the name Singapuri Keysur.
- First cultivated in South East China.

11

SPICES & CONDIMENTS

- Spices are those plants, the product of which are made use as food adjuncts to add aroma and flavour.
- Condiments are also spices, products of which are used as food adjuncts to add taste only.
- Both spices and condiments contain essential oil which provides the flavour & taste. They are used in whole, ground, paste or liquid form mainly for flavouring & seasoning. There are about 63 spices grown in India.

■ Classification of Spices

A Classification based on life cycle –

a) Annual – Coriander, Fennel, Fenugreek

b) Binnial – Onian, Garlic

c) Perennial – Ginger, Turmeric, Black pepper, Clove, Cardamom

B Classification based on growth habit:-

a) Herb – Ginger, Turmeric

b) Shrub – Caper, Juniper

c) Climber – Black Papper

d) Trees – Clove, Cassia, Nutmeg, Cinnamom, All spices.

C.N.	B.N.	Family	2x=	Origin Place	Chemical Content	Part Used	Major Growing (Country / States)	Prepaga-tion by	Yield
Black Pepper	*Piper Nigrum*	Piperacea	128	Tropical Forest of W. Ghat of S.India	Piperine	Berry	Vietnam Kerala (94%)	Runner Shoot Cutting	800-100 kg/ha.
Cardamon (Small)	*Elettaria Cardamon*	Zingiberaceae	24	Tropical Forest of W. Ghat of S.India	X-Terpinyl-acetate 1,8 Cineol	Capsule	Kerala (60%)	Suckers	500 kg/ha.
Cardamon (Large)	*Amomum Subulatum*	Zingibraceae		E. Himalayan		Capsule & Seed	Nepal & Sikkim	Suckers	450 kg/ha.
Ginger	*Zingiber Officinale*	Zingiberaceae	22	S.E. Asia	Pungecy-Oxymethyle pernol (Gingaral)	Rhizome	Kerala (25%)	Seed Rhizomes	15-25 Tonnes/ha.
Turmeric	*Curcuma Longa*	Zingiberaceae		S.E. Asia	Curcumin Yellow Alow	Rhizomes	A.P.	Seed Rhizomes	25-30 Tonnes/ha.
Coriander	*Coriandrum Sativum*	Apiaceae	22	Mediterranean	Linalool (E.O)	Fruit	Rajasthan	Seed 10-15 kg IC 25-30 kg UC 40-42 Kg Veg;eaf	400-500 kg (Rainfall) 600-1200 kg (Irri gated)

Cumin	*Cuminum Cyminum*	Apiaceae	14	Mediterranean	Thymol., cuminol	Fruit	Guj.(I) Raj. (II)	12-15 kg/ha.	7-8 q./ha.
Fenugreek	*Trigonella Foenum Graecum*	Leguminacean	16	S. E. Europe	Diosgenin	Seed	Rajasthan (80%) A.P.	20-25 kg/ha.	9-10 q./ha.
Fennel	*Foeniculum Vulgar*	Apiaceae	22	Mediterranean	Anethole 70% (Sweetness)	Fruit	Gujarat, Rajasthan	10-12 kg/ha.	9-10 q./ha.
Dill	*Anethum Graveolens*	Apiaceae		Eurasia	Carvone	Fruit	Gujarat, Mahara-shtra	5-6 kg/line 10 kg-Broadcast	8-9 q./ha.
Celery	*Apium Graveolens*	Umbelliferae	22	Abyssinia and Asia	D-limonene D-selinene	Leaves & Seed	Punjab	375-500 g / ha	330-560 q./ha green leaves 175 - 275 Kg Seed/ha.
Clove	*Syzygium Aromaticum*	Myrataceae	44	Indonesia	Eugenol (70-90%) Eugenyl-acetate (5-12%)	Unopened Flower bud	T.N., Kerala	By Seed (Mother Clove)	2 kg/tree
Nutmeg	*Myristica Fragrans*	Myristicaceae		Indonesia	Terbein (7-16%)	Dried Kernel Mace-Aril	Kerala, T. N.	By Seed Epicotyl Grafting	800 kg Nutmeg/ha. 100 kg Mace /ha.

Cinamom	*Cinnamomum Verum*	Lauraceae	24	Sri Lanka	Eugenol (70-80%) Cinnamic Aldehyde	Bark	Kerala, T.N.	By Seed	200-300 kg Dried Bark
All Spice	*Pimenta Dioica*	Myrtaceal		West Indies		Fruit			
Vanilla	*Vanilla Planifolia*	Orchidaceae		Mexico-Brazil	Vanillin	Fruit (Beans)	Malagasy Republic	By Stem Cutting Kerala, T.N.	300-400 kg cured bean /ha.
Saffron	*Crocus Stativus*	Iridaceae		India	Cicrocrocin	Stigma	Only in Kashmir	By Carms	5 dried saffron / ha.
Curry Leaf	*Murrya Koenigii*	Rutaceae	18	India	Rutaceae	leaves	Andhra Pradesh	seeds	1000 kg / ha. (From 4th Year)
Ajowan	*Trachysp-ermum Ammi*	Apiaceae	-	Eqypt	Thymol	Seed	-	3-4 kg/ha. Seed	-
Thyme	*Thymus Vulgaris*	Labiatae	-	-	Tymol	Leaf & Flower	-	By Seed & Cutting	-
Rosemarry	*Rosemarinus Afficivalis*	Labiatae	-	Med. Region (S. Europe)	-	Leaves & Flower	-	-	-
Kokum	*Garcinia Indica*	Guttiferae	24	India	B-hydroxy Acetic acid	Fruit	-	-	-
Arafoetida	*Ferulla Foetida*	Apiaceae	-	-	Ferumine	Oleoresin Gum	-	-	-
Juniper	*Juniperous Communis*	-	-	-	-	-	-	-	

C Based on Economic importance.

a) Major Spices: Black pepper, Ginger, Cardamom and Turmeric.

b) Minor Spices: Cumin, Coriander, Fennel, Fenugreek.

c) Seed Spices: Dil, Aniseed, Celery, Bishopweed

D Based on Plant part used –

a) Leaves: Cassia, Curryleaf, Chive.

b) Bark: Cinnamon.

c) Unopened flower buds: Clove.

d) Rhizome: Ginger, Turmeric, Asafoetida (exudate).

e) Seeds: Coriander, Cumin, Fennel, Fenugreek.

f) Fruits : Cardamom, Chilli, Dill, Celery, Cumin, Black Pepper, Allspice, Vanilla (pod), Aniseed, Ajowan

g) Stigma : Saffron.

h) Buds : Onion, Garlic, Shallot, Leek.

i) Kernel: Nutmeg.

j) Root: Horse radish.

k) Herbal Spices: Rosemary, Thyme, Parsely, Horse radish.

■ Botanical Classification:-

I) Monocotyledonous

(1) Family – Liliaceae

1. Meadow Saffron – *Colchicum autumnale*

(2) Family – Amarylidaceae

1. Onion – *Allium cepa*
2. Garlic – *Allium sativum*

(3) Family — Iridaceae

Saffron – *Crocus sativus*

(4) Family – Zingiberaceae

1. Ginger – *Gingeber officinale*
2. Turmeric – *Curcuma longa, C. domestica*
3. Small Cardamom – *Electtaria cardamom*
4. Nepal Cardamom – *Amomum sublatum*
5. Bengal Cadamom – *A. aromaticum*

(5) Family – Orchidaceae

1. Vanilla – *Vanilla planifolia, V. fragrans*

II) Dicotyledenous –

(1) Family Myrtaceae

1. Allspice – *Pimenta dioicea*
2. Clove – *Syzygium aromaticum*
   Eugenia caryophyllus

(2) Family Labiateae

1. Basil – *Ocimum basilicum*
2. Mint – *Mentha arvensis*
3. Peppermint – *Menthe piperita*
   Spearmint – *M. spicata*
   M. viridis
4. Sage - *Salvia officinalis*

(3) Family – Lauraceae

1. Cassia – *Cinnamomum inners*
2. Cinnamon – *C. verum*
3. Tejpat – *C. tamala*

(4) Family – Piperceae

1. Black pepper – *Piper nigrum*
2. Long pepper – *P. longum*

(5) Family – Leguminoseae

1. Fenugreek – *Trigonella foenum graceum*
2. Kasturi methi – *Trigonella corniculata*

(6) Family – Myristicaceae

1. Nutmeg – *Myristica fragrans* / mace

(7) Family – Guttiferae

1. Kokam – *Garcinia Indica* / Garcinia

(8) Family – Umbelliferae (Apiaceae)

1. Coriander – *Coriandrum sativum*
2. Cumin/White cumin – *Cuminum cyaminum*
3. Cumin Black – *Nigella sativa*
4. Fennel – *Foeniculum vulgare*
5. Celery – *Apium graveolens*
6. Dill/Sowa – *Anethum graveolans*
7. Ajowan / Bishop's Weed – *Carcum capticum, Trachyspermum ammi*
8. Parsely – *Petroselinum crispum*
9. Asafoetida – *Ferula asafoetida*
10. Aniseed – *Pimpinella anisum, Anisum vulgare, Anisum officinalis*
11. Caraway – *Apium carvi, Carum carvi, Seseli carvi*

(9) Family – Rutaceae

1. Curry Leaf – *Murraya koenigi*

(10) Family – Capparidaceae

1. Caper – *Capparis Spinosa* –unopened flower bud is used.

General about spices and Condiments –

- Indian share in world trade of spices is around 18%.
- Nearly 90-95% of the total production is consumed locally and the rest is exported.
- India produces spices on 2.0 million hectares with an annual production of about 2.3 million tonnes valued at about 4500 crores, contributing nearly 20% of world's

production, 30% of the trade in terms of quantity but only 10% in terms of value.

- India is the biggest exporter of spices, annually exporting about 1.9 lakh tonnes of different spices and spice products, earning a foreign exchange of about 700 crores.
- India is known as "Home of Spices" because most of the spices are native to our country.
- Spice oil is obtained by steam distillation method.
- Largest area under spices – Rajasthan.
- Largest production of spices – Andhra Pradesh.
- India is the largest producer, consumer and exporter of spices in the world.
- Southern States produce 30% of Cardamom and 25-30% of pepper.
- Dry and arid areas of Gujarat and Rajasthan produce 80% of Coriander and 90% of Cumin produced in the world.
- India – 90% turmeric production of world.
- Jammu and Kashmir – Produces world's best saffron.
- Punjab – 90% celery production of total production of the world.
- Black pepper and Turmeric constitute more than 50% of total export earnings among spices.

INDIA – Production facts

- Black pepper – 54 %
- Black pepper – 26.3% area under cultivation of whole world.
- Ginger – 70%
- Turmeric – 76%

KERALA alone produces

- Black pepper – 94%
- Cardamom – 60%
- Ginger – 25%

RAJASTHAN

- Fenugreek – 80% area
- Fennel
- Coriander

GUJARAT – Cumin is major produced spice.

Andhra Pradesh – Leads in Turmeric, Chilli

- Major Importer and consumer of oleoresin – USA (50%).
- Foot rot, slow decline, Pollu beetle – Important disease and pest of Black pepper.
- Capsule rot, ''Katte'', Kokke Kandu, Thrips – Important disease and pest of Cardamom.
- Rhizome rot – Important disease of Ginger and Turmeric.
- King of Spices – Black pepper.
- Queen of Spices – Small Cardamom.
- Cochin ginger / Indian ginger – Best in the world.
- Clove is the second most important spice next to black pepper in the world.
- Nutmeg, All spice, Juniper- dioecious in nature.
- Turmeric is antioxidant due to phenolic character of curcumin.
- Large Cardamom – Shade loving plant (Seophyte).
- India alone contributes about 10-12% of the total world trade in spices.
- Aril – Commercially important part of Mace.

- Pistil – Commercially important part of Saffron.
- International Organization for Standardization (ISO) approved Indian Vernacular names, common english names as well as scientific names to help identify correct spices because spices in known by different vernacular names in different parts .
- "Spices and Herbs in food industry" – Book written by Lewis (1984).
- Oil of black pepper consists mainly of terpenes.
- Per capita consumption of spice in USA is ½ times of Indian consumption.
- "Ajowan ka Phool" is crystal form of thymol.
- Essential oil in dill is 1.5-4%.
- Male musk deer smell resembles to Ambrette seed.
- ISI - Now BIS (Bureau of Indian Standards).
- ISO - International Organization for Standardization is situated at Budapest, Hungary.
- "Vasco De Gama" was the first to reach Malabar Coast, Kerala (India) and took with him a cargo of spices and other Indian products.
- In the 16th century "Cloves" for instance were among the spices used to preserve food without refrigeration.
- Cloves contain a chemical called eugenol that inhibits the growth of bacteria.
- Clove is still used to preserve "Virginia Ham" food. Likewise later mustered and ground mustared were also found to have preservative qualities.
- Ministry of Food and Agriculture was established in 1951.
- Goverment of India constituted a high level of committee -Spices Enquiry Committee (1951-53) which submitted its report to the Govt. of India in October 1953.

- Indian Central Spices and Cashewnut Committee – 1961.
- CPCRI, Kasargod (Kerala) was established in 1970.
- All India co-ordinated Improvement projects for spices was established in December 1970, but it actually started functioning in 1971
- Spices Development Council, Division of Horticulture and Department of Agriculture & Co-operation – 1966.
- All Spices Development Council – 1966
- Under the aegis of the Directorate of Marketing and Inspection (DMI), Ministry of Agricullture Govt. of India was set up in 1937 under the APGM.
- The first "Agmark" specification was formulated for Curry powder in 1956, for Cardamom and Chilli in 1962 and for Turmeric, Ginger, Coriander, Chilli powder, Onion, Garlic, Ambadi, Cumin, Poppy seed in 1964.
- Central Agmark Laboratory (CAL) – Nagpur (1967).
- CAL is tempo of quality control to assist growers, merchants and exporters of spices in quick compulsory pre shipment inspection and quality control.
- Whole spice grading units – Mumbai, Cochin and Virudhunagar.
- The Central Agmark Laboratory served both as an Apex and National Appellate Laboratory.
- APGM – Agricultural Produce Grading and Marketing Act-1937. Indian Spice quality control.
- Out of 26 spices covered under the Agn ark Grading Scheme the 13 covered under the "Compulsory Grading Scheme" under the Agmark for export purpose are Black pepper, Cardamom, Turmeric, Ginger, Onion, Celery Seed, Chillies, Coriander, Cumin, Fennel, Fenugreek and Curry powder.

- Agmark Organization (DMI) called mother of standardization in county, rendered pioneering services in the quality control of spices for over 50 years.
- CFTRI – Established in 1950 Mysore (Krt).
- First National Symposium on spices – "Spice Role in National economy" in collaboration with the spice export promotion council.
- Spices Export Promotion Council – Cochin (Kerala) was set up under the Ministry of Commerce, Govt. of India.
- Cardamom Board – 1966 (Under ministry of commerce) on 14th April 1966.
- Spices Board – 1986 (under ministry of commerce).
- Spices act – 1986.
- National Research Centre for spices, Peruvannamuzhi, Calicut was established in 1986 but later the name NRC for spices was changed to IISR (Indian Institute of Spices Research) in July 1, 1995.
- NRC on seed spices – Tabiji, Ajmer (Rajasthan).
- International Pepper Community (IPC) – Jakarta (Indonesia).
- At the International level, the three leading producer India, Malaysia and Indonesia formed IPC in March 1972 but started permanent functioning on August 1976 at Jakarta.
- Indian Standards Institution (ISI) is now renamed as "Bureau of Indian Standards (BIS), New Delhi.
- ISG – International Spice Group.
- About 95% pepper is grown in Kerala, 3-4% pepper is grown in Karnatka and rest pepper is grown in Tamil Nadu.
- India is the largest producer and supplier of Ginger and Turmeric.

- India is having a cut throat competition from Guatemala for Cardamom and from Malaysia, Indonesia and Brazil for pepper.
- 275 kg/ha yield of pepper in India is lowest in world as compared to 4,500 kg/ha. in Malaysia.
- Large Cardamom is mainly confined to Sikkim.
- Area under Spices in India – 25 Lakh ha.
- Production of Spices in India – 29.07 Lakh tonnes.
- Chilli share 32% of total spice production.
- Andhra Pradesh – Leading state in chilli production.
- Chilli has oleoresin and natural colouring material.
- Andhra Pradesh 49.1%, Karnataka 12.9% of Total Chilli production.
- Turmeric share 19% of total production of spices mainly grown in Andhra Pradesh, Tamil Nadu and Orissa.
- Contribution of total spice production -

Garlic – 18%	Coriander – 12%
Ginger – 9%	Cumin – 4 %
Chilli – 32%	Turmeric – 19%

- More than 90% of spices produced in the country are used for domestic consumption and rest exported as raw as well as value added products.
- Per capita consumption for spices in India : 2.639 Kg (1999-2000).
- Per capita consumption for spices in USA – 1.448 Kg (1999-2000).
- Major Seed Spices –

  1. Coriander, 2. Cumin, 3. Fenugreek

"BLACK PEPPER" (Kali Mirch)

- King of Spices.
- Branching vine or climbing, perennial evergreen shrub.
- Black Pepper is the dried, mature but unripe berry (fruit) of *piper nigrum.*
- Area under cultivation – 1.36 lakhs ha.
- Production/Annum – 32 Thousand tonnes.
- India accounts 54% of the total area under pepper in the world but share of production is only 26.2%.
- India exports about 41000 tonnes annually.
- India earns a foreign exchange of Rs. 240 crores annually.
- Area of black pepper - Kerala (94%), Karnataka (5%) rest Tamil Nadu, Andhra Pradesh and Assam.
- Fruiting branches – Plagiotropes.
- Topshoot – Orthotropes.
- Hanging Shoot – Geotropes.
- Leaves – Lanceolate.
- Inflorescence – Catkin produced at nodes opposite to upper leaves.
- Flowers – Monoecious or dioecious or hermaphrodite forms occur in different varieties.
- High yielding – More percentage of bisexual flowers (80%), male flower 1 to 19%.
- Fruit - Single seeded berry.
- Fruit takes approximately 6 months to mature after flowering.
- Fruit setting – Normally 50% in cultivated spices.
- Spike shedding occurs to the extent of 14-65%, causes considerable loss.

- Spraying of IAA 50 ppm or planofix 50 ppm and or Zinc 0.5% at the time of berry setting stage reduces spike shedding.
- Indian pepper reaches to 75 countries, North America region being the major importer of Indian peppers.
- NAFED – National Agricultural Co-Operative Marketing Federation of India established – 2nd Oct 1958.
- Pepper Research Station – Penniyur.
- Largest producer of Black pepper – Vietnam.
- Malaysia leads in yield and give about 16 times higher yield than India.
- India average yield – 273 Kg/ha.
- Black pepper – Gynomonoecious.
- First hybridization in black pepper was done in 1954.
- First hybrid of Black pepper released in India – Penniyur – 1.
- Pungent taste is due to – Piperine (Found only in berries).
- Piperine is toxic to houseflies than pyrethrum and a mixture of 0.05% piperine and 0.01% pyrethrum is more toxic than 0.10% pyrethrum alone so it can used as an insecticide.
- Important breeding method – Clonal selection.
- *Piper colubrinum* – Resistant to foot rot, introduced in India from Brazil.
- Main buyers of Indian pepper – USSR, USA, Canada, Italy.
- USSR and USA take about 55 - 60% of total pepper export.
- Consumption period of pepper : 6 - 8 months.

■ Varieties :

- Majority of cultivated types are monoecious.
- In India over 75 cultivars are cultivated.

- Karimunda is the most popular cultivar of pepper among growers of Kerala.

1. Panniyur-1 : F1 hybrid between Uthirankottah x Cheriyakania Kadan, (Released in 1966). Oleoresins – 11.8%, Oil 3.5%, dry recovery -35.3%, piperine – 5.3%. Sensitive to excess shade.
2. Panniyur– 2 (Krishna) : Open pollinated seeding of Balankottah 2159 Kg/ha, Oleoresin – 10.6%, Piperine – 6.2%, Released in 1989.
3. Penniyur– 3 (Shima) : F1 hybrid between Uthirankottah x Cheriyakania Kadan), Released in 1989, Oleoresin- 12.6%, Piperine – 5.2%.
4. Panniyur-4 : Selection from Kuthiravally type II, performs well even under adverse condition, Released in 1989.
5. Panniyur-5 : Open pollinated progeny of Perumkodi, suitable for intercropping in arecanut, oleoresin – 12.33%, piperine-5.3%.
6. Sreekara : Selection from Karimunda, essential oil 7%, tolerant to drought, Oleoresin 13%, released in 1989.
7. Subhakara : Selection from Karimunda (KS 27), released in 1989.
8. Panchami : Selection from Aimpiriyan Coll-856, 8320 Kg green pepper/ha. Late maturing, released in 1991.
9. Pournami : Oleoresin – 13.87% tolerant to root knot nematode (*Meloidogyne incognita*), released in 1991.
10. PLD-2 : Suited to Thiruvananthapuram and Quilon district of Kerala.

■ **Other important varieties –**

- Tellichery, Malabar – Large, attractive, dark reddish, brown to black, very aromatic.
- Malabar Garbled – Alone contribute nearly 90% of the total pepper export from India.

- Lampong and Singapore – Smaller, more shrivelled.
- Arakkulamunda – Regular bearer, early maturity, berries bold.
- Balancotta – Largest leaves among Kerala cultivars, its oil content is one of the highest so very high flavour quality, tolerant to moisture stress and phytophthora wilt.
- Cheriyakaniyakkadan – Small leaves, regular bearer, wilt resistant.
- Cheriakodi – Fruits smallest among all types, bearing in alternate years.
- Daddagya – Fruit largest among mysore type. Esteemed for making white pepper.
- Kalluvally – Regular bearer, tolerant to water stress and disease Kaniakkadam.
- Karimunda – Most popular cultivars grown throughout Kerala, bisexual, Flowering- May June.
- Kottanadan – Most popular in South Kerala, bisexual, highest oleoresin content 17.8%, Pierine – 6.6%, Volatile oil – 2.5-4.2%
- Kuthiravally – Alternate bearer, spike very long and slender (Like horse's tail hence the name Kuthiravally) flowers bisexual, oleoresin 15%, piperine 6% essential oil 4.5%. Produces very high quality pepper.
- Thomanankodi, Aranavalan – Other prominent varieties.
- New released variety
    - Thevam
    - Shakti – Tolerent to fruit rot.
    - Girimunda – Hybrid, for higher elevation
    - Malabar excel- Tolerent to foot rot, hybrid, for higher evaluations.

  ] IISR

- Pepper is a crop of humid tropic (75 - 95 % R.H.).
- It grows successfully between 20° North and South latitude and from sea level upto 1500 m above Mean Sea Level (MSL).
- Crop tolerates temperature between 10° - 40 °C Ideal rainfall – 125-200 cm.
- Western Ghats are ideal place for cultivation.
- Best thrives on virgin soil (Ph 4.5 - 6.0).
- Propagation by – Runner shoot cutting (2 - 3 nodes).
- Punned shoots are separated from vine during – February -March.
- Cuttings from middle 1/3 of the shoot are desirable as they are high yielding.
- Cutting becomes ready for planting in May - June.
- A rapid multiplication technique has been developed by National Research Centre for spices.
- As a mixed crop grown with arecanut, coconut, mango and Jack fruit.
- Pepper – Planting is done with the onset of south west mansoon.
- Erytherina, *Grevelia robusta*, Ailanthus & *Garuga pinnata* are the tree used for climbing of pepper.
- Excessive shading during flowering and fruiting encourages pest infestation.
- Prunning –When the top reach 6 m (March – April).
- Panniyur 1 and Karimunda – Suitable variety for raising bush pepper by planting horizontal side branches or fruiting branches (plagiotropes), cutting with 2-3 nodes are suitable.
- Pepper vine start yielding usually from 3-4th year.
- Flowering – May-June (It takes 6 to 8 months from flowering to ripening stage).

- Harvesting – In plains – November to February.
- In Hills – January to March.
- Harvesting is done when one or two berries on the spike turn bright or red, the whole spike is plucked.
- White pepper is prepared by removing the outer skin and pulp below it before drying the berries.
- Tripod-stand is used for harvesting.
- Recovery of White pepper is about 25% of ripe berries.
- Recovery of Black pepper is about 33% of ripe berries.
- Yield – 800-1000 kg. Black Pepper from 1 hectare of 7-8 years old.
- Full bearing start 7-8th years after planting.
- Decline start after 20-25 years of planting.
- In bush pepper cultivation, all the leaves except the flag leaf are removed.
- Cuttings dipped in 1000 ppm IBA for 45 seconds are used for propogation in bush pepper cultivation.
- White pepper commends a higher market price for use in products as such mayonnaise, where the black speeks of black pepper is not liked.
- Average pepper yield –
  - In India – 273 kg/ha.
  - In Indonesia – 425 kg/ha.
  - In Malaysia – 2000 kg/ha.
  - In Sri Lanka – 431 kg/ha.
- Piperine usually ranges from 4 - 10 %.
- The alkaloid piperine (Melting point 129-130°C) is responsible for bitting taste of pepper.

"CARDAMOM" (Small)

- Known as Queen of spices.
- It is 2nd most important spice of India after Black pepper.
- Cardamom is dried fruit of a tall perennial herbacious plant.
- Area under cultivation – 100,000 ha.
- Annual production – 4000 metric tonnes, 40% of which is exported to more than 60 countries.
- Kerala 60%, Karnataka 31%, Tamil Nadu 9% contribution of total area in India.
- Recently, Guatemala has emerged as a main competitor.
- Inflorescence – Panicle.
- Flower – Bisexual and fragrant.
- Fruit – Capsule.
- Pollination – By honey bees.
- Cardamom flower remains open for 15-18 hours and stigma receptivity and pollen viability are maximum during morning hours.
- Four to fives bee hives/hactare should be maintained for pollination.
- Major producer country – India, Srilanka and Guatemalan.
- Guatemala has overtaken India from second position to first rank; now supplying 62% of the world's requirement of Cardamom with Indian share reduced to 28% only.
- Productivity of India – 60 Kg/ha.
- Guatemala – 300 Kg/ha. (5 times more).
- Aroma is due to α-terpinylacetate and 1, 8 – Cineol.
- Flower initiation – March to April.
- From initiation to full bloom it takes nearly 30 days and from bloom to maturing it takes 5 to 6 months.

- Cardamom found a place on "Queen Sheba" gift box presented to King Solomon.
- Guatemala has become the world's leading cardamom supplier accounting for three quarter of world trade in this commodity.
- In Arabian country beverages, cardamom flavoured coffee called Gawa is generally offered to guests at social and religious functions.
- Varieties are based on fruit size (two varieties).
- *Elettaria cardamomum* var. major (Wild).
- *Elettaria cardamomum* var. minor (Cultivated) is divided into three types -
    - Maysore Type 900-1200 m. MSL
    - Malabar Type 600-900 m. MSL
    - Vazukha-Hybrid (Mysore × Malabar)

New Varieties.

1. Coorg Cardamom Malabar Sel. 1 – Malabar type released by NRCS.
2. ICRI-1 – Malabar type, by ICRI, Myladumpra.
3. ICRI-2 – Mysore type by ICRI, Myladumpra.
4. Mudigree 1 – Malabar type by RRS Mudigree.
5. PV-1 – Malabar type by CRS Pumpadumpra.
6. SKP-14 – Malabar type, By ICRI, Salekpur (Karnataka) Prostrate type of panicle, suited to Karnataka.
   Yield – 260-409 Kg/ha.

- Primary and Secondary nursery is required.
- Malabar cardamom indicates ripening by paling of fruits.

Propagation – Through suckers.

- 1 kg of seed capsule may produce 5000 seedlings.

- Earthing up is done when seedling attain 5-6 leaf.
- In Kerala and Tamil Nadu 18 month's old seedling are used while in Karnataka 10 months old seedlings are used.
- Trashing – Removal of old drying shoots of the plant once in a year with the onset of mansoon under rainfed condition and 2-3 times in high density plantation (HDP) provided with irrigation facility.
- Cardamom is a pseophyte plant, very sensitive to moisture stress.
- Shade regulation is done to –
  1. Regulate soil moisture and temperature.
  2. Provides congenial micro climate.
- Palangi (*Artocarpus fraxinifolius*), Jack, Red Cedar (*Cedirella toona*), Karimaram (*Diospyres ebenum*) and Elangi (*Mimusropes elangi*) are most suitable shade tree.
- Temporary shade trees like *Erythrina lithosperma* and *E. India* are most unsuitable as they compete for nutrients and soil moisture.
- Regular irrigation (25 mm-37.5 mm/Irrigation) helps in initiation of panicales, flowering and fruit set.
- Malabar Cardamom – Drought resistance, mostly suitable for lower hills.
- Mudigree-1 – Tolerant to thrips infestation.
- MCC – Tolerant to azhukal disease.
- Cardamom normally starts bearing after 2 years of planting
- Harvesting is done in October –November, average yield – 500 Kg/Ha.
- Ripe capsules are harvested in order to get maximum green colour during curing.
- Cardamom grading and marketing rules were formed in 1962.

- Malabar type is superior in flavour quality.
- Soaking the freshly harvested green cardamom capsule in 2% washing soda solution for 10 minute prior to drying helps to retain green colour during drying Temperature in Drier – 45 ^{o} – 50° C for 14-18 hours.

  Temperature in Kiln – 50° - 60° C for overnight.
- Summer showers are essential for flowering.
- ECCARD drier (Economical Cardamom Drier) is based on the principle of economically drying using firewood for drying cardamom. It can dry 80-100 kg of Cardamom in 26 hours.
- MELCARD – (Modern Economical Low-Cost Cardamom Drier).
- Cardamom grades/Agmark/ISI specification.
  1. Alleppey Green, coorg green, "Bleached" half bleached, bleached white and mixed Cardamom and Cardamom seed.
  2. Alleppy Green is by far the most important type of Cardamom exported from India, accounting for about 85% of the total cardamom export.
  3. Volatile Oil – Average 8.3%.
- Malabar type – Most pleasant and mellow aroma and flavour character, no well suited for green cardamom, cured fruits are round in shape, 18.5 mm length, and lower cineol content.
- Mysore Type - Harsher aroma and flavour well suited for green cardamom, 21 mm length, and higher cineol content sp.
- Cardamom oil is produced commercially by steam distillation of crushed fruits.
- Best oil is obtained with alleppy green.
- Pyrosis – Excess water in mouth.
- Indian cardamom is low in fat and high is protein, iron, Vitamin – B and C.

- Eating a cardamom once daily with a tablespoon of honey improves eye-sight, strengthens nervous system and keeps one healthy.
- Excessive use of cardamom causes impotency.
- Varities Avinash, Vijetha and PV2 are recommended for release.
- Malabar tupe superior in flavour quality.

"CARDAMOM" (Large)

- Large cardamom also known as Nepal cardamom or greater Indian Cardamom is the dried fruit of *Amomum subulatum* (Seed contain 2-3% essential oil).
- It is shade loving plant (seophyte).
- It is cultivated in Sikkim, Darjeeling and Assam hills.
- India is main producer of large cardamom.
- Sikkim and Darjeeling District of West Bengal produces about 4200 tones of cured cardamom annually.
- About 150 tones are annually exported earning 70 lakhs/ annum.
- Large cardamom is a perennial crop.

Propagation – Rhizomes / Suckers.

- Seed propagation requires 10g. seed for 1 M area.
- Fruit 2.5 cm long ovoid containing 40-50 seeds.
- Average yield – 300 - 1000 Kg/ha from 4-5th year.
- It starts bearing after 3 to 5 years.
- Primary and secondary nursery is required.
- Ideal time for gap filling – May - June.
- Curing is done to retain only about 10-13% moisture.

■ **Varieties –**

1. Bedo – Deep root makes it drought tolerant, tolerant to foorkey and viral infection, 2% essential oil.
2. Bharlangey – 500 kg cured capsule/ha. 15-20 capsules at each spike, 2.5 essential oil.
3. Golsey – Self shaded, require less shade, suited to close spacing, immune to chirkey but not to foorkey disease, susceptible to fungal leaf-spot disease, 2-3% essential oil.
4. Ramla – Moderately tolerant to chirkey but susceptible to foorkey disease, 2-3% volatile oil.
5. Ramsey – Major area under cultivation in Sikkim and Darjeeling, susceptible to Chirkey and foorkey especially at lower altitudes.
6. Sawney – Susceptible to both viral and leaf spot and spike rot disease, Heavy yield under congenial condition.
7. Two viral disease occurs in Large Cardamom.
    1. Chirkey – Transmitted by Corn aphid.
    2. Forkey – Transmitted by Banana aphid.

"GINGER"

- Flower – Bisexual.
- Inflorescence – Spike.
- Fruit – Capsule.
- Ginger is one of the oldest spices with a distinct flavour and pungency.
- India is the largest grower of ginger and also largest producer of dry ginger in the world.
- Major ginger growing states are Kerala, Orissa, Andhra Pradesh, Himachal Pradesh, Meghalaya and West Bengal.

- Kerala – 60% area and 25 % production of our country.
- India account 70% of total world production and 50% in world export.
- India annually exports 5000 tonnes earing Rs. 20 Crores.
- China is our major competitor in the world market.
- India – 53300 ha. area under cultivation with 130000 tonnes of dry ginger production.
- Ginger oleoresin is commercially known as gingerin.
- Ginger of commerce or Adrak is the dried underground stem or rhizomes of Zingiberous, herbaceous plant.
- Standing 3rd to 4th competing with chillies.
- The sundried rhizomes in the trade are known as "Hands" or "races".
- Jamaica and India produce the best quality ginger.
- Ginger in Saudi Arabia is predominantly used for flavouring coffee.
- Major Importing countries from India –

  Saudi Arabia – For fresh and bleached ginger.

  USA, UK, Canada and Singapore – For unbleached dry ginger.
- Indian ginger or Cochin ginger is considered one of the best in the world.
- Ginger rank 3rd in export from India in spices after Black pepper and Cardamom.
- The major bottle necks in boosting exports of Indian ginger are its higher fiber content than that of its main competitor, the Jamaican ginger in world market and its higher cost of production.

Varieties –

1. Rio-de-Janerio – 25-30 tonnes/ha., dry ginger recovery 16-18%, fiber content 5.9%, best for green ginger and

oleoresin content, Introduced from Brazil in Kerala, high moisture requirement.

2. Wynad Manantody – Widely grown in Kerala, best for dry ginger, contributes 50% of total production.
3. China – Suitable for green ginger.
4. Maran – Suitable for dry ginger, most popular cultivar of Assam.
5. Suprabha – Oleoresin 8.9%.
6. Suruchi – Oleoresin 10%.
7. Surari – An X-ray inducted mutant, oleoresin 10.2%
8. Jamaica
9. Bangkok
10. Ernad
11. Chernad
12. Kuruppampadi
13. Tafengiya
14. Himgiri – Tolerant to rhizome rot.

- Pungency is due to gingerol and shogaol.
- Volatile oil – Gingeberene.
- Ginger is always propagated by portions of the rhizomes known as seed rhizomes (Bits).
- Rhizomes length – 2.5-5.0 cm, weight 20-25g. Each having one or two good buds is suitable for sowing. Seed rate (rhizomes) – 1500-1800 kg/Ha.
- Rhizomes sets should be treated with 0.3% dithane M–45 solution for 30 minutes to control fungal diseases.
- Spacing – 30 × 30 cm under bed system of planting.
- Sowing time – April-May.
- Harvesting – For green ginger 5-6 months (Dec.-Jan.).

- For dry ginger – Harvesting is done between 245 to 260 days when leaves turn yellow and gradually start drying up.
- Average yield – 15-25 tonnes/ha.
- For preparing dry ginger the outer skin is removed with split bamboo having pointed ends.
- **Unbleached ginger** – Washing over night – rubbed well to clean – outer skin is removed-dried in sun (for a week) – Rhizomes rubbed together for a week.
- **Bleached ginger** – To get good appearance, Rhizome are peeled and soaked in 2% lime water for 6 hours and dried.
- The yield of dry ginger is 16-25% of the fresh ginger.
- For processing into its products, ginger rhizomes are harvested 5 months after planting.
- The major constraints are non-availability of better genotypes, higher cost of production and high fiber content in varieties.
- There is no flowering and seed set because of chromosomal sterility.
- Polyploidy breeding therefore offered some hope to explore the possibilities of obtaining flowering and seed set.

■ Physical forms of ginger –

- **Hands and fingers** – As dug up from the ground.
- **Black ginger** – Whole rhizomes which have been scaled for 10-15 minutes in boiling water before being scraped and dried.
- The two types of Indian ginger entering international market are Cochin and Calicut ginger.
- Dry ginger is packed in gunny bags.
- Size of gunny bag most commonly used is 122 cm × 67 cm, which is known as Liverpool bag.

- The pungent principle of ginger is oxymethyl phenols.
- Commercial preparation of oleoresin contains gingerol, shogaol and zingerone.
- Freshly prepared oleoresin contains gingerol as the main constituent (approximately 1/3 of weight of oleoresin).
- The mode of ginger consumption are 42-43% dry, 36% green ginger, 17-18% seed and 3-4% spoilage.

■ Physical forms of ginger –

1. "Peeled", Scraped or uncoated – Whole rhizomes from which the cork skin has been cleanly removed without damaging the underlying tissue.
2. "Rough scraped"- Whole rhizome from which the skin has been partially removed usually on flat sides.
3. "Unpeeled" or "Coated" – Whole rhizome which have been dried with the skin intact.
4. "Bleached" or "Limed" – Clean peeled-whole rhizome which have been treated with lime or sulphurous acid to achieve a lighter colour.
5. Splits – Unpeeled rhizome which have been split longitudinally to accelerate drying.
6. Slices – Unpeeled rhizomes which have been sliced to accelerate drying.
7. Ratoons – Second growth rhizomes from plant.

"TURMERIC" (Haldi)

- The spice "turmeric" constitutes boiled, dried, cleaned and polished rhizomes (the underground swollen stem of plant) of *Curcuma longa.*
- The plant is herbaceous, perennial and tropical crop.
- India is by far the largest producer and exporter of turmeric in the world.

- India accounts for 76% of total world production.
- Turmeric occupies about 6% of the total area under spices and condiments in India.
- Andhra Pradesh – Leading turmeric producer state.
- Yellow colour is due to curcumin.
- Rhizome has 1.8 to 5.4% curcumin.
- Rhizome has 2.5 to 7.2% essential oil.
- Rhizomes are called ginger.
- There are four important species of curcuma.
- *C. longa* – Widely cultivated.
- *C. aromatica* – Cochin turmeric / Kasturi Manjal.
- *C. angustifolia* – East Indian Arrow root having plenty of starch in its rhizome.
- *C. amada* – Mango ginger (Which has the taste and flavour of raw mango.)
- In International trade, turmeric is graded into "finger – wholes" and splits.
- Best quality – Fingers.
- Alleppey (fingers) – Well known in trade for its colour.
- Turmeric is antioxidant due to phenalic character of curcumin.

Varieties –

1. CO1- X-ray induced mutant from Erode local, suitable for drought and saline soil, rhizome big size, yield 35 t/ha, duration 270 days, curcumin 32%.
2. BSR 1 – X-ray induced mutant from Erode local, suitable for waterlogged condition, 31 t/ha., curcumin 4.2%.
3. Suguna – Clonal Selection.
4. Suvarna – Clonal Selection.

5. Sudharshana – Clonal Selection.
6. Sugandham – Clonal Selection.
7. Krishna – Clonal Selection moderately resistant to rhizome fly and leaf spot disease.
8. Roma – Less susceptible to diseases, curcumin 9.3%.
9. Suroma – Curcumin 9.3%.
10. Rajendra Sonia.
11. Ranga.
12. Rashmi.
13. Sobha – Oleoresin 15%.
14. IISR Prabha – Oleoresin 15%.
15. IISR Prathiba – Oleoresin 16.2%.
16. Kanti.
17. Allepy turmeric – Best in world.
18. Amba – A variety of Mango ginger.
19. Pant Peetabh – Latest variety of turmeric.

- Highest amount of curcumin has been found in variety Edapalayam (10.9%) and oleoresin in "Konni" (19.2%).
- Propagation – Seed rhizomes (35 - 44 g.) 2500 kg of rhizomes are required for 1 ha., spacing – 30 × 15 cm, 30 × 20cm in beds, 40 - 60 cm × 25 cm (in ridge or furrow), 50 × 15 cm in flat system.
- Turmeric is a Shade loving plant (Seophyte).
- Planting time – April – June.
- Harvesting – 7 - 9 months after planting.
- Harvesting time – January to March.
- Early variety – Matures in 7 - 8 months.
- Medium variety – Matures in 8-9 months.
- Late variety – Matures after 9 months.

- Average yield – 20-25 tonnes of green turmeric. 25-30 tonnes/ha. (HDP)
- Turmeric processing involves three steps curing, polishing and colouring.
- Curing involves boiling of fresh rhizomes in water.
- Colouring is done to give good appearance and better finish to the product.
- Composition of the emulsion required for coating 100 kg of half boiled turmeric is Alum 0.4 kg. Turmeric powder – 2 kg, Castor Seed oil : 0.14 kg, Sodium bisulphate – 30 g., Concentrated HCl-30 ml.
- Dried rhizomes – 5 - 6 % oil.
- About 58% of the oil is composed of turmerones (sesquiterpene ketons) and 9% tertiary alcohol.
- Acetone, alcohol and ethylene dichloride were found suitable for extracting oleoresin from turmeric.
- Oleoresin contains 30-35% curcumin and 15-20% volatile oil.
- Dry recovery of turmeric is 15-30%.
- Trade name – Alleppy, Erode, Duggirala, Nizamabad, Rajapuri and Chudlapah.

"SEED SPICES" (Coriander)

- Term coriander is deried from Greek word "Karis" meaning "bedbug" because of pleasant foetid, bug like odour of green and unriped fruits.
- It is an annual herb, cultivated for its green leaves as well as for its fruits.
- Highly cross pollinated crop.

- It is one of the first seed spice to be used by mankind as early as 5000 BC.
- Flowers are umbel, hermaphrodite, sometimes unisexual or polygamous white or rose coloured.
- Fruits – Schizocarp.
- Essential oil – Linalool (80%), fatty oil (19-21%).
- The main importing countries are those in the gulf region.
- Pakistan, Egypt and Iran are the main competitors of India in international trade.
- Coriander is employed for flavouring liquor particularly gin.
- The odour and taste are due to the essential oil content which varies from 0.1 to 1.0% in dry seed.
- Essential oils are used for flavouring liquors and cocoa preparation.
- Dried fruit – Major ingredient of curry powder.
- Good quality oleoresin is extracted from coriander seed.
- Decydenyde (0.1% volatile oil) is also used in perfumery.
- It is a tropical crop and can be grown throughout the year (except very hot season i.e. March-May) for leaf purpose.
- Black cotton soils with high retentivity of moisture are suitable for its cultivation.
- Coriander is the leading seed spice crop grown in India.
- Coriander is the most heavly consumed herb plant.

■ Varieties –

1. RCr 41 – Highly resistant to stem gall and wilt but moderately tolerant to powdery mildew.
2. RCr 20 – Moderately tolerant to powdery mildew, wilt as well as stem gall.
3. RCr 435, 436

4. RCr 446 – Avg. yield 12q./ha.
5. GC 1- Yellow coloured grain, moderately tolerant to wilt and powdery mildew.
6. GC 2 - Yellow coloured grain, moderately tolerant to wilt and powdery mildew.
7. Sindhu – Resistant to aphid and moderately tolerant to wilt and powdery mildew.
8. Sadhna – Tolerant to white fly and mites.
9. Swathi – Tolerant to white fly, grain mould and wilt.
10. RD 44 (Rajendra Swathi) – Resistant to stem gall and moderately resistant to wilt aphid and weevil suitable for Bihar.
11. Co 1, 2, 3 – Dual purpose.
12. DH-5, Hissar Anand.
13. Pant Haritama – For Punjab.
14. NRCSS 1, 2 – New released varieties.
15. Karan – Small seeded variety suitable for Rajasthan.
16. Lam sel. – For Andhra Pradesh.
17. Merrocon

- Coriander is propagated by seed.
- Seed rate

  10-15 kg – In irrigated condition.

  25-30 kg – In rainfed condition.

  40-42 kg - For leafy vegetable.
- Sowing Time

  Kharif – June – July.

  Rabi – October-November

A spray of GA 100 ppm at five leaf stage increases the male flower while Ethrell and CCC reduce the ratio of bisexual to male flowers.

- GA @ 5-50 ppm hastens and improves flowering when sprayed at 5 leaf stage.
- Harvesting – 90-140 days after sowing (DAS)

Yield –

400-500 kg (Rainfed crop)/ha.

600-1200 kg (Irrigated crop)/ha.

- Drying is done to maintain a moisture level below 13% in seeds.
- Spanish people eat fresh coriander leaves "*cilantro*" and it is also called Chinese parsley.
- Hissar Sugandh – Latest variety.
- IARI Sheetal – 36-3, CIMPO Sel. S-28 and S-52 are recommended for green leaf crop.
- The plants are andromonoecious i.e. both staminate and the perfect flowers are produced on the same plant and even in the same umbellet.

"CUMIN"

- It is an annual herb.
- Flower colour – White or pink.
- Inflorescence – Umbel.
- Volatile oil – 2.5-3.55 produce by steam distillation.
- The seed also contain 17.7% protein, 23.8% fat, 35.5% carbohydrate and 7.7% minerals.
- In India mainly cultivated in Rajasthan and Gujarat.
- It is commercially cultivated in Iran, India, Turkey, China, Morocco, Indonesia and Japan.
- Iran – Major cumin exporter country in the world.
- Fruits are rich in thymol.

- Cumin produces strong and heavy flavour due to an alcohol, cuminol and is slightly bitter in taste.
- It is used in preparation of curry powder.
- The main ingredient of volatile oil is aldehyde.
- It also contains 10% fixed (non-volatile) oil.
- Moderately cool and dry climate is best for cumin cultivation.
- High humidity and frequent rains particularly after flowering period increase incidence of disease like blight and powdery mildew.
- Planned work on collection and improvement of cumin was started by ICAR in 1957 at IARI.
- Aroma is due to cuminol.

■ Varieties –

1. RZ 19 – Tolerant to wilt and blight, released in 1988 developed at RAU, Jobner.
2. RZ 209 – Highly resistant to blight.
3. GC 1 – Tolerant to wilt and blight, 3.6% volatile oil released in 1988.
4. GC 2 – Developed by GAU through pure line selection released in 1991.
5. Vijapur 5 – Popular variety of Gujarat, susceptible to diseases.
6. MC 43
7. S – 404
8. RS 1
9. UC 19
10. UC 194
11. Topolka – Introduce from Bulgaria.

- Seed rate – 12-15 kg/ha.
- Best sowing time – Middle of November.

- Jeeri (*Plantago pumila*) a serious weed of cumin, can be effectively controlled by fluchloralin.
- Harvesting – 120-150 DAS (Days After Sowing).
- Yield – 7 - 8 q/h.
- Cumin is one of the earliest known minor spices used by mankind.
- A drink prepared by cumin in Germany is known as Kummel.
- Black Cumin (Kalongi) – *Nigella sativus* – Rabi season crop var- kala jeera.
- Kalongi has a carbonyl compound "nigellone".

"FENUGREEK"

- Annual herb (both tropical and temperate crop).
- Flower – Papilionaceous (white or yellow).
- Fruits – Legume.
- The species is a typically self- pollinated and destogamous.
- Pollen fertility ranges from 95-98% in the unopened flower bud and 67-80% in open flowers.
- The cultivated forms are diploids $2x = 16$.
- A sterile hexaploid $2x = 44$ has been recorded.
- Anthesis – 9 AM and 6 PM, 11.30 AM (peak period).
- Anther dehiscence – 10.30 am – 5.30 pm, 11.30-12.30 (peak period).
- Stigma become receptive 12 hours before flower opening and remains receptive for about 10 hours after opening of flower.
- Duration of flowering phase – 7-18 days.
- Common Methi – T. *foenum graecum.*
- Kasthuri Methi – T. *corniculata.*

- Kasthuri methi is slow growing type and remains in rosette condition during vegetative phase of growth period.
- Common methi take 5-6 days to germinate while kasthuri takes 7-8 days.
- Fenugreek is third largest seed spice in India after coriander and cumin.
- Fenugreek – Used for curry powder.
- Fenugreek is mainly cultivated in Rajasthan, Gujarat, Madhya Pradesh and Tamil Nadu.
- More than 80% area and production of the country is contributed by Rajasthan alone.
- India is the major producer and exporter of fenugreek.
- Fenugreek seeds contain "diosgenin" which is used as a starting material in the synthesis of sex harmones or preparation of contraceptive pills.
- Diosgenin in seed – 0.40 - 1.26 %.
- Fenugreek is rich in amino acid and trigonelline 150 mg / 100 g.
- The seed also good for elimination of bad breath and body odour and also help to combat dandruff and is a cure for baldness in man.
- Indian women use the seeds power of methi to promote lactation.

Recommended Varieties of fenugreek

S.N.	Variety	Features
1	Pusa Early Bunching	It is quick growing with upright shoots, good yielder, suitable for 2-3 cuttings. Developed at Indian Agricultural Research Institute, New Delhi.

2	Pusa Kesuri	A late flowering variety, rosette type leaves with special fragrance, heavy yiedler and gives 5-6 cuttings. Developed at Indian Agricultural Research Institute, New Delhi.
3	Methi No. 47	Leaves are broad. Succulent and rich in Vitamin C. Released by Maharashtra state.
4	Co-1	A high yielding variety suitable for both greens as well as for seed. Developed at TNAU, Coimbatore.
5	Lam Selection	Plants of medium height, bushy, more branched. Developed by APAU.
6	Hissar Sonali	It is a variety for green leaves. It is developed by HAU, Hissar.

■ Varieties –

1. RMt 1 – Suitable for Rajasthan 1500 kg grain/ha, moderately resistant to root rot and tolerant to powdery mildew.
2. RMt 143 – Moderately resistant to powdery mildew. It is especially recommended for heavier soils of Chittor, Bhilwara, Jhalawar and Jodhpur of Rajasthan.
3. CO 1 – Dual purpose variety suitable for Tamil Nadu.
4. Rajendra Kanti – Moderately resistant to powdery mildew suitable for Bihar.
5. Lam Sel. 1 - Tolerant to root rot, powdery mildew caterpillar and aphid suitable for Andhra Pradesh.
6. Hissar Sonali – Moderately resistant to leaf spot and root rot complex diseases.
7. Pusa early Bunching – Champa Fenugreek.
8. Kasuri selection – Champa Fenugreek.
9. HM 1 – Moderately resistant to leaf spot disease. 100kg/ ha. grain suitable for Haryana.
10. Prabha (NLM).
11. Plume – 20.

- Last week of Oct. – first week of Nov. is best time for sowing
  Seed rate 20-25 kg/ha.
  40 kg – broadcasting method.
- Spacing – 30 × 10 cm.
- Seed should be treated with *Rhizobium* culture.
- Harvesting – 70-165 Days after sowing (DAS), depends on variety.
- Average seed yield – 10-11 q/ha.
  8-10 tonnes of leaves/ha.
- Storage – 7-8% seed moisture having 40% R.H.
- For good yield field inoculation with *Rhizobium meliloti* is useful,
- A spray of ascorbic acid at 250-450 ppm improves plant growth and enhances number and size of pod.
- Rajendra Khushba – Latest released variety of fenugreek.
- Pant Ragini - Latest released variety of fenugreek.

"FENNEL"

- Annual herb.
- The crop is protandrous and cross pollinated.
- Fennel is mainly cultivated in Romania and India.
- In India – Gujarat is major producer followed by Rajasthan and Uttar Pradesh.
- The fruit of fennel yield a volatile oil, the main constituent is anethole.
- Bitterness is due to – Fenchone.
- Sweetness is due to – Anethole (70%).
- In India seeds are used for mastication and chewing alone or with betel leaves and scented soaps.
- Seeds also contain limone.

■ **Varieties –**

1. RF 35 – Moderately tolerant to sugary disease, leaf spot and leaf blight, green seeds.
2. RF 101- Bear large umbel.
3. RF 125 – Yield 17.5 q/ha.
4. Gujarat Fennel 1 – Moderately tolerant to sugary disease and leaf spot, suitable for early sowing, tolerant to drought.
5. CO 1- suitable for drought prone water logged saline and alkaline codition, suitable for hilly areas.
6. PF 35
7. S-7-9
8. Mutant 1,2,3

- Fennel require cool and dry climate for its cultivation.
- Seed Rate – 10-12 kg/ha.
- Sowing time – Mid September to mid October (Best time).
- Chewing type "Lucknawi" Fennel is produced by plucking the umbel 30-40 days after pollination when the size of seed is just half of the full development seed and then dried in shade.
- Fennel matures in 170-180 days, Yield 9-10 q./ha.
- Fennel is most vulnerable to frost damage at the flowering and seed formation stage.
- Control – Spray 0.1% H_2SO_4 and irrigate the crop.

"DILL" (Sowa)

- It is a winter annual herb.
- Flower – Pale yellow.
- Sowa is a long day plant (LDP).
- On distillation an essential oil containing high carvone content is obtained.

- The seed oil of sowa has superior aroma but the oil contains an oxygenated compound "Dill apiole" which is toxic in large doses.
- The herb oil harvested at vegetative stage is rich in phellandrene where as seed oil has high carvone content.
- It is grown in Europe, USA and India.
- In India – Gujarat, Maharastra, Rajasthan and Madhya Pradesh are main producers.
- *Anethum sowa* has longer fruits than the European sp.
- *Anethum panmorium* has strong resemblance to the *A.faniculum*.
- India dill or sowa – *A. graveolens var sowa.*
- Seed Rate –

5 - 6 kg.	Line sowing.
10 kg.	Broadcasting.

Sowing Time – Mid October (Best Time).

Yield – 8 - 9 q./ha.

Varieties- NRCSS-AD 1 (Contains 3.5% essential oil).
NRCSS-AD 2 (Contains 3.5% essential oil).
Haldwani Selection.
Sel-163, NP-179.

- Aroma in herb is due to hexahydrobenzafuran.

"CELERY"

- Latin name : *Apium graveolens* L. var *dulce.*
- "*Graveolens*" means strong scented; "*dulce*" means sweet.
- Family name : Umbelliferae (Apiaceae).
- Diploid (2n = 22) and Biennial.
- Related species used as vegetables : carrot, parsnip and celeriac.

- Origin - Mediterranean and Eurasia.
- It is a winter annual herb.
- Introduced from France.
- Commercially grown in Amritsar, Gurdaspur and Jalandhar district in Punjab, Uttar Pradesh, Himachal Pradesh and Tamil Nadu.
- Harvested organ : petiole, eaten raw or cooked.
- Used as a medicinal herb by the Egyptians and Romans.
- Older varieties required blanching.
- Modern cultivars are grown unblanched.
- Principle consituents of celery seed oil are-
  1. d-limonence (60%)
  2. d-Selinence (10%)

     Sedanonic acid anhydride (0.5%)

     Sedanolide (2.5-3.0%)

     Both are responsible for the aroma of oil.
- Celery has 16-20% fatty oil.
- It is sensitive to very low day temperature (5-10°C) during vegetative growth causing early bolting.
- Mainly grown for salad purpose.
- Days to harvest maturity : 90 to 140 days after planting.

■ **Varieties –** Two type

A. Self blanched or yellow (a) Golden (b) Florida Golden.

B. Green Leaved –

a. Fork Hook Emperor – Late variety.

b. Standard Beared – Early Variety.

c. Wright Grove Giant – High yielding.

RR 185-1 – Improved variety.

Seed rate – 375-500 g./ha.

Sowing Time – August-September

Yield – 300-560 q./ha, green leaves.
175-275 kg/ha. Seed yield.

- **Physiological disorder –**
  1. Pencil Strip – Excess of phosphorus.
  2. Black Heart – 'Calcium' deficiency.
  3. Cracked Stem – 'Boron' deficiency.
- Major insect pests are aster leafhopper, tarnished plant bug, aphids, carrot weevil, variegated cutworms, loopers, celery leaf tier, leafminer and mites.

"CLOVE"

- It is an evergreen tree spice.
- Clove is the dried unopened flower bud of *syzygium aromaticum* (syn. *Eugeneia caryophyllus*).
- In India introduced around 1800 AD by the East India Company in their garden in Courtallam (Tamil Nadu).
- Indonesia, Zanzibar and Madagascar – Major producer.
- Indonesia alone contributes for 66% of world production.
- 2nd most important spice next to black pepper in the world.
- India – Nilgiris, Tirnelveli and Kanyakumari district of Tamil Nadu, Calicut, Kottayam districts of Kerala are major producer.
- Clove is also known as **chicken tongue spice**.
- Clove has alternate bearing habit.
- In Indonesia major part is absorbed for making KRETEK cigarette industry.
- Flower – Hermaphrodrite.
- Fruit – One seeded drupe (SSD).

- Clove is a strictly a tropical plant.
- Propagation – Seed (Known as mother clove).
- Seed become available from June to October and lose their viability with in one week.
- Approach grafting of clove on its own root stock is successful.
- The 18-24 month old seedlings are ready for transplanting in the field (June-July).
- Intercropping with Banana is very good.
- Bearing start after 7-8 years of planting and full bearing start after 15 to 20 years.
- Flowering season –

  Sept. – Oct. (In plains).

  Dec. – Feb. (At high altitude).
- Flowers are produced on young flush.
- Optimum stage of clove picking is when bud changes colour from green to slighty pinkish tinge.
- It takes 4-5 days for drying; well dried cloves will be only about 1/3rd of the weight of the original.
- About 11,000 to 15,000 dried cloves make 1 kg.
- Yield : 4-8 kg/tree from well maintained full grown tree.
- 2 kg/tree average annual yield after 15 years.
- Clove Oil – 16-21% in the buds.
- Oil contains – Eugenol (70-90%) (JRF 2008).
- Eugenyl acetate – 5 to 12%.
- Aroma is due to – Methyl amylketone.
- Varieties – Pevang, Lezipore, Amulya.

"NUTMEG"

- It is an evergreen, conical tree.
- Nutmeg is the dried kernel of seed.
- Mace is dried aril surrounding the seed.
- Nutmeg was introduced in India towards the end of the 18th century.
- Over 15% of the world exports originate from Indonesia.
- Greneda is the second largest exporter of Nutmeg and Mace.
- In India – Kerala, Tamil Nadu and Karnataka are major producers.
- Nutmeg is dioecious tree **(JRF 2008)**
- Fruit – Drupe.
- Pollination – Wind and Insect.
- Propogation by seed.
- About 18-24 months old seedlings are used for transplanting in field.
- Epicotyl grafting technique developed by NRCS, Calicut is found to be promising now.
- Sraight growing shoots are known as orthotropic or chupan.
- Spacing – 8 × 8 M.
- Shade plant – *Erythrina indica*, *E. lithosperma* are planted for Nutmeg.
- Bearing starts in 4 to 5 years when grafted.
- Bearing starts in 7 to 8 years when seedling tree is grown.
- Takes 15 to 20 years for full bearing.
- Harvesting – June-August.
- Fruit take 9 month from flowering to harvesting.
- Freshly pericarp is used for making pickles, jams and jellies.

- Nutmeg fruit weight 60g. of which seed weight 6 to 7g., mace 3 to 4g. and rest pericarp.
- Yield – 2000-3000 fruits/year.
  800 kg nutmeg/ha.
  100 kg mace/ha.
- 10 : 1 (female : male) plants should be planted.
- In established plantation the unproductive male plants can be converted into productive female tree by top working.
- Volatile oil terbein have weedicidal property.
- Pericarp contains 12-14% pectin.

"CINNAMOM"

- True cinnamom also called as Ceylon cinnamom or Sweet Wood, one of the oldest known spices is the dried bark of *C. verum* (syn. C. *zeylanicum*).
- Cassia is popularly known as "Chinese Cinnamom" false cinnamon or bastard cinnamom also one of the oldest known spice.
- Indian Cassia (Tejpet) – *Cinnamomom tamala.*
- Cinnamom bark has 0.5-2% essential oil, eugenol is the main constituent in the leaf oil.
- Cassia bark has 1 to 4.5% essential oil, cinnamaldehyde is the main constituent in the leaf oil.
- The National Conservatory of Cassia Germplasm consists of 30 acessions.
- The oldest cinnamom plantation is the Anjarakandy Estate (250 ha.) in Cannore district of Kerala.
- Srilanka – Largest producer of cinnamom bark with best quality.

- In India grown in Kerala, Tamil Nadu and Karnataka.
- Indonesia, China and Vietnam are the largest producer of Cassia and it is not grown in India.
- Commercial method of propagation – By seed.
- Flowering - January.
- Fruit – June-August
- Shoot harvest – Sep.–Nov. (Ideal for bark extraction).
- Cut 1.0-1.25 m length and 1.25 cm thickness when they are about 2 years old.
- During drying the bark contracts and assumes the shape of quill.
- The quills are graded from 00000 being the finest quality to 0 the coarsest quality.
- The small pieces of bark, left after preparing the quills are graded as quillings.
- The very thin inner pieces of bark are dried as featherings.
- Other grade – Scrapped chips (Bark is scrapped instead of peeling).
- Yield – 200-300 kg dried bark/ha.
- Leaf and bark oil contains eugenol (70-80%) cinnamic aldehyde.

■ Varieties –

1. Navashree: 6-7 shoot/year, high yield 56 kg/ha. in the first 4 year, high bark recovery (40.6%), bark 2.7% oil, Cinnamaldehyde content 73%, in oil bark oleoresin 8%, leaf oil 2.8%, recommended for all cinnamom growing areas.
2. Nithyashree – Released from IISR – Calicut.
3. Konkan Tej
4. Yercaud 13

"ALLSPICE"

- Its flavour is said to resemble a blend of cinnamom, clove and nutmeg.
- The allspice of commerce is the dried immature fruits of the tree *Pimenta dioica* (small evergreen tree).
- Jamaica : Main producer of allspice.
- Flower structurally hermaphrodite but functionally dioecious.
- Fruit – Two seeded berry.
- Flowering – March-June (white colour).
- Edible part – Dried immature fruits (Berries).
- Berries are rich source of vitamin A.
- Male tree – Flower earlier.
- Propagation – By seed.
- After 3 weeks of seed extractions they start reduce viability and loose completely viability after 9 week.
- Yield – 50 - 60 kg dry berries/tree.
- The pimento oil is distilled commercially in Jamaica.
- Oil in berries – 3.3 - 4.5% (Eugenol 65-66%).
- Pimento leaf oil is also distilled only in Jamaica (0.7-2.7% in dried leaf), (0.35 - 1.25 % in fresh leaves)
- Cultar show positive response to flowering and fruiting in all spices.

"VANILLA"

- Malagasy Republic grows 70-80% of the world's crop of Vanilla bean followed by Reunion.
- USA is the largest importer.
- Introduced in India as early as 1835.

- In India grown in Wynad (Kerala) and Nilgiris (T.N.).
- Vanilla is an orchid.
- *V. planifolia* – Produce short thick pods.
- *V. pompana* – Produce largest pods.
- Natural pollination is impossible due to presence of rostellum which prevents contact of anther cap and stigma.
- In its native home, melapone bees and humming birds aid in pollination.
- Propagation – By stem cutting (60-120 cm long vine) having 5-6 internodes.
- Due to peculiar structure of the flower artificial pollination by hand is the rule for fruit setting.
- 85-100% success is obtained by hand pollination.
- Ideal time for pollination – 6 am to 1 pm.
- Pods take 6 weeks to attain full size from fertilization but takes 4 to 10 months to reach full maturity.
- Harvesting – When immature, bean is dark green in colour but when ripe yellowing start from its distal end.
- Oil is called balsom of vanilla (vanillin).
- Vanillin – 2.3 - 2.9 % in beans.
- Vanillin is developed as a result of the enzyme (B-glycosides) action on the glucoside contained in the beans during the process of curing.
- The maxican process yield, 4.15 to 4.40% of vanillin content.
- Vanillin is extracted from the cured beans by hydroalcholic extraction.
- Vanillin content will not be less than 2.5% if properly cured beans are used.
- Caumarin – Cheap substitute for vanilla.
- Yield – 300 - 400 kg cured beans/ha.

"SAFFRON"

- Saffron is as precious as gold not only because of its low production and high demand but also because of its wide range usage in medicine and religious rituals.
- It is used to colour food even at 1 ppm concentration which gives distinct yellow tinge.
- Saffron contains 0.5 to 1.0% essential oil.
- Main principle is cicrocrocin – Responsible for colour.
- Stigma is funnel shaped, used a spice.
- Aroma is due to Saffranol.
- In India it is cultivated only in Kashmir.
- Bitterness is due to – Pecrocrosin.
- Propagation – By bulbs /corms best method (8 tonnes/ha.) in the month of August.
- Flowering – End of October (At this time picking of flower should be done early in morning).
- Shahi Saffron – The finest, purest and most expensive saffron in trade.
- Mogra Saffron and lache saffron – Inferior grade.
- Stigma is used to prepare shahi saffron and mogra saffron and lache saffron are prepared by rest flower parts.
- About 1.5 million flowers on drying give 1 kg of saffron.
- **Yield –** 160 kg fresh flower/hectare.
  5 kg dried saffron/ha.

"CURRY LEAF"

- It plays an important role as a condiment in the culinary preparation of South Indian dishes.
- Self pollinated crop.

- Richest source of calcium.
- Crystalline glucoside ksenigin - Obtained from leaves.
- Muragin – Obtained from flower.

Varieties –

- DWD 1 is sensitive to winter, during winter growth is poor, whereas DWD 2 is winter insensitive (gives high yield).
- Suwasani.
- Senkaampu.
- Mainly propagated through seed.
- Yield : 20 - 25 tonnes of leaves.
- DWD-1 having 5.22% oil where as DWD contains 4.09% oil, both have good aroma.

"PAPRIKA"

B.N. – *Capsicum annum.* F – Solanceae.

- Bright red, sweet to mild pungent, rich in vitamin C.
- Mainly used for adding colour to finished product.
- Colouring matter – Capsanthin (Outer fleshy pericarp).
- Pungent due to – Capsaicin (Inner tissue and seed).
- Spain, Hungary, Morocco, Yugoslavia and USA are the chief producer of paprika.
- Seed rate – 1 kg/ha.
- Yield – 200 kg/ha.

"CHILLI"

- Bird chilli – *Capsicum fruitiscens.*
- India is largest producer of chilli in World, accounting 45% of total area under cultivation.

- Variety suitable for pickle making – Sindhur.
- Variety suitable for capsicin extraction –
  (1) Aparna, (2) Pachas Yellow.
- About 90% capsicin in chilli has been noticed in placenta connecting seed with pericarp.
- Green chillies retard cancer due to natural enzyme asperginase.

"BISHOP'S WEED/AJOWAN"

- Alkaloid – Thymol.
- Thymene (dethymolized oil).
- Xanthotoxin (0.4%).
- Uses – Treatment of vitilago or leucoderma and in making suntan lotin.
- Seed rate – 3-4 kg/ha.
- Variety – Lam sel. 1,2.
- Essential Oil – 3 - 4%.
- Fatty Oil – 26%.
- Sowing Time – October-November
- Harvesting – May-June.
- Yield – 8-10 q/ha.

"THYME"

- Volatile oil 1.08-2.0 %.
- Thymol (45%) – It is responsible for aroma and has germicidal property.
- Propagation – By seed and cutting.

"ROSEMARRY"

- Propagation is done by cuttings (15cm) long.
- Flowering – May-June.
- Harvesting is done after full flowering stage.
- Essential oil – Pinene, Comphene, Cineol, Camphore, Borneol and Bormylacetate.

"KOKUM"

- Dioecious in nature.
- Konkan Amrit is an important variety.

"ASAFOETIDA"

- Hadda is most priced and strongest trade.
- Hingashakta is prepared from this, used in ayurvedic medicines.
- Tapping is done foliage turn green to yellow.

"JUNIPER"

- Dioecious in nature.
- Juniperin – Mixture of tannins and sugar is obtained from juniper.

ANNEXURE

1. **Gametes are usually...........cells. (JRF-2007)**
(a) Diploid (b) Haploid
(c) Germ (d) Polyploid.

2. **In freezing which principle is followed ? (JRF, 2001)**
(a) Condensation (b) Crystallization
(c) Drying (d) All

3. **The economical part in brocoli is termed as ________**
(a) Curd (b) Flower bud
(c) Flower stalk (d) Head

4. **In metaphase stage of cell division chromosomes are arranged on (JRF-2007)**
(a) Equitorial plate (b) Polar plate
(c) Centromere (d) None of these

5. **Which vegetable crop is day neutral? (JRF, 2004)**
(a) Tomato (b) Potato
(c) Sweet potato (d) Spinach

6. **Ripening fruits exit which type of gas ? (JRF, 1999)**
(a) Ammonia (b) Ethylene
(c) Methane (d) All

7. **The vegetable having 70% exchange earnings in export of fresh vegetable is**
(a) Onion (b) Okra
(c) Cucumber (d) Tomato

8. **Monohybrids are derivd by hybridizing two individuals which differ for (JRF-2007)**
(a) Single-character (b) Double character
(c) Quadriple character (d) Multiple character

9. **Mendelian population refers to ....sharing individuals (JRF-2007)**
(a) Different gene pool (b) Mixed gene pool
(c) Same gene pool (d) Both A and B

10. **Quantitative characters are governed by (JRF-2007)**
(a) Monogene (b) Bi-gene
(c) Recessive gene (d) Polygene

11. Cells without cell well are known as (JRF-2007)

(a) Chloroplast (b) Protoplast
(c) Proplastics (d) Chromoplast

12. Hardening is a process in the tissue culture for (JRF-2007)

(a) Cellus formation
(b) Morphogenesis
(c) Making the plants fit for growing under field conditions after tissue culture
(d) Plantlet regeneration

13. Which vegetable crop is commercially used for processing in India? (JRF-2007)

(a) Tomato (b) Ashgourd
(c) A & B both (d) Tapioca

14. Sweet potato is a (JRF-2007)

(a) Modified stem (b) Modified crow
(c) Modified rhizome (d) Modified root

15. Genetically modified vegetable crop recommended for commercial cultivation in India is (JRF-2007)

(a) Potato (b) Brinjal
(c) Tomato (d) None of these

16. Ideal processing potato cultivator should have (JRF-2007)

(a) High TSS and low sugars
(b) High dry matter & low sugars
(c) High sugars (d) High acidity

17. Cassava is used for the manufacture of (JRF-2005)

(a) Pectin (b) Sugar
(c) Sago (d) Latin

18. Which crop gives higher protein recovery per unit area? (JRF-2007)

(a) Pea (b) Gherkin
(c) Carrot (d) Potato

19. The long day type variety of onion is ....

(a) Early Grano (b) Pusa White flat
(c) Brown Spanish (d) Arka Pitamber

20. **Which of the following is a stem vegetable? (JRF-2007)**
(a) Colocasia (b) Amaranth
(c) Sweet potato (d) Knolkhol

21. **Muriate of Potase is called (JRF-2005)**
(a) K_2SO_4 (b) KCI
(c) K_2O (d) $KMNO_4$

22. **Softness of Pea is measured by (JRF, 2002)**
(a) Derometer (b) Refractometer
(c) Alanometer (d) Tendrometer

23. **Which vegetable is rich in Calcium ? (JRF, 2004)**
(a) Amaranthus (b) Cowpea
(c) Potato (d) Brinjal

24. **For inactivation of microbial growth which method is followed (JRF, 2001)**
(a) Freezing (b) Sulphuring
(c) Blanching (d) None

25. **Tomato variety which can with stand high as well as low temperature is (JRF-2007)**
(a) Pusa Naveen (b) Hissar Lalit
(c) Pusa Sheetal (d) Pusa Sadabahar

26. **In tomato, red colour is due to (JRF-1998)**
(a) Anthocynin (b) Lycopene
(c) Xanthophylls (d) None

27. **How many types of gametes (2^n) will be produced by F_2 in a dihybrid cross ?**
(a) 4 (b) 2
(c) 9 (d) 16

28. **'Blossom end rot' of tomato is due to deficiency of (JRF, 2002)**
(a) Mg (b) Bo
(c) Mo (d) Ca

29. **Which phenotypic ratio is expected in F_2 in case of complementary gene action?**
(a) 12 : 3 : 1 (b) 13 : 3
(c) 15 : 1 (d) 9 : 7

30. **Pungency of Garlic is due to** **(JRF, 2002)**
(a) Allyl propyl disulphide (b) Diallyl disulphide
(c) Isothiduinate (d) None

31. **Which of the following has the highest cation exchange capacity ?** **(JRF-2007)**
(a) Sandy loam soil (b) Loam soil
(c) Loamy sand soil (d) Clay soil

32. **Which of the following is single seeded vegetable ?** **(JRF, 2001)**
(a) Cucumber (b) Leek
(c) Chow-chow (d) Tomato

33. **Potato contain how much % of protein** **(JRF, 1999)**
(a) 1.8 (b) 2.2
(c) 1.4 (d) 3.2

34. **Cryo - preservation is associated with** **(JRF, 2001)**
(a) Liquid nitrogen (b) Liquid oxygen
(c) Liquid potassium (d) Liquid carbon di-oxide

35. **The process of bringing a wild species under human management is referred to as**
(a) Acclimatization (b) Introduction
(c) Selection (d) Domestication

36. **The crop which can be grown in acidic soils is (JRF,1997)**
(a) Spinach (b) Cauliflower
(c) Watermelon (d) French bean

37. **Which of the following is the oldest known spice ?**
(a) All spice (b) Nutmeg
(c) Black pepper (d) Cassia

38. **Bush pepper is propagated from**
(a) Runners (b) Laterals
(c) Top shoots (d) Hanging shoots

39. **The biting taste of black pepper is because of**
(a) Terpenine (b) Piperine
(c) Pipermenthene (d) Pentanol

40. Essential oil in Dill is

(a) 0.1 - 2.6% (b) 0.1 - 1.0%

(c) 1.5 - 4.0% (d) None

41. Cat face is a disorder of (JRF-2007)

(a) Cho-cho (b) Pointed gourd

(c) Tomato (d) Cauliflower

42. Highest production of spice is in which state ?

(a) Himachal (b) Andhra Pradesh

(c) Madhya Pradesh (d) Uttar Pradesh

43. Arka Navneet is the variety of (JRF, 2001)

(a) Brinjal (b) Tomato

(c) Okra (d) None

44. Linalool is a/an important ingredient of

(a) Fennel (b) Coriander

(c) Cumin (d) Aniseed

45. Origin of ginger is

(a) Mediterranean region (b) Coriander

(c) South East Asia (d) North West Asia

46. Terbein is the important constituent of

(a) Ginger (b) Cinnamon

(c) Nutmeg (d) Curry leaf

47. Origin of Small cardamom is

(a) Sri Lanka (b) Indonesia

(c) India (d) China

48. TSS of watermelon is (JRF, 2002)

(a) 10-12° Brix (b) 14-16° Brix

(c) 18-22° Brix (d) 20-22° Brix

49. The red colour of chilli fruit at ripening stage is due to the pigment (JRF-2007)

(a) Capsaicin (b) Capsicutin

(c) Capsanthin (d) Lycopene

50. Which of following is associated with vegetable research? (JRF, 2001)

(a) Swaminathan (b) G. Kallo
(c) T K Bose (d) Randhawa

51. Massing and Selfing technique of breeding was first time suggested in (JRF-2009)

(a) Papaya (b) Turnip
(c) Allium spp. (d) Mango

52. The stalk of the ovule is called : (JRF-2007)

(a) Pedicel (b) Petiole
(c) Panicale (d) Micropyle

53. The first step in genetic improvement of plant species is

(a) Domestication (b) Germplasm collection
(c) Selection (d) Hybridization

54. Seed rate of cucumber per hectare is (JRF, 1999)

(a) 2.5-3.5 kg (b) 5-6 kg
(c) 1-2 kg (d) 6-8 kg

55. Daria cultivation is followed in which crop ? (JRF, 2001)

(a) Tomato (b) Root crops
(c) Cucurbits (d) Bulb crops

56. Commercially used growth retardent is (JRF, 2004)

(a) Paclobutrazol (b) BA
(c) Both (d) GA3

57. Louis de Vilmorin (1856) used the concept of progeny test for the improvement of

(a) Pea (b) Sugarbeet
(c) Potato (d) Maize

58. Capacity of refrigeration is measured by (JRF,1999)

(a) Gram (b) Ton
(c) Pascal (d) Lux

59. Pollination and fertilization before anthesis is called (JRF-2007)

(a) Cleistogamy (b) Chasmogamy
(c) Homogamy (d) Dichogamy

60. **Oleoresin in extracted from** **(JRF-2007)**
(a) Colocasoa (b) Chilli
(c) Brinjal (d) Onion

61. **Bottleneck is a disorder of** **(JRF, 2002)**
(a) Cucumber (b) Bottle gourd
(c) Carrot (d) Rose

62. **Johansen (1903) proposed the pure line theory on the basis of seed size in crop**
(a) Potato (b) Sweet potato
(c) French bean (d) Dolichos bean

63. **In preservation of food how much quantity of colour is permitted ?** **(JRF, 2001)**
(a) 250 ppm (b) 300 ppm
(c) 200 pmm (d) 500 pmm

64. **............ state has monopoly in Black pepper production.**
(a) Karnataka (b) Kerala
(c) Tamilnadu (d) AP

65. **Head office of NBPGR is located at** **(JRF-2007)**
(a) Raipur (b) Hyderabad
(c) New Delhi (d) Bhubaneshwar

66. **Flowering stimulus is mostly perceived by** **(JRF-2007)**
(a) Shoot apex (b) Leaves
(c) Buds (d) Stems

67. **Sree Dhara is a var. of ________**
(a) Sweet Potato (b) Cassava
(c) Coleus (d) None

68. **The crop showing greatest inbreeding depression** **(JRF,1997)**
(a) Onion (b) Lettuce
(c) Tomato (d) Lima bean

69. **Ipomea batatas requires photoperiod for flowering is** **(JRF-2009)**
(a) 7 hrs. (b) 9.5 hrs.
(c) 13 hrs. (d) 11 hrs.

70. **Which is prerequisite for genetic improvement of a vegetable crop ?**
(a) Male sterility (b) Genetic variation
(c) Sexual reproduction (d) Apomixis

71. **The growth regulator responsible for shoot elongation is (JRF-2006)**
(a) BAP (b) GA_3
(c) 2, 4-D (d) IAA

72. **Blossom end rot is caused due to deficiency of (JRF-2007)**
(a) Fe (b) Cu
(c) Zn (d) Ca

73. **In mustard which type of incompatibility exist (JRF, 1999)**
(a) Sporophytic (b) Gametophytic
(c) Both (d) None

74. **Hybrid RS-841 is a rootstock for ....**
(a) Pointed gourd (b) Watermelon
(c) Muskmelon (d) Cucumber

75. **The enhancing effect of coconut water in the promotion of growth and differentiation of excised tissue and organ has been attributed to the presence of (JRF-2006)**
(a) Cytokinin (b) Auxins
(c) BA (d) ABA

76. **Betalains is a pigment in________**
(a) Beet (b) Bougainvillea
(c) Cactus (d) All

77. **Which growth regulator is substitute of low chilling treatment ? (JRF, 2004)**
(a) Auxins (b) Cytokinin
(c) Gibrillin (d) ABA

78. **Polyphenol oxydase enzyme inactivate at the temperature of (JRF-2007)**
(a) 80-90 °C (b) 100 °C
(c) 50 °C (d) above 100 °C

79. **Auxanometer is used to measure (JRF-2006)**
(a) TSS (b) Transpiration rate
(c) Acidity (d) Growth of the plant

80. **Why does blanching is practiced in vegetable? (JRF, 2000)**
(a) Micro organism removal (b) In-activate enzyme
(c) For softening the vegetable (d) All

81. **Selection provides opportunity to isolate the promising genotypes from a**
(a) Homogenous Population (b) Homozygous Population
(c) Heterogeneous Population (d) Heterozygous Population

82. **Which is an antioxidant ? (JRF-2009)**
(a) Vitamin-C (b) Vitamin-A
(c) Sugar (d) Ethylene

83. **Earthing of potato crop is done**
(a) 20 days after sowing (b) 40 days after sowing
(c) 50 days after sowing (d) None of the above

84. **.......... colour light favours stem elongation which unfavourably resemble etiolated plants (JRF-2006)**
(a) Blue (b) Green
(c) Ultra-violet (d) Red

85. **Seed potato takes about..........days to germination after sowing. (JRF-2007)**
(a) 5 -10 (b) 10 - 15
(c) 15 - 20 (d) 30

86. **Auxin is synthesized in the plant at (JRF-2006)**
(a) Stem (b) Apex
(c) Root hairs (d) Leaves

87. **For keeping the sugar content low, potato tubers should be stored at a temperature of**
(a) 5 °C (b) 10 °C
(c) 15 °C (d) None of the above

88. **For seed production of cucumber, maintain following m. isolation between two cultivars**
(a) 800 m (b) 1000 m
(c) 1200 m (d) 1600 m

89. Indian Farming is a publication of (JRF-2007)
(a) IARI, New Delhi (b) CAZRI, Jodhpur
(c) ICRISAT, Hyderabad (d) ICAR, New Delhi

90. Heterosis in bottle gourd was reported by
(a) I.D. Tyagi (b) G.S. Gill
(c) V.S. Seshadri (d) U.K. Verma

91. The hybrids of crop, mostly preferred in India are (JRF-1997)
(a) Brinjal (b) Cucumber
(c) Onion (d) Tomato

92. The staple vegetable in our food is
(a) Potato (b) Brinjal
(c) Chillies (d) Tomato

93. Akashin, a physiological disorder in radish is due to deficiency of
(a) B (b) Ca
(c) Mn (d) Cu

94. ELISA is ....linked test (JRF-2007)
(a) Bacteria (b) Virus
(c) Enzyme (d) Protozoa

95. In the plains of northern India, most of the varieties of pea are suitable for sowing from
(a) Mid-August to Sep. (b) After mid-November
(c) October-November (d) None of the above

96. Drought tolerant onion cultivar is
(a) Hissar-2 (b) Pusa Ratnar
(c) Pusa Red (d) Baswant 780

97. In case of chilli the perennial variety is
(a) Pusa Jawala (b) Arka Lohit
(c) Pusa Sadabahar (d) Arka Abir

98. In case of brinjal the phomopsis blight resistant variety is
(a) Pusa Ankur (b) Pusa Bhairav
(c) Arka Sheel (d) Black Beauty

99. Central Agricultural Research Institute is located at (JRF-2008)

(a) Kolkata (b) Port Blair
(c) Vishakhapatnam (d) Delhi

100. Molecular markers are used for determining.........variation in plants. (JRF-2007)

(a) Genetic (b) Phenotypic
(c) Cytoplasmic (d) Endoplasmic

101. Optimum temperature and humidity for storage of sweet potato tubers are

(a) 5°C & 95-100% RH (b) 10°C & 90-95% RH
(c) 15°C & 85-90% RH (d) None of the above

102. Pear shaped fruit variety of tomato is

(a) Punjab Kesri (b) Punjab Chhuhara
(c) PNR-7 (d) Punjab Tropic

103. Which of the following is a man made vegetable ? (JRF-2008, 2009)

(a) Clusterbean (b) Artichoke
(c) Rutabaga (d) Winged bean

104. Following seed rate is recommended for sowing one hectare watermelon (JRF, 2002)

(a) 1 - 1.5 kg (b) 2 - 2.5 kg
(c) 4.5 - 5.0 kg (d) 5.5 - 6.5 kg

105. *Curcuma longa* is botanical name of (JRF-2009)

(a) Zinger (b) Turmeric
(c) Cardamom (d) Black pepper

106. Restriction fragments length polymorphism (RFLP) is a (JRF-2007)

(a) Marker technique (b) Selection method
(c) Mutation technique (d) Gene slicing technique

107. Alternate form of a single gene at same loci is known as (JRF-2009)

(a) Diallel (b) Chromosome
(c) Allel (d) Plastid

108. Homozygosity and homogeneity is maximum in (JRF, 1999).
(a) Pure line (b) Inbred line
(c) Land races (d) Synthetic varieties

109. *Brassica oleracia* var-*acephala* is the botanical name of (JRF-2009)
(a) Cauliflower (b) Cabbage
(c) Kale (d) Broccoli

110. Mass selection is practiced on the basis of
(a) Genotype (b) Phenotype
(c) Environment (d) All of Above

111. ……… is the source of Bt gene (JRF-2007)
(a) Fungi (b) Bacteria
(c) Virus (d) Mycoplasma

112. *Cinnamomum tamala* is the botanical name of
(a) Cinnamon (b) Tejpat
(c) All spice (d) None

113. Major all spice producing country in the world is
(a) India (b) Sri Lanka
(c) Jamaica (d) America

114. Curry leaf is a
(a) Self-pollinated crop (b) Cross-pollinated crop
(c) Often cross-pollinated (d) None

115. The concept of centre of origin of crop plants was proposed by :
(a) JR Harlan (b) NI Vavilov
(c) JG Mendal (d) C. Linnaeus

116. To retain green colour of capsule of cardamom what is used
(a) Washing soda 2% (b) Calcium carbide
(c) Ethylene (d) All

117. The main function of Cytokinin is (JRF, 2004)
(a) Cell enlargement (b) Cell division
(c) Cell elongation (d) All

118. Family of clove is
(a) Myrtaceae (b) Apiaceae
(c) Alliaceae (d) None

119. The varieties prefixed with Arka are released from

(a) IIHR (b) IARI

(c) CPRI (d) CTCRI

120. Most expensive spice crop of the world is

(a) Saffron (b) Coriander

(c) Turmeric (d) Celery

121. Downy mildew of muskmelon controlled by systemic fungicide (JRF-2005)

(a) Ridomi MZ (b) Blitox

(c) Indofil M45 (d) None of these

122. Plant introduction consist of

(a) Procurement and quarantine

(b) Multiplication and distribution

(c) Cataloguing and evaluation

(d) All of above

123. ....is used as a donor for brinjal fruit and shoot borer (JRF-2006)

(a) Solanum nigrum (b) S. melongena

(c) S. Gilo (d) S. muricatum

124. Delphinidin a plant pigment is responsible for development of .... colour (JRF-2009)

(a) Red (b) Blue

(c) Orange (d) Pink

125. Origin of tomato is from

(a) Peru Ecuador (b) Mexico

(c) Chile (d) India

126. Pionsette is the popular variety of (JRF-2005)

(a) Longmelon (b) Muskmelon

(c) Watermelon (d) Cucumber

127. Which part of the world is considered as center of diversity for tomato ?

(a) Mexico (b) Peru

(c) Chile (d) India

128. **Which is the anti gibberellin factor ?** **(JRF, 2004)**
(a) Cycocel (b) BA
(c) Cultar (d) IBA

129. **Which one is the correct scientific name of cauliflower ?** **(JRF-2005)**
(a) *Brassica oleracea var capitata*
(b) *Brassica oleracea var gummifera*
(c) *Brassica oleracea var botrytes*
(d) *Brassica oleracea var italica*

130. **Vegetative reproduction in vegetable crops may occur through**
(a) Rhizome, corm, stem cuttings and grafting
(b) Tuber, bulb and rhizome
(c) Runner, stolon, sucker and meristem culture
(d) All of the above

131. **All spice, *Pimenta dioca*, belongs to family**
(a) Myrtaceae (b) Rutaceae
(c) Myrsticaeae (d) Solanaceae

132. **Family of sweet potato is** **(JRF, 2002)**
(a) Euphorbiacae (b) Aizoiaceae
(c) Agavaceae (d) Convolvulaceae

133. **Small cardamom is propagated by**
(a) Suckers (b) Seed
(c) Rhizomes (d) Root cuttings

134. **Which one of the following is used for production of Virus free plants ?** **(JRF-2006)**
(a) Nodes (b) Meristem
(c) Anther (d) Ovule

135. **Which of the following is the leading producer of all spice?**
(a) India (b) Sri Lanka
(c) Jamaica (d) Indonesia

136. **Inflorescence of small cardamom is**
(a) Panicle (b) Spadix
(c) Spike (d) Raceme

137. **In tissue culture .... is used to induce both callus growth and root initiation** **(JRF-2006)**
(a) Auxin (b) GA
(c) ABA (d) Cytokinin

138. **Cavity spot disorder of carrot is caused due to deficiency of :** **(JRF-2007)**
(a) Ca (b) Fe
(c) S (d) Cu

139. **Browning of media causing explants drying is mainly due to :** **(JRF-2006)**
(a) Tannins (b) Abscissic acid
(c) Benzoic acids (d) Gibberellins

140. **Which is correct about the cells present in embryosac?**
(a) Egg cell, two antipodal cells, a secondary nucleus and two synergids
(b) Three antipodal cells, two synergids, one egg cell and one 2n secondary nucleus
(c) One egg cell, two synergids and two secondary nucleus
(d) A 2n secondary nucleus, one egg cell and three antipodal cells

141. **Which cucurbit have highest Vitamin A ?** **(JRF, 2001)**
(a) Bitter gourd (b) Cucumber
(c) Pumpkin (d) All

142. **Pusa Kesar is a variety of**
(a) Summer Squash (b) Turnip
(c) Winter Squash (d) Carrot

143. **Artificial pollination is essential in** **(JRF-2006)**
(a) Cinnamon (b) Clove
(c) Vanilla (d) Nutmeg

144. **Pusa Yamdagini is a variety of**
(a) Carrot (b) Turnip
(c) Beetroot (d) Radish

145. **Which of the following varieties are mutants of tomato?**
(a) CO1 (b) Pusa lal Meeruti
(c) S-12 (d) All the above

146. Plug trays are related to **(JRF-2007)**
(a) Management nursery (b) Fruit harvesting
(c) Vegetable harvesting (d) All are true

147. Parbhani Kranti is variety of
(a) Tomato (b) Okra
(c) Brinjal (d) Sweet potato

148. Pre harvest spray of which of the following is done to reduce storage loss of onion ? **(JRF-2006)**
(a) Maleic hydrazide (b) Gibberellic acid
(c) Indole butyric acid (d) Ethephon

149. Which of the following in chilli variety is a result of hybridization ?
(a) G1 (b) G2
(c) G3 (d) G5

150. "Capsaicin" is chemically
(a) $C_{17}H_{27}NO_3$ (b) $C_{11}H_{22}NO_3$
(c) $C_{18}H_{28}NO_3$ (d) None of these

151. Mosaic in chilli is transmitted by
(a) Hopper (b) White fly
(c) Aphid (d) None of these

152. Sex form present in betelvine is
(a) Monoecious (b) Dioecious
(c) Andromonoecious (d) None

153. In cardamom, mainly pollination occurs due to
(a) Honey bee (b) House fly
(c) Air (d) Water

154. Katte disease mainly problematic in
(a) Cinnamon (b) Cumin
(c) Cardamom (d) Pepper

155. Poisonous baiting is done to control (JRF-2008)
(a) Borer (b) Whitefly
(c) Thrips (d) Fruitfly

156. Which of the following is a recognized grading standard of tomato ?

(a) Super B (b) Super C
(c) Super A (d) Super D

57. Which of the following tomato varieties are resistant to nematodes ?

(a) Nemamukt (b) SI-120
(c) Nemared (d) All of these

158. 'Arka Alok', a tomato variety is resistant to (JRF-2008)

(a) Wilt (b) Nematode
(c) Fruit borer (d) High temperature stress

159. "Muscilage substance" of okra used for clearing sugarcane juice is extracted from

(a) Flower (b) Leaf
(c) Stem and root (d) Fruit

160. Which of the following turnip varieties is not the result of hybridization ?

(a) Pusa Swarnima (b) Pusa Kanchan
(c) Pusa Sweti (d) Pusa Chandrima

161. In cryopreservation, seed are immersed in liquid nitrogen at .... °C (JRF-2006)

(a) 100 (b) –196
(c) –100 (d) 0

162. The chilli hybrid CH-1 is bred using the mechanism of ....

(a) Self-incompatibility (b) GMS
(c) CGMS (d) Exerted stigma

163. The optimum temperature for seed germination in case of pea is ....

(a) 5 °C (b) 10 °C
(c) 20 °C (d) 22 °C

164. Florida Butter is an important variety of ....

(a) Broad bean (b) Lima bean
(c) Indian bean (d) Winged bear

165. Which organization is related to agricultural marketing ? (JRF, 2004)

(a) NAPHED (b) APHEDA
(c) NABARD (d) All

166. Pusa Kranti is a variety of

(a) Tomato (b) Okra
(c) Brinjal (d) Sweet potato

167. Highest potato producing state in India is : (JRF-2009)

(a) Uttar Pradesh (b) Madhya Pradesh
(c) Bihar (d) West Bengal

168. Which of the following is most commonly used rooting hormone ? (JRF-2006)

(a) IBA (b) NAA
(c) IAA (d) GA

169. Use of rootstock is common in (JRF-2007)

(a) Cole crops (b) Cucurbits
(c) Tuber crops (d) Leafy vegetable crops

170. Sugar Baby, watermelon has been introduced in India from

(a) USA (b) Phillipines
(c) Denmark (d) Israel

171. .... is used as donor parent for salt tolerance in Tomato breeding (JRF-2006)

(a) Lycpersicon (b) *L. hirsutum*
(c) *L. cheesmanii* (d) *L. pimpinellifolium pennellii*

172. Hill soils are deficient in (JRF-2007)

(a) Phosphorus (b) Potassium
(c) Nitrogen (d) Calcium

173. Bolting of onion takes place if temperature goes below (JRF, 2004)

(a) 10 °C (b) 15 °C
(c) 20 °C (d) 25 °C

174. In which of the following underground root is edible ? (JRF, 2002)

(a) Potato (b) Ginger
(c) Onion (d) Radish

175. To breed for disease resistance which method is used ? (JRF, 1999)

(a) Back cross (b) Mass selection

(c) Pure line selection (d) All

176. .... is used as a source of YVM resistance in Okra breeding (JRF-2006)

(a) *Abelmoschus angulosus* (b) *A. crinitus*

(c) *A. manihot sp. tetraphyllus* (d) *A. moschatus*

177. Man made vegetable is (JRF-2009)

(a) Rutabaga (b) Cauliflower

(c) Kale (d) Cabbage

178. The bitterness is colocasia corms is felt due to the presence of : (JRF-2005)

(a) Calcium carbonate (b) Capscicin

(c) Calcium oxalate (d) Sodium chloride

179. In self-pollinated crops, cross pollination varies from

(a) 5 - 10 % (b) 20 - 30 %

(c) 30 - 50 % (d) 50 - 70 %

180. Which of the following variety of tomato have potato leaf marker ?

(a) Pusa Ruby (b) Pusa Red Plum

(c) Pusa Divya (d) Pusa Hybrid-1

181. Okra hybrids are very popular as F1, all are resistant to YVMV governed by

(a) Single dominant gene (b) Single recessive gene

(c) Two recessive gene (d) All

182. Heretosis can be fully exploited in the form of (JRF-2005)

(a) Composites (b) Hybrids

(c) Multilines (d) Synthetics

183. While applying basal dose of DAP, it should never mixed with : (JRF-2007)

(a) Zn (b) Cu

(c) Fe (d) Mg

184. Greening in potato is due to **(JRF,1998)**

(a) High temperature (b) High light intensity

(c) Low temperature (d) All

185. Storing of seed at freezing temperature **(JRF, 2004)**

(a) Orthodox (b) Recalcitrant

(c) Ex-situ conservation (d) In situ conservation

186. Common blight of bean is caused by **(JRF-2009)**

(a) Virus (b) Phytoplasma

(c) Bacteria (d) Fungus

187. Commonly used carbon energy source in plant tissue culture media is **(JRF-2006)**

(a) Sucrose (b) Glucose

(c) Lactose (d) Galactose

188. Anther mutant cultivar of Tomato is **(JRF, 2002)**

(a) Pusa Divya (b) Pusa Gaurav

(c) Pusa Ruby (d) None

189. A vegetable crop showing less than 5 per cent cross-pollination is considered as

(a) Cross-pollinated (b) Self-pollinated

(c) Often self-pollinated (d) Often cross-pollinated

190. Name the country which ranks first in vegetable production? **(JRF-2007)**

(a) India (b) USA

(c) China (d) Japan

191. The most commonly used solidifying agent in plant tissue culture media is **(JRF-2006)**

(a) Agar (b) Sucrose

(c) Starch (d) Maltose

192. Oxalates are present more in **(JRF, 2001)**

(a) Leafy vegetables (b) Root vegetable

(c) Bulb crops (d) None

193. Which phenotypic ratio is expected in F2 in case of masking gene action ?

(a) 12 : 3 : 1 (b) 13 : 3

(c) 15 : 1 (d) 9 : 7

194. **Potato variety more suitable for chips making is** **(JRF-2002)**
(a) Kufri Jyoti (b) Kufri Naveen
(c) Kufri Sheetman (d) Kufri Chipsona

195. **Phytochrome P 730 stimulates flowering of** **(JRF-2006)**
(a) Short day plants (b) Long day plants
(c) Day neutral plants (d) Dark

196. **Temporarily self-incompatibility may be over come through**
(a) Irradiation and bud pollination
(b) Bud pollination and heat shock
(c) Grafting and by removal of stigmatic surface
(d) All of the above

197. **Complete genetic heterozygosity & heterogenity is found in** **(JRF, 2002)**
(a) F_1 hybrids (b) Pureline
(c) Multilines (d) Synthetics

198. **In a typical monohybrid cross, genotypic ratio in F2 will be :**
(a) 1 : 2 : 1 (b) 1 : 1 : 1
(c) 3 : 1 : 1 (d) 2 : 1 : 1

199. **In a DNA molecule, thymine always pairs with (JRF-2009)**
(a) Adenine (b) Guanine
(c) Cystocine (d) Urecil

200. **The optimum temperature for Potato tuberization is** **(JRF-2006)**
(a) 7°C (b) 17°C
(c) 14°C (d) 24°C

201. **Which one of the following is a yellow vein mosaic resistant variety of okra ?** **(JRF-2009)**
(a) Pusa Sawani (b) Harbhajan
(c) Arka Anamika (d) Pusa Makhmali

202. **Early December is the improved variety of** **(JRF-2007)**
(a) Pea (b) Cauliflower
(c) Cabbage (d) Carrot

203. Which one of the following states has maximum coefficient of variation in rainfall ? (JRF-2005)

(a) Assam (b) Maharashtra

(c) Rajasthan (d) West Bengal

204. Which of the following is due to Mg deficiency ?

(a) Browning (b) Hollow stem

(c) Chlorosis (d) Whiptail

205. "Golden Flake" is due to excess Ca oxalate and low K : Ca ratio in

(a) Tomato (b) Carrot

(c) Potato (d) Cabbage

206. In Knol Khol, fleshy edible part is an enlargement of (JRF-2005)

(a) Stem (b) Root

(c) Leaf (d) Bud

207. "Pencil strip" is physiological disorder of celery due to

(a) Phosphorus (b) N_2 deficiency

(c) K calcium (d) Ca deficiency

208. All cole crops are reportedly derived from

(a) Colewort (b) Cabbage

(c) Polewort (d) Cauliflower

209. Which of the following vegetable belong to family- composite? (JRF, 2002)

(a) Spinach (b) Sorrel

(c) Endive (d) Chive

210. Botanically pumpkin is (JRF-2007)

(a) Berry (b) Pepo

(c) Catkin (d) Drupe

211. 'Chirkey' and 'Foorkey' are major viral diseases of

(a) Small Cardamom (b) Large Cardamom

(c) Nutmeg (d) Black Pepper

212. Round shape variety of Radish is (JRF-2009)

(a) Japanese White (b) Rapid Red White Tipped

(c) Pusa Himani (d) Chinese Pink

213. **Which one of the following spices is known as 'King of Spices' ?**
(a) Ginger (b) Cardamom
(c) Turmeric (d) Black Pepper

214. **The ability of cell to generate into a whole plant is known as " (JRF, 2001)**
(a) Regenaration capacity (b) Totipotancy
(c) Autolysis (d) None

215. **Which of the following spices is known as 'Queen of Spices' ?**
(a) Ginger (b) Cardamom
(c) Turmeric (d) Black Pepper

216. **India is the major importer of**
(a) Ginger (b) Garlic
(c) Black Pepper (d) Coriander

217. **Which operation for temporary suppression of self-incompatibility reaction is most successful with both sporophytic and gametophytic systems ?**
(a) Bud pollination (b) Removal of stigma
(c) Grafting (d) Irradiation

218. **White rust of crucifer is caused by (JRF,1998)**
(a) Crostopus candida (b) Albugo candida
(c) Peranosporra (d) Cleviceps

219. **Diploid apagomy is found in (JRF-2009)**
(a) Allium spp. (b) Knol-khol
(c) Okra (d) Cluster bean

220. **Cleistogamy conditions of flower promotes**
(a) Self-pollination (b) Cross-pollination
(c) Apomixis (d) Male sterility

221. **Hybrid Onion programme originated in 1925 by (JRF-2006)**
(a) T.W. Whitaker (b) B. D. Dowker
(c) V. Swarup (d) H.A. Jones

222. **Seedless watermelon is (JRF-2007)**
(a) Triploid (b) Hexaploid
(c) Tetraploid (d) Diploid

223. Pro trays are related to **(JRF-2006)**
(a) Post harvest (b) Nursery
(c) Transport (d) Processing

224. About 95 to 99% portion of plant tissues are made of: **(JRF-2009)**
(a) C, H and O (b) N, P & K
(c) Ca, Mg and C (d) Cu, Zn and Fe

225. The varieties prefixed with Pusa are released from
(a) IIHR (b) IARI
(c) CPRI (d) CTCRI

226. Cross-pollination in plants is encouraged by
(a) Monoecy (b) Dicliny
(c) Dichogamy (d) All above

227. International Institute of Horticulture is situated in **(JRF-2006)**
(a) Brazil (b) Italy
(c) USA (d) None of these

228. Which phenotypic ratio is expected in F2 in a monohybrid cross ?
(a) 1 : 2 : 1 (b) 1 : 1 : 1
(c) 3 : 1 (d) 2 : 1 : 1

229. Photophosphorylation is **(JRF, 1999)**
(a) Hill reaction
(b) Formation of ATP in presence of light
(c) Formation of NADP in presence of light
(d) None

230. Dormancy of seed can be due to **(JRF-2007)**
(a) Impermeability of seed coat
(b) Requirement of water
(c) Presence of ABA in the seed
(d) All of these

231. Which of the following is four angled Bean? **(JRF, 2002)**
(a) Winged Bean (b) Cluster Bean
(c) Lima Bean (d) Broad Bean

232. Which phenotypic ratio is expected in F2 in case of polymeric gene action ?

(a) 12 : 3 : 1 (b) 9 : 6 : 1
(c) 15 : 1 (d) 9 : 7

233. .... is the progenitor form of the cultivated cucumber (JRF-2006)

(a) *Cucumis sativus* var. hardwickii
(b) *Cucumis melo*
(c) *Cucumis anguria*
(d) *Citrus lanatus*

234. Stamenless mutant is reported in : (JRF-2009)

(a) Cabbage (b) Brinjal
(c) Tomato (d) Carrot

235. _________ is barrier of self pollination in vanilla

(a) Labellum (b) Rostellum
(c) Anther cap (d) Column

236. Sweet potato is a (JRF-2007)

(a) Short day plant (b) Long day plant
(c) Day neutral (d) Intermediate

237. Fruit of brassica is called as (JRF, 1999)

(a) Pome (b) Berry
(c) Siliqua (d) Drupe

238. How many genotypes will be possible in F2 (3n) in case of a trihybrid cross ?

(a) 16 (b) 9
(c) 27 (d) 81

239. Most suitable radiation for disinfection of food is (JRF, 2001)

(a) UV rays (b) Gamma rays
(c) Alpha rays (d) All

240. In India who has introduced climate controlled green house cultivation (JRF-2006)

(a) ICAR (b) Mahyco seed Co.
(c) Indo-American Hybrid Seed Co. (d) Suton & Sons

241. Fruits and shoot borer is a serious pest of : (JRF-2009)
(a) Cabbage (b) Brinjal
(c) Bitter gourd (d) Tomato

242. The main constituent of cinnamon bark is ________
(a) Eugenol (b) Cinnamaldehyde
(c) Eugeny l acetate (d) Limone

243. The cleistogamous vegetable is : (JRF-2009)
(a) Garden pea (b) Lettuce
(c) Capsicum (d) Tomato

244. ________ var. of tomato sets fruits at low temp.
(a) Pusa Ruby (b) Pant Bahar
(c) Hissar Arun (d) Pusa Sheetal

245. Riceyness is a physiological disorder associated with: (JRF-2009)
(a) Cabbage (b) Cauliflower
(c) Broccoli (d) Chinese cabbage

246. Seed rate of cauliflower is (JRF, 1997)
(a) 100 g/ha (b) 200 g/ha
(c) 350-600 g/ha (d) 1000 g/ha

247. .... is a self blanched var. of cabbage
(a) Pusa Shubhra (b) Pusa Sharad
(c) Pusa Him Jyoti (d) Pusa Snowball

248. Male sterile line is maintained by (JRF, 1999)
(a) 'Crossing with heterozygous
(b) Male fertile line
(c) Both of the above
(d) None of the above

249. Which of the following is a small seeded variety of coriander
(a) Karan (b) Swathi
(c) Sindhu (d) Sadhna

250. Pant C-1 is resistant to
(a) Bacterial blight (b) TMV
(c) CMV (d) All of these

251. **Which of the following vegetable is known as 'God sent Vegetable' ?** **(JRF-2009)**
(a) Pea (b) Tomato
(c) Winged bean (d) Brinjal

252. **The most common technique used in the incorporation of multiple disease resistance (MDRs) is**
(a) Recombinant cross (b) Pure line selection
(c) Mass selection (d) Back cross

253. **In green vegetables the limiting amino acid is (JRF-2008)**
(a) Tryptophan (b) Methionine
(c) Lysine (d) Arginine

254. **Okra is a**
(a) Self-pollinated (b) Often cross-pollinated
(c) Cross-pollinated (d) None of these

255. **The safe seed storage moisture content should be between:** **(JRF-2009)**
(a) 8-9 % (b) 10-12 %
(c) 13-14 % (d) 15-16 %

256. **The smallest perfect F2 population (4n) will be in case of a dihybrid cross**
(a) 4 (b) 8
(c) 27 (d) 16

257. **Which spice is referred as "King of spices" ?**
(a) Black pepper (b) Small cardamom
(c) Large cardamom (d) Clove

258. **In plant spacing 2x2, how much plant can be accommodated?** **(JRF, 2001)**
(a) 2000 (b) 2500
(c) 3000 (d) 3500

259. **Tomato is rich source of :** **(JRF-2009)**
(a) Vitamin A (b) Iron
(c) Phosphorus (d) Vitamin C

260. **In a monohybrid cross, in case of incomplete dominance, phenotypic ration in F2**
(a) 1 : 2 : 1 (b) 1 : 1 : 1
(c) 3 : 1 (d) 2 : 1 : 1

261. Self sterile variety of black pepper
(a) Balankotta (b) Both a and c
(c) Kalluvalli (d) None of these

262. Pusa Chikni is a variety of
(a) Amaranth (b) Bottle gourd
(c) Smooth gourd (d) Brinjal

263. In tomato blossom end rot is due to (JRF, 1999)
(a) Ca deficiency (b) Low Temp
(c) High temp (d) Very high humidity

264. Pusa Early Bunching is a variety of
(a) Amaranth (b) Bottle gourd
(c) Fenugreek (methi) (d) Brinjal

265. ________ element is necessary for root growth
(a) N (b) P
(c) K (d) Zn

266. The Indian Institute of Vegetable Research is located at (JRF, 2001)
(a) Ranchi (b) Delhi
(c) Varanasi (d) Bangalore

267. Jade Cross is a variety of
(a) Cauliflower (b) Broccoli
(c) Cabbage (d) Brussels sprout

268. Crossing over occurs during
(a) Leptotene (b) Pachytene
(c) Zygotene (d) Diplotene

269. .... acid is present in Kokum
(a) Citric (b) Malic
(c) Acetic (d) Hydroxy citric

270. Which vegetable crop is propagated by stem cutting ?
(a) Raddish (b) Carrot
(c) Cucumber (d) Sweet potato

271. Sauerkraut can be made from : (JRF-2009)
(a) Broccoli (b) Cabbage
) Chinese cabbage (d) Spinach

272. **Number of all possible crosses in a diallel cross (excluding reciprocals) are**
(a) n (n–1) (b) n (n–1) /2
(c) n (n–1) / 3 (d) $n^2/2$

273. **Arka Mohini is a variety of**
(a) Tomato (b) Okra
(c) Brinjal (d) Sweet Pepper

274. **Damping off is caused by** **(JRF, 1999)**
(a) Pythium spp (b) Fusarium
(c) Alternaria (d) All

275. **Kihara's technique is related with which of the following crop ?** **(JRF-2006)**
(a) Onion (b) Watermelon
(c) Tomato (d) Muskmelon

276. **Improved Japanese is a variety of**
(a) Cauliflower (b) Broccoli
(c) Cabbage (d) Brussels sprout

277. **................ is called as Chinese Potato.**
(a) Coleus (b) Canna indica
(c) Giant taro (d) WI Arrowroot

278. **Golden Acre is a variety of**
(a) Cauliflower (b) Broccoli
(c) Cabbage (d) Brussels sprout

279. **A serious disease of onion commonly found in seed crop is :** **(JRF-2009)**
(a) Late blight (b) Purple blotch
(c) Red rot (d) Leaf curl virus

280. **Gerkhins belong to which genus** **(JRF, 2001)**
(a) Cucurbit (b) Cucumis
(c) Citrulus (d) None

281. **Number of single crosses in full diallel are**
(a) n (n–1) (b) n (n–1) /2
(c) 2n (n–1) /2 (d) n^2

282. The average daily requirement of vegetables for human adult is (JRF-2006)

(a) 130 g (b) 180 g
(c) 285 g (d) 510 g

283. Black rot is a most common disease of : (JRF-2009)

(a) Tomato (b) Potato
(c) Celery (d) Cabbage

284. Soft rot of vegetables is caused by (JRF, 1999)

(a) Verticillium (b) Erwinia carotovora
(c) Fusarium (d) None

285. Heterosis is the result of which type of gene effects ?

(a) Over-dominance (b) Dominance
(c) Additive (d) Epistatic

286. The popular variety of radish for summer and rainy season in North Indian plains is (JRF-2009)

(a) Pusa Himani (b) Pusa Chetki
(c) Japanese White (d) White Icicle

287. Ideal storage temperature for brinjal is (JRF, 2002)

(a) 14-15° C (b) 10° C
(c) 0° C (d) 20° C

288. In Sweet potato root system is (JRF 2005)

(a) Deep (b) Medium
(c) Medium deep (d) Shallow

289. The type of germination in garden pea is : (JRF-2009)

(a) Epigeal (b) Hypogeal
(c) Hypo-epigeal (d) Epi-hypogeal

290. Sporophytic self incompatibility system is present in (JRF-2006)

(a) Cabbage (b) Tomato
(c) Chilli (d) Okra

291. Multistyle flowers are common in (JRF-2005)

(a) Okra (b) Brinjal
(c) Chillies (d) Lima Bean

292. Sweet potato is an autopolyploids and has the ploidy level of :

(a) 3 x (b) 4 x

(c) 5 x (d) 6 x

293. The proportion of additive genetic variance to phenotypic variance is called (JRF-2006)

(a) Heterobeltiosis (b) Heritability

(c) Dominance relationship (d) Linkage relationship

294. Nitrogen deficiency in plants leads to- (JRF, 2003)

(a) Excessive (b) Profuse flowering

(c) Lush green colour (d) Chlorosis

295. The most useful insect in seed production of vegetable is: (JRF-2009)

(a) Bumble bees (b) Honey bees

(c) Wasps (d) Thrips

296. Breeder seed is the progeny of (JRF-2005)

(a) Registered seed (b) Nucleus seed

(c) Foundation seed (d) Certified seed

297. Number of all possible three way crosses among n lines should be

(a) n (n-1) (n-2) /2 (b) n (n-1) (n-2) /3

(c) n (n-1) (n-2) /4 (d) n (n-1) (n-2) (n-3) /3

298. Self-pollination increases

(a) Homozygosity (b) Heterozygosity

(c) Homogeneity (d) Heterogeneity

299. Curd initiation and development in cauliflower is mainly influenced by (JRF-2009)

(a) Temperature (b) Photoperiod

(c) Relative humidity (d) None of these

300. Which of following is hybrid variety of black pepper ? (JRF, 2001)

(a) Penyur - 3 (b) Penyur - 2

(c) Penyur - 4 (d) None of the above

301. Pollen sterility in tomato can be identified by using **(JRF-2006)**

(a) Acetocarmine (b) $KMnO_4$
(c) Methylene blue (d) Phenolphthalein

302. In purelines variety genetic variation may arise over a period of time due to

(a) Mechanical mixture (b) Out-crossing
(c) Mutation (d) All above

303. Clove belong to family **(JRF-1997)**

(a) Leguminoceae (b) Myrtaceae
(c) Stirculiaceae (d) Rosaceae

304. Purity of pureline varieties can be maintained through regular

(a) Pedigree selection (b) Mass selection
(c) Recurrent selection (d) Pure line selection

305. Tapioca belongs to the family **(JRF-2005)**

(a) Araceae (b) Euphorbiaceae
(c) Rosaceae (d) Dioscoreaceae

306. Bolting problem in onion occur if temperature goes below **(JRF, 1999)**

(a) 25° C (b) 20° C
(c) 15° C (d) 10° C

307. Swarna rekha is the var of **(JRF, 2001)**

(a) Pointed gourd (b) Bitter gourd
(c) Bottle gourd (d) None

308. Pureline selection can be applied for the improvement of

(a) Local varieties (b) Introduced varieties
(c) Old pureline varieties
(d) All the above in self-pollinated crops

309. In male sterility, the pollen is **(JRF-2005)**

(a) Abortive (b) Shrivelled
(c) Non functional (d) All of these

310. In Brinjal, only those flowers set fruits which have **(JRF-1997)**

(a) Short (b) Long style
(c) Medium (d) Both (b) & (c)

311. In a random mating population, gene and genotypic frequency remains same generation after generation provided that there is no

(a) Selection (b) Migration
(c) Mutation (d) Random drift
(e) All of above

312. Browning in cauliflower is due to (JRF, 1997)

(a) Cu (b) Bo
(c) Ca (d) Mo

313. The gain under selection for quantitative traits is impeded by :

(a) Non-additive gene action (b) Low selection intensity
(c) Low heritability of the trait
(d) All of the above

314. In which crop the adventitious root is present? (JRF, 1999)

(a) Sweet potato (b) Turnip
(c) Radish (d) Carrot

315. Effect of self-fertilization on the frequency of one of the two homozygote (AA/aa) at a single locus (A/a) after n generation of selfing will be calculated by formulae

(a) $2^{n-1/2n+1}$ (b) $2^{n/2n+1}$
(c) $2^{n-1}/2^n$ (d) $2^{n+1}/2^n$

316. Vanilla grows best at a height of (JRF,1997)

(a) 1 m (b) 2 m
(c) 3 m (d) 0.5 m

317. Which of the following is hybrid variety of black pepper?

(a) Penniyur-1 (b) Penniyur-2
(c) Penniyur-4 (d) None

318. Which of the following state is the largest producer of Black pepper ?

(a) Kerala (b) Andhra Pradesh
(c) Tamil Nadu (d) Karnataka

319. Which of the following state is the largest producer of Cardamom ?

(a) Kerala (b) Andhra Pradesh
(c) Tamilnadu (d) Karnataka

320. **Major constraint in vegetative propagation of cardamom is**
(a) Katte disease (b) Capsule Borer
(c) Thrips (d) None

321. **Which of following is responsible for sourness of radish? (JRF, 2000)**
(a) Isothiosinate (b) Anthocynins
(c) Auxins (d) All

322. **India is second largest producer of vegetables after (JRF, 1999)**
(a) Brazil (b) China
(c) Africa (d) Japan

323. **Part of onion we eat is called**
(a) Corm (b) Bulb
(c) Rhizome (d) Tuber

324. **The pigment present in watermelon is (JRF, 2002)**
(a) Carotenoid (b) Anthocynin
(c) Lycopene (d) Both b & c

325. **Food storage organs of ginger and turmeric are called**
(a) Rhizome (b) Tuber
(c) Corm (d) Root

326. **Pungency of onion is due to the presence of (JRF-2005)**
(a) Isothiocynate (b) Allyl propyl disulphide
(c) Solanin (d) Betanin

327. **.......... cultivation is mostly concentrated in peri-urban areas. (JRF-2007)**
(a) Fruit (b) Vegetable
(c) Medicinal crop (d) Plantation crop

328. **First time cytoplasmic genetic male sterility is used in (JRF, 2000)**
(a) Onion (b) Pea
(c) Ground nut (d) Carrot

329. **Population of cross-pollinated vegetable crops (composites and synthetics) are**
(a) Homozygous and homogenous
(b) Homozygous and heterogeneous

(c) Heterozygous and heterogeneous
(d) Heterozygous and homogenous

330 Mass selection is based on the
(a) Progeny test (b) Maternal parent only
(c) Paternal parent only (d) Both parents

331. Diara method of cultivation is common in (JRF-2005)
(a) Cucurbits (b) Cole crops
(c) Root crops (d) Tuber crops

332. Proportion of completely homozygous plants after m generation of selfing when n number of genes are segregating
(a) $[2^m - 1/2^m]^n$ (b) $[2^n - 1/2^n]^m$
(c) $[2^m - 1/2^{m+1}]^n$ (d) $[2^m - 1/2^{m+1}]^n$

333. IIHR headquarter is located at (JRF,1997)
(a) Kasargod (b) Calicut
(c) Kottayan (d) Bangalore

334. VK-1/ Priya is a cultivar of ....
(a) Snake gourd (b) Bitter gourd
(c) Ridge gourd (d) None

335. The ETL of aphid population in case of Potato is _________
(a) 20/100 leaves (b) 30/100 leaves
(c) 10/100 leaves (d) 40/100 leaves

336. Propagation of curry leaves is generally done by ________
(a) Suckers (b) Seed
(c) Both (a) and (b) (d) Stem cutting

337. Which is non-pungent vegetable when uncooked? JRF, 2004)
(a) Lettuce (b) Radish
(c) Onion (d) Chilli

338. The diploid chromosome number of cluster bean is
(a) 2n = 22 (b) 2n = 24
(c) 2n = 14 (d) 2n = 16

339. The origin of Dolichus bean is
(a) China (b) India
(c) Mexico (d) Mediterranean region

340. Chromosome number of cucumber is : (JRF-2005)

(a) 2n = 24 (b) 2n = 14
(c) 2n = 12 (d) 2n = 20

341. Origin of water melon is

(a) Africa (b) India
(c) Iran (d) None

342. Viral disease of potato are spread by (JRF-2005)

(a) Jassids (b) Aphids
(c) White flies (d) Mites

343. Which of the following cucurbits is also used as ornamental type ?

(a) Summer squash (b) White squash
(c) Pumpkin (d) Winter squash

344. Chromosome number of okra is (JRF-2005)

(a) 2n = 22 (b) 2n = 130
(c) 2n = 88 (d) 2n = 20

345. Little leaf of brinjal is transmitted by (JRF-2008)

(a) Aphids (b) Jassids
(c) Hopper (d) Whitefly

346. Shoot and fruit borer of okra belongs to

(a) *Earias vitella* (b) *Amrasca bigttula bigttula*
(c) *Tetranychus spp* (d) *Heliothis armigera*

347. Carrot is a good source of (JRF-2005)

(a) Vitamin A (b) Riboflavin
(c) Vitamin C (d) Protein

348. Which of the following cucurbits is propagated by stem cuttings ?

(a) Kakrol (b) Kartoli
(c) Little gourd (d) All of these

349. The amount of solanin/tuber considered as unfit for consumption is

(a) 5 mg (b) 10 mg
(c) 20 mg (d) 25 mg

350. Leafy vegetables are usually rich source of **(JRF-2008)**
(a) Iron (b) Protein
(c) Carbohydrate (d) Carotene

351. In which of the following vegetable all parts are edible ?
(a) Winged bean (b) Limabean
(c) Broadbean (d) Carrot

352. Sree Shilpa is an first hybrid of
(a) Greater yam (b) Elelhant foot yam
(c) Lesser yam (d) White yam

353. *Cucumis ficifolia* is a type of species
(a) Perennial (b) Biennial
(c) Annual (d) None of these

354. Self-incompatibility can be effectively utilized in .... crops. **(JRF-2006)**
(a) Tomato (b) Cole crops
(c) Bhendi (d) Peas

355. Phytic acid a toxic substance is present in
(a) Leafy vegetables (b) Peas and beans
(c) Cucurbits (d) Root vegetables

356. All spice belongs to family
(a) Umbelliferae (b) Apiaceae
(c) Myrtaceae (d) None

357. Simple recurrent selection is most suitable for characters with
(a) High heritability (b) Low heritability
(c) Moderate heritability (d) All of the above

358. In tomato Hybrid Hisar Anmol the donor for leaf curl virus is .... **(JRF-2006)**
(a) *Lycopersicon peruvianum*
(b) *L. pimpinellifolium*
(c) *L. esculentum* (d) *L. hirsutum L sp. glaberatum*

359. Potato is a **(JRF, 2002)**
(a) Allohexaploid (b) Autohexaploid
(c) Auto tetraploid (d) Allotetraploid

360. The difficulties encountered in the production of interspecific hybrids may be broadly classified into

(a) Failure of zygote formation and hybrid sterility
(b) Failure of zygote development and F1 seedling development
(c) Failure of zygote formation and development
(d) All of the above

361. Self incompatibility is overcome by (JRF-2006)

(a) Bud pollination (b) Surgical techniques
(c) Irradiation (d) All of these

362. Ramsey and Golsey are important variety of

(a) Cinnomon (b) Small Cardamom
(c) Ginger (d) Large Cardamom

363. Nepal cardamom is botanically known as

(a) *Amomum aromaticum* (b) *Elettaria cardamom*
(c) *Amomum sublatum* (d) None of these

364. In cucumber production of male flowers can be induced by the application of (JRF-2006)

(a) Boron (b) Ethephon 600 ppm
(c) GA 2000 ppm (d) Auxin

365. The chromosome number of *Coriandrum sativum* is

(a) 2n = 11 (b) 2n = 33
(c) 2n = 22 (d) None of the above

366. Dill is a native of

(a) India (b) Euroasia
(c) Mediterranean (d) South America

367. Whiptail of cauliflower is due to deficiency of (JRF, 1999)

(a) Mo (b) Zn
(c) Co (d) Cu

368. Leading cardamom growing state in India is

(a) Tamilnadu (b) Andhra Pradesh
(c) Kerala (d) Karnataka

369. Nutmeg plants are (JRF-2008)

(a) Monoecious (b) Andromonoecious
(c) Androdioecious (d) Dioecious

370. *Sauropus androgynous*, chekurmanis belongs to
(a) Euphorbiaceae (b) Solanaceae
(c) Convolvulaceae (d) Polygonaceae

371. Type of self-incompatibility observed in cole crops is
(a) Sporophytic (b) Both (a) and (c)
(c) Gametophytic (d) None of theses

372. The active principle of clove oil is (JRF-2008)
(a) Macine (b) Oleum
(c) Aldehyde (d) Eugenol

373. Indian accouts for ... % of turmeric production in the word
(a) 80 (b) 70
(c) 76 (d) 90

374. Type of self-incompatibility observed in tomato is
(a) Sporophytic (b) Both (a) and (c)
(c) Gametophytic (d) None of these

375. Part used for consumption of clove is
(a) Seed (b) Bark
(c) Flower (d) Unopened bud

376. Plant part of saffron used as spice
(a) Seed (b) Bark
(c) Flower stigma (d) Rhizome

377. Almost all the vegetable crops belong to the plant sub community
(a) Bryophyte (b) Spermatophyte
(c) Thalophyte (d) None of these

378. Commercial part of all spice is
(a) Whole tree (b) Leaves
(c) Bark only (d) Fruits only

379. Origin of clove is
(a) Indonesia (b) Africa
(c) India (d) None

380. Bacterial wilt resistant variety of Brinjal (JRF, 2004)
(a) Arka Abha (b) Arka Vardan
(c) Arka Gaurav (d) All

381. **Bush pepper is produced from :** **(JRF, 2003)**
(a) Runners (b) Laterals
(c) Top shoots (d) Hanging shoots

382. **Efficiency of mass selection principally depends on the**
(a) Heritability of trait
(b) Gene frequency
(c) Number of genes controlling the trait
(d) All of the above

383. **Food Corporation of India (FCI) was established in year** **(JRF, 2002)**
(a) 1950 (b) 1955
(c) 1960 (d) 1965

384. **Arka Abir is the variety of** **(JRF, 2001)**
(a) Brinjal (b) Cucurbit
(c) Tomato (d) Chilli

385. **Sterility in distant hybrids may be due to the**
(a) Cryptic structural hybridity
(b) Cytoplasm of alien species
(c) Lack of chromosome pairing
(d) All of the above

386. **Which vegetable is dioecious in nature and propagated by stem cutting ?** **(JRF, 2002)**
(a) Spinach (b) Sweet potato
(c) Tobacco (d) Pointed gourd

387. **Which of following is responsible for cracking in tomato?** **(JRF, 2000)**
(a) B deficiency (b) Ca deficiency
(c) S deficiency (d) Mn deficiency

388. **The FPO licence is given by** **(JRF, 2001)**
(a) Govt of India (b) State Govt.
(c) Agri University (d) None

389. **Development of a variety usually takes**
(a) 4 – 6 years (b) 7 – 8 years
(c) 10 – 12 years (d) 15 – 20 years

390. Sree Dhara is a variety of
(a) Sweet potato (b) Coleus (Chinese potato)
(c) Colocasia (d) Cassava

391. Which of the following vegetable produces maximum amount of nutrient/unit area/unit time ?
(a) Potato (b) Carrot
(c) Tomato (d) Onion

392. Little leaf of brinjal is transmitted by
(a) Fly (b) Leaf hopper
(c) Whitefly (d) None of these

393. Delayed harvesting in radish may cause
(a) Pithiness (b) Forking
(c) Deformed roots (d) All

394. Jobner Green is a high yielding variety of
(a) Spinach (b) Indian Spinach
(c) Amaranthus (d) Onion

395. Which of the following varieties of Brussels sprout is dwarf one ?
(a) Long Island (b) Catskill
(c) Amager Market (d) Danish Prize

396. "Rutin" and Vitamin-P are associated with which vegetable crop ?
(a) Chilli (b) Tomato
(c) Brinjal (d) Potato

397. Yellow colour of onion is due to **(JRF, 2002)**
(a) Qucertin (b) Myrcetin
(c) Allicin (d) Malvelin

398. For simply inherited characters, selection may be started from :
(a) F_4 (b) F_3
(c) F_2 (d) F_1

399. PK-ft-16 is the variety of **(JRF, 2001)**
(a) Chilli (b) Tomato
(b) Brinjal (d) Cucurbits

400. Segregation for different characteristics does not fit Mendelian ration in F_2 from wide crosses due to

(a) Irregular meiosis (b) Impaired recombination
(c) Reduced survival of recombinant gametes and zygotes
(d) All of the above

401. Most important factor in a hybridization programme is

(a) Choice of parents for hybridization
(b) Emasculation
(c) Handling of segregating generations
(d) Pollination

402. Browning due to boron deficiency is common in- **(JRF, 2003)**

(a) Radish (b) Cauliflower
(c) Onion (d) Tubers

403. ________ is called as stress hormone

(a) Auxine (b) Gibberellins
(c) Cytokinins (d) ABA

404. ________ crop/crops is/are suitable for greenhouse cultivation

(a) Sweet pepper (b) Cucumber
(c) Tomato (d) All

405. Which vegetable crop is rich source of niacin? (JRF, 2004)

(a) Spinach (b) Cowpea
(c) Raddish (d) Sweet potato

406. 'Hortscience' is published by

(a) ICAR (b) Horticultural Society of India
(c) ASHS (d) Govt. of India

407. The ________ is done to obtain virus free plants

(a) Pollen culture (b) Cell culture
(c) Embryo culture (d) Meristem culture

408. ________ is the site of B-oxidation in cell.

(a) Chloroplast (b) Vacuole
(c) Mitochondria (d) Ribosomes

409. Edible part in jerusalem artichoke is- **(JRF, 2003)**
(a) Seed (b) Cutting
(c) Layers (d) Tubers

410. Which vegetable has high calorie value ? **(JRF, 2004)**
(a) Fenugreek (b) Amaranthus
(c) Potato (d) Tomato

411. Early blight of tomato is caused by **(JRF-2005)**
(a) Fungus (b) Virus
(c) Bacterium (d) None of these

412. _______ is 1st tropical variety of cabbage developed for cultivation under high temperature conditions at IARI, New Delhi.
(a) Pusa Sambandh (b) Pusa Mukta
(c) Pusa Drum Head (d) Pusa Ageti

413. Yellow anther type in case of chilli is ________
(a) Bhagy Lakshmi (b) Jwalamukhi
(c) Andhra Jyoti (d) Bhaskar

414. Vegetable of 20th century is- **(JRF, 2003)**
(a) Cluster bean (b) Amaranthus
(c) Winged bean (d) Chikurmains

415. Variety resistant to cyst nematodes in Potato is __________
(a) Kufri Swarna (b) Kufri Sutlej
(c) Kufri Giriraj (d) Kufie Asoka

416. Tomato belongs to the species **(JRF-2005)**
(a) *Peruvianum* (b) *Esculentum*
(c) *Hirsutum* (d) *Pimpinellifolium*

417. The mechanism of protogyny is observed in _______
(a) Onion (b) Cabbage
(c) Carrot (d) Radish

418. Which is called as a protective food ? **(JRF, 2004)**
(a) Vegetable (b) Fruits
(c) Fruits & vegetable (d) Cereals

419. In combination breeding, which is most important consideration for choice of parents ?

(a) *Per se* performance (b) Combining ability
(c) Genetic variability (d) Heritability

420. Protein is found in abundance in **(JRF-2005)**

(a) Beans (b) Cauliflower
(c) Cabbage (d) Bittergourd

421. Sulphur containing food is stored in **(JRF, 2001)**

(a) AR cans (b) SR cans
(c) Both (a) and (b) (d) None of the above

422. Main objective of emasculation in breeding programme is

(a) Prevention of selfing
(b) Encourage crossing
(c) Prevention of out crossing
(d) Hybrid development

423. Large cardamom is propagated through- **(JRF, 2003)**

(a) Seed (b) Cutting
(c) Layer (d) Sucker

424. Which vegetable is rich source of Vit. C (>100 mg/100g) **(JRF, 2004)**

(a) Tomato (b) Cabbage
(c) Turnip (d) Radish

425. Most widely accepted method of emasculation in vegetable crops is

(a) Hand emasculation (b) Alcohol treatment
(c) Suction method (d) Hot water treatment

426. Pre-cooling is meant to **(JRF-2005)**

(a) Improve fruit quality (b) Remove field heat
(c) Fresh the object (d) None of these

427. Better principle in bitter gourd is **(JRF, 2003)**

(a) Cucurbitacine (b) Momordicin
(c) Solanin (d) None

428. Arka Pragati is a variety of **(JRF, 2002)**

(a) Okra (b) Onion
(c) Palak (d) Lettuce

429. Cleistogamy favours **(JRF, 1999)**
(a) Self pollination (b) Cross pollination
(c) Both (d) None

430. Incompatibility is common is family **(JRF,1998)**
(a) Malvaceae (b) Rosaceae
(c) Liliaceae (d) Poaceae

431. According to ISI tomato has been specialized into **(JRF, 2002)**
(a) 4 grades (b) 3 grades
(c) 2 grades (d) Only one grade

432. Spice crop largely exported from India **(JRF, 1997)**
(a) Black pepper (b) Cardamom
(c) Cumin (d) Turmeric

433. Highest hybrid varieties in vegetable are grown in **(JRF, 2002)**
(a) Tomato (b) Cabbage
(c) Brinjal (d) Carrot

434. Good storage require **(JRF, 1998)**
(a) High temp & high humidity
(b) Low temp & high humidity
(c) High temp & low humidity
(d) None of the above

435. For handling wide crosses which is most suitable breeding scheme ?
(a) Backcross method (b) Pedigree
(c) Diallel selective mating (d) Biparental mating scheme

436. Common blight of bean is caused by **(JRF, 2002)**
(a) Virus (b) Physiological
(c) Bacteria (d) Mycoplasma

437. Resistance to which disease was transferred from *Solanum demissum* into potato ?
(a) Late blight (b) Early blight
(c) Frost (d) Virus X

438. Little leaf of brinjal is caused due to (JRF,1998)
(a) Virus (b) Mycoplasma
(c) Bacteria (d) Fungi

439. Which group of crops show severe inbreeding depression?
(a) Cross-pollinated crops and Vegetative propagated crops
(b) Apomixis
(c) Self-pollinated crops
(d) Seed propagated crops

440. Clones generally degenerates due to
(a) Viral infection (b) Bacterial infection
(c) Mutations (d) All of the above

441. Heterobeltiosis is estimated over the (JRF, 2002)
(a) Wild parent (b) Better parent
(c) Popular variety (d) Popular hybrid

442. Genetic variation within a clone may arise due to the
(a) Somatic mutations (b) Mechanical variation
(c) Segregation and recombination
(d) All of the above

443. In cucumber common sex expression is (JRF-2005)
(a) Monoecism (b) Dioecism
(c) Andremonoecism (d) None of these

444. Mosaic virus in cucumber spreads by (JRF-2005)
(a) Seeds (b) Aphids
(c) Jassids (d) Whitefly

445. Daria method of cultivation is followed in which vegetable crop ? (JRF, 1998)
(a) Cucurbits (b) Cole crop
(c) Tubers (d) Root crops

446. Which amino acid is precursor of ethylene ? (JRF, 1999)
(a) Cystine (b) Purine
(c) Methionine (d) Proline

447. In radish which substance is present ? (JRF, 2001)
(a) Allyl propyl disulphide (b) Diallyl disulphide
(c) Isothiocynate (d) None

448. Sweet potato belongs to the family **(JRF-2005)**

(a) Convolvulaceae (b) Solanaceae

(c) Euphorbiaceae (d) Araceae

449. Diploid apogamy is found in **(JRF, 2002)**

(a) Allium spp (b) Perithelium spp

(c) Tobacco (d) Rice

450. .... is commonly used for propagation of bush pepper **(JRF-2006)**

(a) Geotrophic shoot (b) Orthotropic shoot

(c) Plagiotropic shoot (d) Runner

451. Little leaf of brinjal is caused by **(JRF-2005)**

(a) Bacteria (b) Fungi

(c) Mycoplasma (d) Zn deficiency

452. Gynoecious lines are common in **(JRF,1997)**

(a) Water melon (b) Musk melon

(c) Long melon (d) Cucumber

453. The percentage of recovery of black pepper is about .... of the ripe berries **(JRF-2006)**

(a) 20 (b) 23

(c) 48 (d) 33

454. Leguminous leafy vegetable is **(JRF,1998)**

(a) Cowpea (b) Fenugreek

(c) Spinach (d) Palak

455. Which of the following is protandrous ? **(JRF, 2002)**

(a) Onion (b) Tomato

(c) Cauliflower (d) French bean

456. Origin of Brinjal is **(JRF, 1997)**

(a) Africa (b) America

(c) India (d) China

457. In Nutmeg, male and female plants are planted in the ratio of : **(JRF-2006)**

(a) 1 : 10 (b) 1 : 15

(c) 1 : 13 (d) 1 : 5

458. Aijowan ka phool is a
(a) Crystal form of Thymol (b) Liquid form of Thymol
(c) Crystal form of Cineol (d) Crystal form of Linalool

459. Bareja system is the term associated with
(a) Nutmeg (b) All spice
(c) Betelvine (d) Black pepper

460. The common name of *Ammomum subulatum* is (JRF-2006)
(a) Large cardamom (b) Nepal cardamom
(c) Great Indian cardamom (d) All the above

461. Leading producer of large cardamom is
(a) Indonesia (b) China
(c) Sri Lanka (d) India

462. Which of the following potato varieties is resistant to late blight ? (JRF-2005)
(a) Kufri Badshah (b) Kufri Sinduri
(c) Kufri Bahar (d) Kufri Chandramukhi

463. Large cardamom is a
(a) Sun loving plant (b) Shade loving plant
(c) Both a and b (d) None

464. Which dioecious vegetable is propagated by stem cutting ? (JRF, 1999)
(a) Yams (b) RUBP
(c) Pointed gourd (d) All

465. India accounts for _ _ _ _ _ % of total area under black pepper.
(a) 56 (b) 55
(c) 54 (d) 50

466. Anti-diabetic factor in bittergourd is (JRF-2006)
(a) Lipid (b) Protein
(c) Pigments (d) None of these

467. Triploid variety is commercially used in (JRF, 2004)
(a) Muskmelon (b) Cucumber
(c) Watermelon (d) All

468. Indian Cassia is **(JRF-2006)**
(a) *Cinnamomum cassia* (b) *C. burmannii*
(c) *C. loureirii* (d) *C. tamala*

469. Genetic analysis in clonally propagated crops is very difficult due to the following
(a) Reduced flowering (b) Perennial life cycle
(c) Sterility (d) All of the above

470. Large cardamom is commercially propagated through
(a) Rhizomes (b) Tubers
(c) Suckers (d) Cormels

471. Rajendra Swati is a variety of **(JRF-2006)**
(a) Cumin (b) Fenugreek
(c) Fennel (d) Coriander

472. Black pepper starts yielding from .... year onwards
(a) 2nd year (b) 3rd year
(c) 2½ year (d) 4th year

473. The F_1 hybrid in Brinjal is ....
(a) Pusa Uttam (b) Pant Rituraj
(c) Arka Kusumakar (d) Arka Navneet

474. Per capita consumption of potato in India/year is _________
(a) 15 kg (b) 20 kg
(c) 30 kg (d) 50 kg

475. Protandrous condition favour cross pollination in
(a) Cabbage (b) Onion
(c) Spinach (d) Bottlegourd

476. Daily requirement of vegetable per capita/day is (JRF, 2004)
(a) 150 gram (b) 200 gram
(c) 285 gram (d) 400 gram

477. Vegetable included in solanaceous fruit group (JRF-2005)
(a) Potato (b) Tomato
(c) Sweet Potato (d) Okra

478. Rhubarb is propagated through ....
(a) Stem cutting (b) Tubers
(c) Division of crowns (d) Root suckers

479. Which of the following is precursor for auxin synthesis ? **(JRF, 2002)**

(a) Methionine (b) Tryptophon
(c) Adenine (d) None

480. Fenugreek is a diploid with chromosome number

(a) 2n = 14 (b) 2n = 16
(c) 2n = 12 (d) 2n = 18

481. Black pepper starts yielding from .... years onwards

(a) 2nd (b) 3rd
(c) 4th (d) 5th

482. Botanical name of bitter gourd is **(JRF, 1998)**

(a) Cucumis melo (b) Momordica
(c) Citrullus (d) None

483. Konkan Tej is the variety of

(a) Cardamom (b) Clove
(c) Cinnamon (d) Nutmeg

484. Pusa Parvati is a variety of French bean developed through.

(a) Clonar selection (b) Mass selection
(c) Mutation breeding (d) None

485. Oil of black pepper consist mainly of

(a) Phenol (b) Alkaloid
(c) Terpenes (d) Tetracycline

486. Indian institute of Vegetable Research is located at **(JRF, 2004)**

(a) New Delhi (b) Allahabad
(c) Varanasi (d) Mumbai

487. Pusa Komal is the variety of **(JRF, 2001)**

(a) French bean (b) Cluster bean
(c) Cow pea (d) Goa bean

488. Brinjal belongs to the species **(JRF-2005)**

(a) *Melongena* (b) *Khasianum*
(c) *Sisymbrifolium* (d) None of these

489. Adhunik and Panchali are varieties of ________

(a) Chilli (b) Okra
(c) Greater yam (d) Bitter gourd

490. Which of following originated in India ?
(a) Pea (b) French bean
(c) Cow-pea (d) Cluster bean

491. Main constraint in commercial cultivation of spices is (JRF, 1997)
(a) Lack of planting material (b) Lack of market
(c) Lack of knowledge (d) All

492. Seed rate for raising one hactare of radish is (JRF-2005)
(a) 1-5 kg (b) 2-3 kg
(c) 4-5 kg (d) 7-8 kg

493. A *chimera* can be maintained through
(a) Cuttings, budding and graftings
(b) Somatic embryogenesis
(c) Apomixis
(d) Adventious shoot regeneration

494. In plant spacing 2 × 2, how much plant can be accommodated? (JRF, 2001)
(a) 2000 (b) 2500
(c) 3000 (d) 3500

495. Mint is botanically known as
(a) *Mentha piperita* (b) *M. arvensis*
(c) *M. spicata* (d) None of the above

496. Following variety of black pepper posses resistance to nematodes and is high yielding released in the year 1991 by NRCS
(a) Sreekara (b) Subhakara
(c) Pournami (d) Panchami

497. Following variety of Black Pepper is one of the open pollinated progeny of Balankotta
(a) Panniyur - 1 (b) Panniyur - 2
(c) Panniyur - 3 (d) None of the above

498. The somatic chromosome number of cumin are
(a) 2n = 11 (b) 2n = 12
(c) 2n = 14 (d) None of the above

499. Kufri Safed is a selection from
(a) Phulwa (b) Kufri Sindhuri
(c) Darjeeling Red Round (d) Uptodate

500. Whiptail in Cauliflower is caused due to the deficiency of (JRF-2005)
(a) Potassium (b) Boron
(c) Molybdenum (d) Copper

501. Which type of heterosis is of practical importance in breeding ?
(a) Average heterosis (b) Heterobeltiosis
(c) Economic heterosis (d) Luxuriance

502. The state leads in vegetable production (JRF, 1997).
(a) Bihar (b) Punjab
(c) U.P. (d) W. B.

503. In radish the system of pollination is (JRF-2005)
(a) Self pollination (b) Cross pollination
(c) Often Cross-pollination (d) None of these

504. Heterosis can be fixed by
(a) Vegetative propagation (b) Apomixis
(c) Balanced lethal system (d) All of the above

505. Which variety of tomato can be sown throughout year ? (JRF, 2000)
(a) Pusa Ruby (b) Pusa Sadabahar
(c) Sioux (d) None of these

506. Pusa Deepti is the variety of (JRF, 2001)
(a) Brinjal (b) Pumpkin
(c) Melon (d) None

507. Heterosis can be fully exploited in the form of (JRF, 2002)
(a) Hybrids (b) Composites
(c) Synthetics (d) Multilines

508. General combining ability can be estimated from performance of :
(a) Parents (b) F_1s
(c) Backcross (d) F_2s

509. Most commonly used male sterility system for hybrid seed production in vegetables is

(a) Cytoplasmic genetic male sterility
(b) Genetic male sterility
(c) Photoperiod sensitive genetic male sterility
(d) Cytoplasmic male sterility

510. Which of following bears yellow colour flower? (JRF, 2001)

(a) Bottle gourd (b) Luffa
(c) Cucumber (d) All

511. Heterosis may be fixable if it is mainly due to

(a) Additive × dominance effects
(b) Dominance × dominance effects
(c) Additive and additive × additive effects
(d) All of the above

512. "Saponine" is present in

(a) Spinach (b) Asparagus
(c) Tomato (d) All of these

513. Intervarietal hybridization is also called

(a) Interspecific hybridization (b) Intergeneric hybridization
(c) Intraspecific hybridization (d) Distant hybridization

514. Metaxenia is observed in (JRF, 2001)

(a) Root crops (b) Tuber crops
(c) Bulb crops (d) Cucurbits

515. Dioscorine is present in

(a) Yam (b) Elephants foot yam
(c) Cassava (d) Colocassia

516. Peas and beans contain which of the following toxic substances ?

(a) Phytic acid (b) Anti vitamin E factor
(c) Trypsin inhibitor (d) All of these

517. Which of the following toxic substances is present in celery

(a) Apiin (b) Sinigrin
(c) Cucurbitacin (d) Choline esterase

518. "Cheratin" is present in
(a) Bitter gourd (b) Watermelon
(c) Bottle gourd (d) Muskmelon

519. Which of the following is a commercial source of Levulose?
(a) Queensland arrowroot (b) Jerusalem Artichoke
(c) Colocasia (d) West Indian Arrowroot

520. Economic or useful heterosis can be estimated by

(a) $$\frac{\text{Mean value of } F_1 - \text{Mean value of commercial cultivar}}{\text{Mean value of commercial cultivar}} \times 100$$

(b) $$\frac{\text{Mean value of } F_1 - \text{Mean value of mid parent}}{\text{Mean value of mid parent}} \times 100$$

(c) $$\frac{\text{Mean value of } F_1 - \text{Mean value of best parent}}{\text{Mean value of best parent}} \times 100$$

(ঘ) $$\frac{\text{Mean value of } F_1 - \text{Mean value of } F_2}{\text{Mean value of } F_1} \times 100$$

521. The most effective method for the transfer of oligogenic character is (JRF, 2002)
(a) Pedigree breeding (b) Bulk breeding
(c) Back cross breeding (d) None

522. The large cardamom is propagated by (JRF, 2001)
(a) Seeds (b) Suckers
(c) Cutting (d) All

523. Which type of male sterility system is most convenient to use
(a) Genetic (b) Cytoplasmic
(c) Cytoplasmic-genetic (d) Temperature sensitive genetic

524. Term heterosis was given by
(a) East (1908) (b) Shull (1914)
(c) Powers (1908) (d) Davenport (1910)

525. Modern technique of inbred evaluation is based on
(a) Testing of GCA and SCA
(b) Phenotypic evaluation
(c) Prediction of double cross performance from single cross data

(d) All of the above

526. Term mutation was first introduced in 1900 by
(a) Muller (b) Bateson
(c) Hugo de Vries (d) Gaul

527. For population improvement in any crop, best method is (JRF, 2002)
(a) Pureline selection (b) Pedigree selection
(c) Mass selection (d) Recurrent selection

528. Generally the frequency of spontaneous mutations is
(a) 10^{-6} (b) 10^{-4}
(c) 10^{-5} (d) 10^{-7}

529. The pungency of chilli is due to (JRF, 2001)
(a) Capsanthin (b) Capsacin
(c) Isothiocynate (d) None

530. Mutagenic changes are the results of
(a) Deletion (b) Addition
(c) Substitution (d) All of the above

531. Chillies are good source of (JRF, 1997)
(a) Protein (b) Fat
(c) Carbohydrates (d) Vitamin

532. Which vegetable is underground stem ? (JRF, 1999)
(a) Potato (b) Radish
(c) Carrot (d) Okra

533. Which type of mutations is used in crop improvement
(a) Lethal (b) Sublethal
(c) Vital (d) Subvital

534. Most effective wavelength of ultraviolet rays is
(a) 254 nm (b) 280 nm
(c) 325 nm (d) 354 nm

535. Largest producer of cauliflower is (JRF, 1997)
(a) China (b) Kenya
(c) India (d) Australia

536. Mutagen treatment reduces
(a) Seed germination and fertility
(b) Growth rate
(c) Vigour (d) All of the above

537. In radish which substance is present ? (JRF, 2001)
(a) Allyl propyl disulphide (b) Diallyl disulphide
(c) Isothiocynate (d) None

538. Ratio of beneficial to useless mutations is one in
(a) 500 (b) 800
(c) 1200 (d) 1500

539. Krishna is the variety of (JRF, 1997)
(a) Tomato (b) Brinjal
(c) Chilli (d) Potato

540. Which is the drought resident crop ? (JRF, 1999)
(a) Sweet potato (b) Okra
(c) Cauliflower (d) Potato

541. Dominant mutations can be selected in
(a) M_1 generation (b) M_2 generation
(c) M_3 generation (d) M_2 and M_3 generation

542. A cross between an inbreed line and an open pollinated variety is called as (JRF, 2002)
(a) Test cross (b) Back cross
(c) Poly cross (d) Top cross

543. Mutation breeding can be used to
(a) Improve specific character
(b) Induce desirable mutant alleles
(c) Improve polygenic traits
(d) All of the above

544. Which plant part can be used for mutagenic treatments?
(a) Seeds (b) Pollen grains
(c) Cuttings (d) All of the above

545. Pink colour of onion is due to (JRF, 1998)
(a) Qucertin (b) Anthocynin
(c) Carotene (d) None

546. Generally what concentration of Cochicine is used for seed treatment ?

(a) 0.05 % (b) 0.20 %
(c) 0.50 % (d) 1.0 %

547. Which planting material is most conveniently and effectively used for mutagenic treatment ?

(a) Seeds (b) Pollen grains
(c) Cuttings (d) All of the above

548. Whiptail of cauliflower is caused due to deficiency of (JRF, 2004)

(a) Mo (b) Cu
(c) Zn (d) B

549. Which growth regulator is commonly used as weedicide? (JRF, 1997)

(a) ABA (b) IBA
(c) NAA (d) 2, 4-D

550. Night temperature requirement for tomato is (JRF, 1999)

(a) 5-10 °C (b) 15-20 °C
(c) 25-30 °C (d) 10-15 °C

551. Mutagenesis serves which function

(a) Parents for hybridization
(b) Direct release as a variety
(c) Material for genetic studies
(d) All of the above

552. Wintering is done to (JRF, 1999)

(a) Expose plants to cold temperature
(b) It is alternative to root pruning in rose
(c) It protects plant against cold
(d) All

553. Which operation is essential during M1 generation ?

(a) Prevention of out crossing
(b) Prevention of mechanical mixture
(c) Selection of superior types
(d) Both (a) and (b)

554. Spontaneous mutation used mostly for **(JRF, 1999)**
(a) Ornamentals (b) Fruits
(c) Vegetables (d) Spices

555. Most commonly used agent for chromosome doubling is
(a) Nitrous oxide (b) Colchicine
(c) Gibrellic acid (d) Alcohol

556. Purple blotch is problem in **(JRF, 2004)**
(a) Okra (b) Tomato
(c) Onion (d) Turmeric

557. Clubrot of cauliflower is caused by **(JRF, 2002)**
(a) Xanthomonas compestris
(b) Plasmodiophora brassiceae
(c) Fruining crotovera (d) None of the above

558. The highest/maximum productivity of potato is in _______
(a) U.P. (b) Bihar
(c) Punjab (d) W.B.

559. Pungency in onion is due to **(JRF, 1998)**
(a) Allyl propyl disulphide (b) Di-allyl disulphide
(c) Capsacin (d) None

560. .... ranks 1st in total vegetable production in India.
(a) U.P. (b) Bihar
(c) Maharashtra (d) W.B.

561. Vegetable crop mostly affected by thrips **(JRF, 2004)**
(a) Tomato (b) Okra
(c) Chilli (d) Onion

562. A disease having regular occurrence at a particular area is called as **(JRF, 1999)**
(a) Epidemic (b) Endemic
(c) Sporadic (d) None

563. .... ramls first in both area and production of Coriander in India.
(a) Gujarat (b) U.P.
(c) M.P. (d) Rajasthan

564. Chemical formula of colchine is

(a) $C_{22}H_{25}O_6N$ (b) $C_{25}H_{22}O_6N$
(c) $C_{22}H_{27}O_8N$ (d) $C_{22}H_{25}O_6N_2$

565. Ricyness in cauliflower is due to (JRF, 1997)
(a) Growing in rice field (b) Unfavourable temperature
(c) Mo deficiency (d) None

566. Nullisomics produce which type of gametes ?
(a) n + 1 (b) n – 1
(c) n (d) n + 1 – 1

567. Meristem culture can be used for
(a) Clonal propagation
(b) Recovery of virus free stocks
(c) Germplasm exchange and conservation
(d) All of the above

568. The quick freezing is prefect over slow freezing, due to (JRF, 2001)
(a) Formation of large crystal
(b) Formation of small crystal
(c) To save time (d) None of the above

569. Little leaf of Brinjal is caused by (JRF, 2004)
(a) Bacteria (b) Fungus
(c) Mycoplasma (d) Virus

570. Isolation distance in Brinjal for seed production is (JRF, 2002)
(a) 50-100 m (b) 10-20 m
(c) 200-300 m (d) 500 m

571. Which one of the following is used to enhance more fruit set ? (JRF,1997)
(a) NAA (b) 2, 4-D
(c) 2,4,5-T (d) Ethylean

572. Phomopsis blight is serious disease of (JRF, 2004)
(a) Tomato (b) Brinjal
(c) Chilli (d) Onion

573. Which is a rich source of Vitamin C ? (JRF, 2001)
(a) Cabbage (b) Tomato
(c) Chilli (d) All

574. Green manure should be incorporated at its (JRF, 2002)
(a) Early tender stage (b) Fully matured stage
(c) Flowering stage (d) Drying stage

575. Long term storage, which is most commonly used
(a) Liquid nitrogen (b) Liquid oxygen
(c) Carbon dioxide (d) All these

576. The fruit of tomato is (JRF, 2002)
(a) Pone (b) Stone
(c) Berry (d) Drupe

577. Colchicine is soluble in
(a) Alcohol (b) Cold water
(c) Chloroform (d) All of the above

578. Most commonly used cryoprotectant is
(a) DMSO (b) Glycerol
(c) Proline (d) Sorbitol

579. ____________ is the major pest of Brinjal (JRF, 1997)
(a) Fruit fly (b) Thrips
(c) Fruit & shoot borer (d) Mealy bug

580. Acridity in colocasia is due to (JRF, 1998)
(a) Calcium oxalate (b) Potassium
(c) Sulphur (d) None

581. Heterosis can be estimated by

(a) $\frac{\text{Mean value of } F_1 - \text{Mean value of commercial cultivar}}{\text{Mean value of commercial cultivar}} \times 100$

(b) $\frac{\text{Mean value of } F_1 - \text{Mean value of mid parent}}{\text{Mean value of mid parent}} \times 100$

(c) $\frac{\text{Mean value of } F_1 - \text{Mean value of best parent}}{\text{Mean value of best parent}} \times 100$

(d) $\frac{\text{Mean value of } F_1 - \text{Mean value of } F_2}{\text{Mean value of } F_1} \times 100$

582. Fruit ripening hormone is (JRF, 2002)
(a) Auxin (b) GA
(c) Cytokinin (d) Ethylene

583. NBRI Is situated at _______
(a) New Delhi (b) Culcutta
(c) Lucknow (d) Panpur

584. Indian cardamom export is set back by_______country.
(a) Malaysia (b) Indonesia
(c) Guatemala (d) Chana

585. The productivity of vegetable in India is _______
(a) 11.2 % (b) 14.4 %
(c) 12.8 % (d) 15 %

586. Bitterness in cucumber is due to _______
(a) Glucosides (b) Acids
(c) Alkaloides (d) Metaxenia

587. Which one of the following is highly cross-pollinated ? (JRF-1997)
(a) Okra (b) Bitter gourd
(c) Chilli (d) Cowpea

588. Effect of selfing on the frequency of both the homozygotes (AA + aa) with reference to a single locus (A/a) after n generation of selfing will be calculated by formulae
(a) $2^{n-1}/2^{n+1}$ (b) $2^{n}/2^{n+1}$
(c) $2^{n-1}/2^{n}$ (d) $2^{n+1}/2^{n}$

589. Multistyled flowers are common in (JRF,1998)
(a) Brinjal (b) Okra
(c) Chilli (d) Limabean

590. Which of the following crop show protandry ?
(a) Carrot (b) Tomato
(c) Chilli (d) Brinjal

591. Pusa Ruby is a variety of (JRF, 2002)
(a) Potato (b) Brinjal
(c) Tomato (d) Chilli

592. .... coloured onion var. are suitable for dehydration
(a) Red (b) Yellow
(c) White (d) All

593. ________ **is the seed rate for pole var. of French bean**
(a) 60 kg/ha (b) 80 kg/ha
(c) 40-45 kg/ha (d) 25-30 kg/ha

594. Variety of chilli resistant to leaf curl virus (JRF, 1997)
(a) Bhaskar (b) Pusa Deepti
(c) Jwalamukhi (d) Pusa Jawala

595. The metal nutrient present in the structure of chlorophyll is (JRF, 2002)
(a) Fe (b) Zn
(c) Ca (d) Mg

596. What is the optimum dose of a mutagen for effective treatment ?
(a) LD_{50} (b) Close to LD_{50}
(c) Lower than LD_{50} (d) Higher than LD_{50}

597. Major constraint in vegetative propagation of cardamom is (JRF, 1997)
(a) Katte disease (b) Thrips
(c) Capsule Borer (d) None

598. Effect of selfing on the frequency of heterozygote (Aa) at a single locus (A/a) after n generation of selfing will be calculated by formulae
(a) $2^{n-1}/2^{n+1}$ (b) $2/2^{n+1}$
(c) $2^{n}/2^{n-1}$ (d) $2/2^{n}$

599. Most serious disease of vegetable seedlings (JRF, 1998)
(a) Root rot (b) Damping off
(c) Wilt (d) None

600. The vegetable which is richest source of Vit. A is ________
(a) Turnip leaves (b) Carrot
(c) Pumpkin yellow (d) Drumstick leaves

601. The per capita consumption of vegetable/availability of vegetable in India is
(a) 300 g (b) 285 g
(c) 175 g (d) 225 g

602. Arka Suman is a variety of ....
(a) French bean (b) Cowpea
(c) Garden pea (d) Dolichus bean

603. The diploid chromosome number of cluster bean is ....
(a) 2n = 22 (b) 2n = 24
(c) 2n = 14 (d) 2n = 1

604. The crop which is not transplanted (JRF,1997)
(a) Tomato (b) Brinjal
(c) Chilli (d) None of the above

605. Standard unit used in expressing refrigeration capacity is (JRF, 2002)
(a) Calorie (b) Ampere
(c) Ton (d) Joule

606. Carrot is a rich source of (JRF, 1998)
(a) Vitamin A (b) Vitamin B
(c) Vitamin C (d) Vitamin D

607. The optimum size of potato tuber for planting is _________
(a) < 25g (b) 30-40 g
(c) > 50g (d) 100 g

608. The edible portion of Jerusalem artichoke is (JRF, 2001)
(a) Leaf stalk (b) Tuber
(c) Fruits (d) Bulb

609. .... is called as multivitamin plant
(a) Amaranthus (b) Spinach
(c) Chekurmanis (d) Asparagus

610. Family of sweet potato is (JRF,1998)
(a) Solanaceae (b) Convolvulaceae
(c) Mavaceae (d) Compositae

611. In garlic, bulbils develop naturally from modified
(a) Flower (b) Root
(c) Stem (d) Fruit

612. For packing of spinach, cauliflower and sweet potato which type of can is used (JRF,1998)
(a) AR cans (b) SR cans
(c) Plain (d) All of the above

613. In flowering plants each anther has
(a) One pollen sac (b) Two pollen sacs
(c) Three pollen sacs (d) Four pollen sacs

614. Male sterility is common in (JRF,1998)
(a) Carrot (b) Bean
(c) Okra (d) None

615. CH-1 variety of chilli is developed at (JRF, 2001)
(a) TNAU, Coimbatore (b) PAU, Ludhiana
(c) IIHR, Bangalore (d) IARI, New Delhi

616. Which of the following is not the hybrid variety of tomato? (JRF, 1998)
(a) Karnataka (b) Naveen
(c) Rashmi (d) Pusa Gaurav

617. Potato is rich source of (JRF, 1999)
(a) Fat (b) Starch
(c) Vitamin (d) Mineral

618. Which of the following insect in non polyphagous ? (JRF, 2000)
(a) Lady bird beetle (b) White grub
(c) Termite (d) Grass hopper

619. Asexually propagated crops are
(a) Highly heterozygous (b) Generally polyploids
(c) Mostly perennials
(d) Shows all characteristics of a, b & c

620. Which one of the following is used as growth retardant? (JRF,1997)
(a) GAB (b) NAA
(c) Ethophon (d) MH

621. In which variety of cauliflower seed production is not possible in north Indian plains ? (JRF, 1999)
(a) Pusa deepali (b) Pusa subhra
(c) Pusa synthetic (d) Pusa snowball -1

622. Storage temperature of cucumber is (JRF, 1999)
(a) 0 °C (b) 4 - 5 °C
(c) 6 - 7 °C (d) 10 °C

623. **The term epistasis is given by** **(JRF, 2001)**
(a) Mendel (b) Halman
(c) Bateson (d) Hull

624. **Lady bird beetle attack on family** **(JRF, 2000)**
(a) Cucurbitaceae (b) Solanaceae
(c) Lilliaceae (d) None

625. **Inbreeding is deleterious in** **(JRF, 1999)**
(a) Tomato (b) Maize
(c) Onion (d) Cucumber

626. **The growth regulator related with abscission of leaves is** **(JRF, 2002)**
(a) ABA (b) Ethylene
(c) GA (d) Cytokinin

627. **Which of the following hormone is most effective in induction of bolting in rosette plant ?** **(JRF, 2002)**
(a) Auxin (b) GA_3
(c) Cytokinin (d) Ethylene

628. **Genetic assortative mating is closest form of**
(a) Outcrossing (b) Inbreeding
(c) Panmixis (d) Open pollination

629. **Irish famine was caused by** **(JRF, 1999)**
(a) Phytopthora (b) Alternaria
(c) Fusarium oxysporum (d) None of the above

630. **Carrot belong to family** **(JRF, 2001)**
(a) Compositae (b) Umbelliferae
(c) Leguminoceae (d) None of the above

631. **Phytic acid a toxic substance present in________**
(a) Leafy vegetable (b) Peas and beans
(c) Cucurbits (d) Root vegetable

632. **The vegetable crop that tops in the share of total vegetable production**
(a) Brinjal (b) Tomato
(c) Potato (d) Cauliflower

633. Late blight resistant variety of potato is (JRF, 1999)
(a) Kufri badshah (b) Kufri sindhuri
(c) Kufri megha (d) None

634. In India Saffron (*Cross sativus*) cultivated only in ....
(a) Punjab (b) Kashmir
(c) Assam (d) Tamilnadu

635. The spacing for celery is ....
(a) 15 × 20 (b) 15 × 40
(c) 20 × 40 (d) 25 × 40

636. Allspice (*Pimenta dioca*) belongs to family ....
(a) Umbelliferae (b) Apiaceae
(c) Myrtaceae (d) None

637. Which is non bulb forming member of onion family ? (JRF, 2001)
(a) Leek (b) Garlic
(c) Onion (d) None

638. East India Company introduced cabbage, cauliflower etc. from :
(a) Mediterranean (b) China
(c) Afghanistan (d) Central Asia

639. In 1883 late blight of potato was inadvertently introduced in India from
(a) Europe (b) Mexico
(c) USA (d) Peru

640. The world collection of sweet potato germplasm is maintained at :
(a) Peru (b) China
(c) New Zealand (d) Austalia

641. The fruits & vegetables are known as (JRF, 2001)
(a) Nutritive food (b) Energy food
(c) Protective food (d) None

642. Fruit of Portulaca (a summer leafy vegetable) is (JRF, 2002)
(a) Capsule (b) Berry
(c) Siliqua (d) None

643. **Anther culture is ordinarily used for production of**
(a) Diploid plants (b) Haploid plants
(c) Polyploid plants (d) Triploid plants

644. **Saffron belongs to the family** **(JRF-2006)**
(a) Orchidaceae (b) Iridaceae
(c) Lauraceae (d) Apiaceae

645. **The browning occur in** **(JRF, 2001)**
(a) Cauliflower (b) Brinjal
(c) Okra (d) None

646. **Eugenol is the main constituent in the leaf oil of** **(JRF-2006)**
(a) Cassia (b) Cinnamon
(c) Bal leaf (d) Curry leaf

647. **Optimum night temperature for fruit set in tomato is** **(JRF, 2002)**
(a) 8 °C (b) 13 °C
(c) 17 °C (d) 21 °C

648. **The 2n number of cucumber is** **(JRF, 2001)**
(a) 14 (b) 24
(c) 38 (d) 18

649. **Daily requirement of vitamin –D for a healthy adult is** **(JRF, 2002)**
(a) 200 I.U. (b) 400 I.U.
(c) 600 I.U. (d) 800 I.U.

650. **In 1900, potato moth came in to India from**
(a) Italy (b) Ireland
(c) USA (d) New Zealand

651. **Botanical name of mango ginger is** **(JRF-2006)**
(a) *Curcuma longa* (b) *Curcuma amada*
(c) *Mangifera indica* (d) *Curcuma aromatic*

652. **Plant exploration would be most successful when it is targeted at**
(a) Farmer's field (b) Primary centre of origin
(c) A microcenter (d) Secondary centre of origin

653. Recessive bud mutation is most likely to expressed in
(a) An inbred (b) A clone
(c) A pureline (d) A hybrid

654. Kufri Red is a selection from
(a) Phulwa (b) Kufri Sindhuri
(c) Darjeeling Red Round (d) Up to date

655. Rupali variety of tomato is developed by (JRF, 2001)
(a) Indo America Hybrid Seed company
(b) Ankur seeds
(c) Mhycho seeds (d) None

656. Selfing reduces heterozygosity in each generation by the factor :
(a) 1/4 (b) 1/3
(c) 1/2 (d) 1/8

657. Which of following hormone is used as a herbicide ? (JRF, 2001)
(a) 2,4-D (b) NAA
(c) ABA (d) IAA

658. Predominant sex form of *Cucumis melo* is: (JRF-2009)
(a) Gynoecious (b) Monoecious
(c) Bisexual (d) Andromonoecious

659. Rossette formation is been due to the deficiency of (JRF, 1999)
(a) Mn (b) Zn
(c) Cu (d) Mo

660. How many types of gametes will be expected in F_1 (2^n) in a trihybrid cross ?
(a) 4 (b) 8
(c) 27 (d) 16

661. Protogyny is prevalent in : (JRF-2009)
(a) Cabbage (b) Onion
(c) Cucumber (d) Spinach

662. Saffron belongs to family
(a) Fabaceae (b) Lauraceae
(c) Iridaceae (d) Myrtaceae

663. Allepy and Duggirala are the trade names of

(a) Ginger (b) Garlic
(c) Tamarind (d) Turmeric

664. *Trachyspermum ammi* is the botanical name of

(a) Hing (b) Aijowan
(c) Zeera (d) Parsley

665. Cinnamon belongs to the family

(a) Lauraceae (b) Fabaceae
(c) Iridaceae (d) Araceae

666. Important constituent in the Bishop's weed is

(a) Cineol (b) Limone
(c) Thymol (d) Phenol

667. In cole crops which mechanism is commonly prevailing for hybrid seed production ?

(a) Male sterility (b) Self-incompatibility
(c) Apomixis (d) All above

668. Spinach belongs to family ....

(a) Compositae (b) Spinaceae
(c) Umbeliferae (d) Chenopodiaceae

669. Vitamin-C is commonly known as (JRF, 2002)

(a) Citric acid (b) Acetic acid
(c) Lactic acid (d) Ascorbic acid

670. .... is leading supplier of chilli during off-season.

(a) Andhra Pradesh (b) Maharashtra
(c) Madhya Pradesh (d) Himachal Pradesh

671. Which country is the leading producer of Vanilla ?

(a) Madagascar Republic (b) Sri Lanka
(c) India (d) Indonesia

672. All India Co-ordinated Potato Improvement Project was started in ....

(a) 1960 (b) 1982
(c) 1975 (d) 1971

673. In carrot cross-pollination is encouraged by which mechanism ?
(a) Protogyny (b) Protoandry
(c) Self-incompatibility (d) Male sterility

674. Potato contains following amount of vitamin-C (mg/100g of edible portion) (JRF, 2002)
(a) 16 (b) 17
(c) 100 (d) 25

675. Which of the following hormone increase senescence ? (JRF, 2001)
(a) Ethylene (b) NAA
(c) IAA (d) ABA

676. Failure of embryo development in interspecific crosses may be due to
(a) Chromosome elimination (b) Genotypic disharmony
(c) Cytoplasmic incompatibility (d) All of the above

677. The present (2010) DG of ICAR is ....
(a) Dr. R.S. Paroda (b) Dr. S. Ayyappan
(c) Dr. Mangala Rai (d) Dr. G. Kaloo

678. Which of the following is class one preservative ? (JRF, 2001)
(a) Silver nitrate (b) KMS
(c) Sugar (d) Vinagar

679. India stands 1st in the production of _______
(a) Tomato (b) Cabbage
(c) Onion (d) Okra

680. .... is the sexually sterile diploid crop
(a) Chekurmanis (b) Rhubarb
(c) Garlic (d) Beet

681. .... var. of cabbage is resident to black leg
(a) Pusa Mukta (b) Pusa drum head
(c) Golden acre (d) None

682. Male sterile line in onion is isolated from_______
(a) Pusa Red (b) Pusa Ratnar
(c) Pusa Madhavi (d) Arka Bindu

683. Fufu is important product made from_________

(a) Bottlegourd (b) Muskmelon
(c) Calocasia (d) Yam

684. Sree Harsha is a var. of _______

(a) Yam (b) Cassava
(c) Calocasia (d) Sweet Potato

685. NARS refers to : (JRF, 2003)

(a) National Ayurvedic Research System
(b) National Agriculture Research System
(c) National Agriculture Research Scheme
(d) National Agriculture Review System

686. Late blight resistant variety of Potato (JRF, 2004)

(a) Kufri jyoti (b) Kufri giriraj
(c) Kufri chipsona (d) All

687. The cultivated turmeric is (JRF, 2001)

(a) Tetraploid (b) Triploid
(c) Diploid (d) None

688. For measurement of TSS which instrument is used ? (JRF, 2001)

(a) Dendrometer (b) Thermometer
(c) Refractrometer (d) None

689. .... is also called as 'four angled bean'

(a) Broad bean (b) French bean
(c) Winged bean (d) Indian bean

690. Mudigree is a var. of _______

(a) B. Pepper (b) Cardamom
(c) Cinnamon (d) Clove

691. CIPHET is situated in (JRF, 2000)

(a) Amritsar (b) Nagpur
(c) Hyderabad (d) Ludhiana

692. .... has antioxidant property

(a) Black pepper (b) Cardamom
(c) Kokum (d) Turmeric

693. __________ is a thermosensitive vitamin.

(a) Vit. A (b) Vit. E
(c) Vit. C (d) Vit. D

694. Large Cardamom is widely cultivated in________

(a) Kerala (b) Karnataka
(c) Tamilnadu (d) Sikkim

695. Which variety of cauliflower is suitable for South Indian condition ? (JRF, 2004)

(a) Pusa subhra (b) Pusa early synthetic
(c) K-3 (d) All of the above

696. Isolation distance of Brinjal for certified seed production is ....

(a) 200 m (b) 5 m
(c) 100 m (d) 50 m

697. State accounting maximum area & production of cardamom is- (JRF, 2003)

(a) Tamilnadu (b) Kerala
(c) Karnataka (d) Meghyala

698. FPO was promulgated in the year

(a) 1949 (b) 1955
(c) 1961 (d) 1963

699. .... is a veg. of 20th century

(a) Clusterbean (b) Broad bean
(c) Winged bean (d) Chekurmanis

700. Mushrooms are an excellent source of ...., a mineral that the body needs to produce red blood cells

(a) Mn (b) Cu
(c) Fe (d) Cl

701. The fruit in mustard is ________

(a) Siliqua (b) Hespericium
(c) Pepo (d) Lomeninum

702. Pellagra is due to deficiency of _________

(a) Niacin (b) Folic acid
(c) Biotin (d) Pyridoxine

703. A cross in which order of male & female parent is reversed is called as (JRF, 2002)

(a) Reciprocal cross (b) Back cross
(c) Test cross (d) Top cross

704. .... instrument is used to measure growth of plant.

(a) Fathometer (b) Galvanometer
(c) Crescograph (d) Anemometer

705. Sulphur fungicide can be freely used on all vegetable except one- (JRF, 2003)

(a) Peas & beans (b) Root vegetable
(c) Cucurbits (d) Okra

706. .... vitamin is also called as Tocopherol.

(a) K (b) D
(c) B_6 (d) E

707. Horticultural Society of India (HST) was established in ....

(a) 1942 (b) 1947
(c) 1954 (d) 1950

708. Which variety of Onion is suitable for kharif season ? (JRF, 2004)

(a) Pusa red (b) Pusa madhri
(c) N-53 (d) Early Grano

709. In our diet, staple vegetable is (JRF, 2000)

(a) Brinjal (b) Potato
(c) Tomato (d) Chilli

710. .... been is also known as Goa bean.

(a) Hyacinth bean (b) French bean
(c) Winged bean (d) Dolichus bean

711. .... is yellow coloured variety of Onion.

(a) Hisar-11 (b) Arka Kalyan
(c) Punjab-48 (d) Early Grano

712. Time & temperature for vegetable processing in canning depend upon (JRF, 2001)

(a) Amount of microbes present in food
(b) Texture of product

(c) Nutrients
(d) None

713. 'Pusa Kanchan' is variety of ________
(a) Tapioca (b) Colocasia
(c) Turnip (d) Onion

714. Type of inflorescence in Beet root is ________
(a) Raceme (b) Cyme
(c) Spike (d) Spadix

715. Which is not a legume crop ? (JRF, 2003)
(a) Pea (b) beans
(c) Wheat (d) Fenugreek

716. In onion cross-pollination is encouraged by which mechanism?
(a) Protogyny (b) Protoandry
(c) Cleistogamy (d) Heterostyly

717. Genetic variability can be created through
(a) Somatic hybridization (b) Hybridization
(c) Mutation (d) All above

718. Nantes is the variety of (JRF, 2004)
(a) Radish (b) Carrot
(c) Onion (d) Turnip

719. The amount of solanin/tuber of potato considered as unfit for consumption is
(a) 5 mg (b) 10 mg
(c) 20 mg (d) 25 mg

720. The origin of sweet potato is ....
(a) Asia (b) Mexico
(c) S. America (d) Africa

721. Object of precooling is (JRF, 2000)
(a) To fresh the product (b) To improve the quality
(c) Both a and b (d) To remove field heat

722. Sree Shilpa is first hybrid of ________
(a) Greater yam (b) Elephant foot yam
(c) Lesser yam (d) White yam

723. The main aim of hybridization in plants is

(a) Improvement in one or more quantitative trait

(b) Use of F1 as a hybrid

(c) Transfer of one or more quantitative trait

(d) All of the above

724. In vegetable improvement programme which is the most commonly used ?

(a) Interspecific hybridization

(b) Intergeneric hybridization

(c) Intervarietal hybridization

(d) Distant hybridization

725. NPK requirement in legumes is- **(JRF, 2003)**

(a) 3 : 1 : 0 (b) 4 : 2 : 1

(c) 3 : 1 : 1 (d) 1 : 2 : 2

726. Which of the following is responsible for controlling of fruit drops ? **(JRF, 2000)**

(a) Auxin (b) Gibberellins

(c) Cytokinin (d) Ethylene

727. Seed rate per hectare of cluster bean is

(a) 10 - 20 g (b) 5 - 10 kg

(c) 40 - 50 kg (d) 30 - 40 kg

728. Self incompatibility results due to **(JRF-2008)**

(a) Genetic causes

(b) Morphological causes

(c) Physiological and Biochemical causes

(d) All of the above

729. How can one protect the stem-end of radish root from becoming green ?

(a) By thinning (b) By earthing

(c) By staking (d) All of the above

730. Plant regulator which improves fruit set in tomato is

(a) Gibberellic acid (b) Maleic hydrazide

(c) 2,4-Dichlorophenoxy acetic acid

(d) Ethephon

731. Pusa Chetki variety of radish is most suitable for sowing

during

(a) March-August (b) October-November
(c) After November (d) None of the above

732. The carrot seed required for sowing one ha is

(a) 1 - 2 kg (b) 4 - 5 kg
(c) 10 - 15 kg (d) 15 - 20 kg

733. Squash melon is believed to have originated in

(a) China (b) India
(c) Malaysia (d) None of the above

734. Systemic fungicide used for the control of downy mildew of muskmelow is

(a) Indofil M-45 (b) Ridomil MZ
(c) Blitox (d) None of the above

735. The best time for transplanting cabbage seedling in the field is

(a) Morning (b) Afternoon
(c) Evening (d) Night

736. *Trichosanthes dioca* is native of (JRF-2008)

(a) China (b) India
(c) Iran (d) Peru

737. Types of vegetable gardens base on the purpose for which they have been developed are

(a) Two (b) Six
(c) Four (d) None of the above

738. The seed rate for one ha planting of *palak* is

(a) 2 - 5 kg (b) 5 - 9 kg
(c) 15 - 20 kg (d) 25 - 30 kg

739. The usual weight of seed piece of potato is

(a) 30 - 40 kg (b) 40 - 50 kg
(c) 50 - 60 kg (d) None of the above

740. First intergeneric hybrid between radish and cabbage was made by (JRF-2008)

(a) Hull (1945) (b) Rimpu (1890)

(c) Andrew Knight (1800) (d) Karpenchenko (1927)

741. **Vegetables are rich source of**
(a) Carbohydrate (b) Vitamins
(c) Proteins (d) None of the above

742. **Which of the following is the early variety of cabbage ?**
(a) Pride of India (b) Early Drum Head
(c) Chieftain (d) None of the above

743. **Vegetables for processing are subjected to drying after**
(a) Blanching (b) Without blanching
(c) Sulphuring (d) Any other means

744. **Carrot roots become longer**
(a) Above 10°C (b) Below 10°C
(c) Above 25°C (d) Below 25°C

745. **India Institute of Spices Research is situated in**
(a) Cochin (b) Trivandrum
(c) Calicut (d) Dharwad

746. **Rio-de-Janerio is an improved variety of**
(a) Turmeric (b) Garlic
(c) Ginger (d) Large Cardamom

747. **Jade cross is a F1 hybrid of ________.**
(a) Brussels sprout (b) Spouting brocoli
(c) Knol-khol (d) Cabbage

748. **The variety resistant to black leg in cabbage is**
(a) Pusa Mukta (b) Golden
(c) Pusa Drum head (d) Pusa sambandh

749. **How much area is suitable for kitchen gardening for a family of five ?** **(JRF, 2004)**
(a) 50 m^2 (b) 100 m^2
(c) 200 m^2 (d) 250 m^2

750. **Akashin a disorder in Radish is due to deficiency of :**
(a) Manganese (b) Boron

(c) Magnesium (d) Galcium

751. The single seeded cucurbit is

(a) *L. acutangula* (b) *L. cylindrica*
(c) *C. moschata* (d) *S. edule*

752. The variety suitable for both cold and hot sector in Tomato is

(a) Pusa sheetal (b) Pusa sadabahar
(c) Pusa Hybrid-1 (d) Arka sambandh

753. Onion is stored at a temperature of (JRF, 2004)

(a) 4-5 °C (b) 25 °C
(c) 6-8 °C (d) 10 °C

754. Completely tuberization in potato is stopped at temperature of : (JRF, 2004)

(a) 20 °C (b) 25 °C
(c) 30 °C (d) 23 °C

755. Which is not a cytokinin ? (JRF, 2004)

(a) BA (b) Zeatin
(c) Etheral (d) Kinatin

756. The F_1 hybrid in case of Muskmelon is

(a) Pusa Madhuras (b) Pusa Rasraj
(c) Arka Rajhans (d) Arka Jeet

757. Largest importer of cauliflowers in the world is

(a) Netherlands (b) Germany
(c) England (d) Switzerland

758. The origin of Dolichus bean is

(a) China (b) India
(c) Mexico (d) Mediterranean region

759. The book HORTI. SCIENCE is written by (JRF, 2004)

(a) Willium gadd (b) Swaminathan
(c) Jennick (d) Randhawa

760. .... is the selective herbicide for broad leaved plants

(a) Paraquat (b) Diquat
(c) 2, 4-D (d) None

761. The vegetable in which all parts of plant can be used as edible part is

(a) Lima bean (b) Winged bean
(c) Radish (d) Broad bean

762. The capacity of a cell to develop into a complete plant is called............ (JRF-2007)

(a) Tissue culture (b) Fecundity
(c) Totipotency (d) Haploidy

763. Maximum life of Vanilla is (JRF,1997)

(a) 10 years (b) 15 years
(c) 30 years (d) 50 years

764. A pureline may be used

(a) As a parent for hybridization
(b) For mutation studies
(c) As a variety (d) All the above

765. The ornamental type in case of cucurbits is

(a) Summer squash (b) White squash
(c) Pumpkin (d) White squash.

766. Herkogamy favours (JRF, 2002)

(a) Self pollination (b) Cross pollination
(c) Mutation (d) Wide hybridization

767. Navashree and Nithyashree are varieties of

(a) Cardamom (b) Cinnamom
(c) Black Pepper (d) None

768. Fenugreek seed contain

(a) Allicin (b) Diosgenin
(c) Ajmacline (d) All

769. Commercial part of mace is

(a) Dried kernel (b) Seed
(c) Flower (d) Dried aril

770. Major bottlenecks in breeding clonally propagated crops are

(a) Perennial life cycle, reduced flowering and fertility
(b) Long life cycle and difficulty in genetic analysis
(c) Severe inbreeding depression

(d) All of the above

771. Which is not the variety of Brinjal ? **(JRF, 2004)**
(a) Pant samrat (b) Pant bahar
(c) Pant rituraj (d) Pusa ankur

772. Male musk deer smell resembles to
(a) Ambrette seed (b) Basil
(c) Celery (d) Davana

773. India's share in world trade of spice is
(a) 8 % (b) 18 %
(c) 28 % (d) 38 %

774. Chemical present in turmeric is
(a) Cineol (b) Zingiberene
(c) Cucurmine (d) Linalool

775. Which of the following spice is native to India ?
(a) Cardamom (b) Kokum
(c) Saffron (d) All the above

776. Vanilla belongs to family
(a) Orchidaceae (b) Guttiferae
(c) Fabaceae (d) Iridaceae

777. Curry leaf belongs to family
(a) Rosaceae (b) Solanaceae
(c) Rutaceae (d) Myrtaceae

778. Nutmeg belongs to family
(a) Myrtaceae (b) Lauraceae
(c) Myrsticaceae (d) Piperaceae

779. Nucleus was first discovered by **(JRF-2008)**
(a) Benda (1877) (b) Camillo Golgi
(c) Robert Brown (1882) (d) Flemming (1882)

780. Root system of sweet potato is basically
(a) Shallow (b) Deep
(c) Medium (d) None of the above

781. Origin of fenugreek is
(a) Europe (b) America
(c) Asia (d) Australia

782. Seed production of which crop is not done on plains ?
(a) Early cauliflower (b) Cabbage
(c) Onion (d) Carrot

783. The plant part used in case of Kokum is
(a) Fruit (b) Leaves
(c) Rhizome (d) Bark

784. Feathering and scrapped chipps are important grades of
(a) Black pepper (b) Clove
(c) Ginger (d) Cinnamon

785. In case of potato, the Economic Threshold Level (ETL) of aphid population is
(a) 20/100 leaves (b) 30/100 leaves
(c) 40/100 leaves (d) 50/100 leaves

786. The leading state in production of Fenugreek is
(a) Rajasthan (b) Gujarat
(c) Madhya Pradesh (d) Uttar Pradesh

787. The leading ginger producing country in the world is
(a) China (b) Srilanka
(c) Italy (d) India

788. Shape of head of cabbage variety 'Pride of India' is
(a) Flat (b) Pointed
(c) Round (d) None of the above

789. Plant part used in asfoetida
(a) Gum (b) Bark
(c) Oleoresin (d) Both (a) and (c)

790. Ginger and turmeric are native to
(a) South East Asia (b) Sri Lanka
(c) Indo Burma (d) Indonesia

791. Nutmeg and clove are native to
(a) India (b) Africa
(c) Indonesia (d) America

792. Saffron is native to
(a) India (b) America
(c) Indonesia (d) Africa

793. Chemical constituent present in pepper is
(a) Piperine (b) Curcumine
(c) Cineol (d) Eugenol

.794. "Cicrocrocin" is present in
(a) Saffron (b) Sweet flag
(c) Clove (d) Dill

795. Which of the following varieties is an introduction of ginger ?
(a) Rio de Janerio (b) Suprabha
(c) Surabhi (d) Suruchi

796. Which spice is called the "queen of spices" ?
(a) Black pepper (b) Small cardamom
(c) Large cardamom (d) Clove

797. "Yercaud-1" is an improved variety of
(a) Cinnamon (b) Clove
(c) Nutmeg (d) Kokum

798. "Konkan Amrit" is an improved variety of
(a) Kokum (b) Vanilla
(c) Juniper (d) Saffron

799. 90% of the pungency in chili is present in which part of fruit ?
(a) Seed (b) Placenta
(c) Pericarp (d) None of these

800. Vanilla is propagated through
(a) Stem cutting (b) Corm
(c) Root cuttings (d) Tubers

801. Which of the following is a dioecious spice ?
(a) Nutmeg (b) Juniper
(c) Kokum (d) All of these

802. Which of the following is a mutant ginger variety ?
(a) Surari (b) Surbhi
(c) Rio de Janerio (d) Suruchi

803. Leafleass mutants arc produce in which crop ?

(a) Brinjal (b) Onion

(c) Okra (d) Pea

804. Which cucumber spps. have resistant against nematodes ?

(a) *Cucumis anguria* (b) *Cucumis ficifilia*

(c) *Cucumis metuliferus* (d) All of these

805. Little leaf resistant spps. of Brinjal ?

(a) *S. torum* (b) *S. nigrum*

(c) *S. gilo* (d) *S. khasianum*

806. Characteristics flavor of cole crops is due to volatile compounds.

(a) Diallyl-disulphide (b) Dimethyl-trisulphide

(c) Isothiocynate (d) Diphenylamine

807. The genetics basis of pureline selection was first explained by : (JRF-2006)

(a) C. A. Brim (b) J.B. Harrington

(c) C.O. Gardner (d) W. Johannsen

808. International Seed Testing Association [ISTA] is situated at:

(a) Colambo (Srilanka) (b) New Delhi (India)

(c) Norway (Switzerland) (d) Rome (Italy)

809. New Zealand spinach (*Tetragonia expansa*) belongs to family:

(a) Chenopodiaceae (b) Solanaceae

(c) Apiaceae (d) Aizoaceae

810. In plant breeding nearly homozygous lines are produced by continued self-fertilization accomplish by selection for ....generations (JRF-2006)

(a) 3 - 4 (b) 5 - 6

(c) 7 - 8 (d) 9 - 10

811. Crop having max/highest productivity in India

(a) Cabbage (b) Cassava

(c) Potato (d) Okra

812. Lima bean (Butter bean) is

(a) Self-pollinated (b) Cross-pollinated
(c) Often-cross-pollinated (d) Highly cross-pollinated

813. Gametophytic self-incompatibility is found in

(a) Beet (b) Cabbage
(c) Turnip (d) Tomato

814. In Onion, ... is used as maintainer for the male sterile line (JRF-2006)

(a) Smsms (b) Nmsms
(c) NMsMs (d) NMsms

815. Bacterial blight resistant variety of cowpea.

(a) Pusa Rituraj (b) Pusa Barsati
(c) Pusa Komal (d) Arka Suman

816. Pusa Red Plum variety of tomato has highest vitamin-C content and it is developed through

(a) Intervarietal crosses (b) Interspecific crosses
(c) Intergeneric (d) Intraspecific crosses

817. Sree Dhara is a variety of

(a) Elephant foot yam (b) Sweet potato
(c) Topioca (d) Coleus

818. Pusa snowball -1 variety of cauliflower of which seed production is only possible in

(a) Plains (b) Hills
(c) Both (d) None

819. Which type of flower in brinjal have maximum fruit cutting?

(a) Long styled (b) Medium styled
(c) Pseudo short-styled (d) Short styled

820. In private sector first F_1 hybrid of tomato "Karnatka Hybrid" and of sweet pepper "Bharat" was released for commercial cultivation in 1973 by a private seed company.

(a) IAHS- Banglore (b) Nath
(c) MAHYCO (d) Beejosheetal

821. Which was the first example of successful breeding for disease or pest resistance ?

(a) Pusa Sawani – YVMV

(b) Pusa Komal – Bacterial blight
(c) Pusa Shubra – Black rot
(d) Pusa Drum Head – Black leg

822. Little leaf of Brinjal is transmitted by
(a) Hopper (b) Aphids
(c) Jassids (d) White fly

823. Selfing & massing technique of breeding (Improvement in a cultivar can be affected by selfing followed by massing) was suggested by
(a) Clarke (1943) (b) Jones & Mann
(c) Pirovano (1972) (d) None

824. Which of the following is a long day type onion variety?
(a) Nasik Red (b) Brown Spanish
(c) Udaipur 102 (d) Pusa Red

825. Karpenchenko (1997) made first intergeneric hybrid between:
(a) Radish & Cabbage (b) Cabbage & cauliflower
(c) Radish & cauliflower (d) Radish & Turnip

826. Pusa Parvati, an improved cultivar of French resistant to mosaic & powdery mildew disease and suitable for hills was developed through
(a) Selection (b) Hybridization
(c) Mutation (d) Polyploidy

827. Pusa Kirti & Pusa Kiran are the variety of
(a) Spinach (b) Palak
(c) Lettuce (d) Amaranthus

828. Asparagus is a
(a) Monoecious (b) Dioecious
(c) Andromonoecious (d) Hermis

829. India ranks in vegetable production
(a) First (b) Second
(c) Third (d) Fourth

830. "Akashin" a physiological disorder observed in radish is due to deficiency of
(a) N (b) Ca
(c) K (d) B

831. August & September are varieties of

(a) Knol-khol (b) Cabbage
(c) Radish (d) Kale

832. Share of vegetable is about ....% of the total cultivable land in India.

(a) 1.2 (b) 2.8
(c) 3.6 (d) 4.8

833. India contribute..........% in world vegetable production

(a) 5.38 (b) 23.5
(c) 13.48 (d) 31.31

834. The productivity of vegetable in India is

(a) 11.2 % (b) 14.4 %
(c) 12.2 % (d) 15.1 %

835. USA stands at which position in total vegetable production ?

(a) 1st (b) 3rd
(c) 2nd (d) 4th

836. Which of the following is leading vegetable crop of India in production ?

(a) Cabbage (b) Potato
(c) Tomato (d) Brinjal

837. Which is the proper stage of tapping asfoetida

(a) Foliage just appear (b) Foliage turns green to yellow
(c) Foliage lush green (d) Foliage senescenc

838. Somatic hybrids of *Solanum melongena* and *S. sisimbrifolium* was made possible by

(a) Embryo rescue (b) Protoplast fusion
(c) Micropropagation (d) Anther culture

839. Transgenics are produced in tomato for

(a) Modified ripening (b) Resistance to herbicides
(c) Resistance to virus (d) All the above

840. "Newleaf" is transgenic of which crop commercialized by Monsanto since 1989 ?

(a) Potato carrying 3A a gene
(b) Cabbage carrying 3A a gene
(c) Tomato carrying 3A a gene

(d) Brinjal carrying 3A a gene

841. Colour contents of tomato are controlled by

(a) Two genes (b) Four genes
(c) Three genes (d) Multiple gene

842. Which of the following is primary center for cassava ?

(a) Central Asia (b) Africa
(c) Near East (d) Central America

843. Global diversity of vegetables crops consists of about

(a) 200 spp. (b) 400 spp.
(c) 300 spp. (d) 500 spp.

844. Which of the following is leading vegetable crop of India in productivity ?

(a) Cabbage (b) Tapioca
(c) Tomato (d) Brinjal

845. Arka Harit is a variety of

(a) Amaranth (b) Bottle gourd
(c) Bitter guard (d) Tomato

846. Which of the following is the leading export commodity of India ?

(a) Fresh onion (b) Fresh cucumber
(c) Fresh potato (d) Fresh tomato

847. Inbreeding depressions is much more which of the following crops ?

(a) Winter cauliflower (b) Broccoli
(c) Summer cauliflower (d) Brussels sprout

848. Mitosis takes place in which of the following organs ? (JRF-2008)

(a) Reproductive organs (b) Somatic organs
(c) Anthers (d) Both (a) and (b)

849. Which of the following is leading vegetable crop of India in area

(a) Cabbage (b) Potato
(c) Tomato (d) Brinjal

850. **Introduction of genes coding for *Bacillus thuringenesis* (Bt.) insecticidal crystal protein confer resistance to**
(a) Insects (b) Bacteria
(c) Diseases (d) Viruses

851. **Strains of Bt produce toxins, which confer resistance against**
(a) Diptera (b) Coleopteran
(c) Lepidoptera larvae (d) All the above

852. **Which of the following is primary center of diversity for bottle gourd ?**
(a) Chinese-Japanese (b) Hindustani Center
(c) Indo-Chinese (d) Mediterranean region

853. **Pusa Mukta is a variety of**
(a) Cauliflower (b) Broccoli
(c) Cabbage (d) Brussels sprout

854. **Which of the following tomato varieties are good for processing ?**
(a) Arka Shaurabh (b) Arkas Ashish
(c) Arka Ahuti (d) All the above

855. **Colour blindness genes are located on** **(JRF-2008)**
(a) Y-chromosomes (b) X-chromosomes
(c) Autosomes (d) All of the above

856. **N-53 is a variety of**
(a) Amaranth (b) Bottle gourd
(c) Onion (d) Brinjal

857. **Which of the following centers is not primary center of diversity for any vegetable crops ?**
(a) Chinese (b) South America
(c) North America (d) Central America

858. **Which of the following is primary center of diversity for cucumber ?**
(a) Chinese-Japanese (b) Hindustani Center
(c) Indo-Chinese (d) Mediterranean region

859. **Pusa Katki is a variety of**
(a) Cauliflower (b) Bottle gourd
(c) Bitter guard (d) Brinjal

860. Arka Nishant is a variety of
(a) Amaranth (b) Bottle gourd
(c) Radish (d) Carrot

861. Improved Japanese, cauliflower is introduced in India from
(a) USA (b) Phillipines
(c) Denmark (d) Israel

862. Which of the following is primary center of diversity for yams ?
(a) Chinese-Japanese (b) Hindustani Center
(c) Indo-Chinese (d) Mediterranean region

863. Angur Lata is a variety of
(a) Amaranth (b) Bottle gourd
(c) Tomato (d) Brinjal

864. Which of the following is component of National Gene Bank ?
(a) National seed repository (b) Cryobank
(c) National tissue culture (d) All the above

865. Development of embryo from egg cell without fertilization is called as (JRF-2008)
(a) Apospory (b) Autogamy
(c) Parthenogenesis (d) Apogamy

866. Pusa Barsati, cowpea is an improvement over introduction from :
(a) USA (b) Phillipines
(c) Denmark (d) Israel

867. Pusa Nasdar is a variety of
(a) Amaranth (b) Bottle gourd
(c) Ridge gourd (d) Brinjal

868. Which of the following is primary center of diversity for radish ?
(a) Chinese-Japanese (b) Hindustani Center
(c) Indo-Chinese (d) Mediterranean region

869. Cryopreservation (cryobank), can be adopted for preservation of which type of seed ?
(a) Orthodox (b) Both (a) and (c)
(c) Recalcitrant (d) None

870. Pusa Jyoti is a variety of
(a) Amaranth (b) Bottle gourd
(c) Spinach (d) Brinjal

871. Which of the following is primary center for spinach ?
(a) Central Asia (b) Africa
(c) Near East (d) Central America

872. Pusa Swarnima is a variety of
(a) Summer Squash (b) Turnip
(c) Winter Squash (d) Carrot

873. Pusa Alankar is a variety of
(a) Summer Squash (b) Turnip
(c) Winter Squash (d) Carrot

874. Golden Acre, cabbage is introduced from
(a) USA (b) Phillipines
(c) Denmark (d) Israel

875. CH-1, hybrid developed at IARI, New Delhi belongs to
(a) Tomato (b) Brinjal
(c) Chilli (d) Capsicum

876. In onion, the type of self sterility is
(a) Genic (b) Cytoplasmic-genic
(c) Cytoplasmic (d) None of the above

877. Bonneville pea is introduced from
(a) USA (b) Phillipines
(c) Denmark (d) Israel

878. Application of ethrel @ 200-300 ppm at two true leaf stages and four leaf stages suppresses male flowers in
(a) Bottle gourd (b) Squashes
(c) Pumpkins (d) All the above

879. Which of the following crops is wind pollinated ?
(a) Spinach (b) Tomato
(c) Cucumber (d) Watermelon

880. Arkel, pea is introduced from
(a) France (b) Phillipines
(c) Denmark (d) Israel

881. Arka Mamik and Arka Jyoti are the sister seedlings of a cross between
(a) Tetra-2 × Pusa Rassal (b) IIHR-21 × Crimson Sweet
(c) IIHR-20 × Crimzon Sweet (d) None of the above

882. "Sylvia" is edible podded variety of
(a) Pea (b) Broad bean
(c) Lima Bean (d) French bean

883. Pusa Sanyog, is a F-1 hybrid of
(a) Cucumber (b) Muskmelon
(c) Watermelon (d) Longmelon

884. Hayes and Jones (1916) first reported heterosis in
(a) Cucumber (b) Muskmelon
(c) Muskmelon (d) Longmelon

885. Onion variety suitable for kharif season is
(a) N-53 (b) Pusa Red
(c) Early Graino (d) Pusa Madhvi

886. The first F1 hybrid to be released from IARI, New Delhi in 1973 was
(a) Pusa Meghdoot in bottle gourd
(b) Pusa Sanyog in cucumber
(c) Pusa Alankar in summer squash
(d) None of the above

887. Area (%) under hybrid is maximum in
(a) Tomato (b) Brinjal
(c) Cabbage (d) Okra

888. The often cross pollinated vegetable crop is (JRF-2008)
(a) Pea (b) Tomato
(c) Amaranth (d) Limabean

889. Palam Samridhi is an improved variety of
(a) Cabbage (b) Broccoli
(c) Cauliflower (d) Brussels sprout

890. Kt-1 (Pusa Deepti) developed at IARI, Regional station Katrain belongs to
(a) Tomato (b) Brinjal
(c) Chilli (d) Capsicum

891. Crossing over takes place during (JRF-2008)
(a) Leptotene (b) Zygotene
(c) Pachytene (d) Diplotene

892. Hybrid seed production in cucumber involves
(a) Use of gynoecious lines (b) Both a & c
(c) Use of monoecious lines (d) None of the above

893. Sex expression in cucurbits is manipulated by
(a) $AgNO_3$ (b) GA_3
(c) Etherl (d) All the above

894. Pinching of male flowers is effective in
(a) Bottle gourd (b) Cucumber
(c) Water melon (d) Muskmelon

895. Which of the following is used as maintainer line in onion breeding ?
(a) Line A (b) Line D
(c) Line B (d) Line C

896. In a DNA molecule, thymine always pair with (JRF-2008)
(a) Adenine (b) Guanine
(c) Cytosine (d) Urecil

897. Arka Pitambar and Arka Kirtiman are hybrids of
(a) Okra (b) Capsicum
(c) Onion (d) Chilli

898. Which of the following monoecious line is used for production of Pusa Rasraj ?
(a) M 1 (b) M 3
(c) M 2 (d) M 4

899. **Non-lobing leaf character is used as single recessive marker gene for hybrid seed production in**
(a) Watermelon (b) Muskmelon
(c) Cucumber (d) None of the above

900. **Bitterness in cucumber is due to effect of external pollens and is called**
(a) Alkaloids (b) Metaxenia
(c) Acids (d) Glucosides

901. **Nutritional quality breeding deals with genetic improvement in : (JRF-2008)**
(a) Vitamins (b) Oil
(c) Protein (d) All of the above

902. **Hybrids in cabbage developed by using cytoplasmic male sterility are suitable for**
(a) Hills (b) Plateau
(c) Plains (d) Islands

903. **Type of incompatibility in cabbage and cauliflower is**
(a) Sporophytic (b) Neither a nor c
(c) Gametophytic (d) Both a and c

904. **Welsh and Grimbell in 1947 first reported male sterility in :**
(a) Radish (b) Onion
(c) Cucumber (d) Carrot

905. **Regeneration of a plant from a single cell in culture medium is called as (JRF-2008)**
(a) Protoplast culture (b) Meristem culture
(c) Organ culture (d) Cell culture

906. **The number of chromosome is reduced to half in plants obtained by (JRF-2008)**
(a) Anther culture (b) Ovule culture
(c) Shoot tip culture (d) Both (a) and (b)

907. **CH-1, A hybrid in chilli is result of a cross between**
(a) MS-12 × Punjab Lal (b) MS-13 × Punjab Lal
(c) MS-1 × Punjab Lal (d) MS-11 × Punjab Lal

908. Which of the following is used as androcides in okra ?

(a) GA_3 (b) MH (0.4 - 0.5 %)
(c) NAA (d) None of the above

909. In India, male sterile line have been isolated from which of the following varieties

(a) Niphad-2-4-1 (b) Pusa Red
(c) Nasik White Globe (d) All of the above

910. Crucifer vegetables are value for (JRF-2008)

(a) Magnesium containing compounds isothicynates
(b) Sulphar containing compounds glucosinolates
(c) Molybdenum containing compounds isoflavonones
(d) Boron containing compounds vanillin

911. Pinching of male flower is effective in hybrid seed production of :

(a) Bottle gourd (b) Ash gourd
(c) Bitter gourd (d) None of the above

912. Which of the following organisms are the most destructive in vegetable crops ?

(a) Fungus (b) Virus
(c) Bacteria (d) None of the above

913. Chimera can be maintained by (JRF-2008)

(a) Somatic embryogenesis
(b) Apomixis
(c) Adventitious shoot formation
(d) Cutting and Grafting

914. Which of the following tomato varieties are resistant to root knot nematodes ?

(a) Hissar Lalit (b) Arka Vardhan
(c) Sel. 120 (d) All of these

915. Which of the following tomato varieties are resistant to bacterial wilt ?

(a) Arka Abha (b) Arka Alok
(c) Arka Shresha (d) All of the above

916. 2, 4-D is a **(JRF-2008)**
(a) Fungicide (b) Auxin
(c) Growth hormone (d) Insecticide

917. Which of the following tomato varieties are resistant to fruit borer ?
(a) Pusa Uphar (b) BT-10
(c) BT-1 (d) None of the above

918. Which of the following species of *Lycopersicon* are resistant to leaf curl virus ?
(a) *L. peruvianum* (b) *L. glandulosum*
(c) *L. pimpinellifolium* (d) All of these

919. Which of the following are tolerant to salt stress in tomato ?
(a) *L. peruvianum* (b) *L. pennellii*
(c) *L. cheesmani* (d) All of these

920. Which of the following are phomopsis blight resistant varieties of brinjal ?
(a) Pant Samrat (b) Pusa Purple Cluster
(c) Pusa Bhiarav (d) All of these

921. Arka Sheel and Manjari Gota are resistant to
(a) Little leaf (b) Root knot rematodes
(c) Aphid (d) None of these

922. Long day type variety of garlic is
(a) Agrifound White (b) Agrifound Parvati
(c) Yamuna Safed (d) G-282

923. Which of the following brinjal varieties is resistant to aphid ?
(a) Arka Sheel (b) Manjari Gota
(c) Annamalai (d) Pusa Bhairav

924. Which of the following chilli varieties is resistant to leaf curl ?
(a) Pusa Jawala (b) Pant C-1
(c) Pusa Sadhabahar (d) All of these

925. Which of the following are resistant to powdery mildew in chilli ?
(a) Arka Lohit (b) Pusa Jawala
(c) Pant C-1 (d) Pusa Sadhabahar

926. Arka Manik watermelon is resistant to
(a) Powdery mildew (b) Anthracnose
(c) Downey mildew (d) All of these

927. Pusa Mukta is black rot resistant variety of
(a) Knol Khol (b) Cabbage
(c) Cauliflower (d) Tomato

928. Poinsette cucumber is resistant to
(a) Powdery mildew (b) Angular leaf spot
(c) Anthracnose (d) All of these

929. Black leg is an important disease of which of the following vegetable crops ? (JRF-2008)
(a) Cauliflower (b) Tomato
(c) Winged bean (d) Cabbage

930. *Allium fistulosum* is resistant to
(a) Neck rot (b) Onion smut
(c) Pink rot (d) All of these

931. Pusa Sadhabahar, a chilli variety is resistant to
(a) CMV (b) Leaf curl
(c) TMV (d) All of these

932. Lima bean is a
(a) Self-pollinated (b) Often cross-pollinated
(c) Cross-pollinated (d) None of these

933. In which of the following state potato is generally not grown ?
(a) Kerala (b) Tamil nadu
(c) Uttar Pradesh (d) Andhra Pradesh

934. Type of inflorescence in carrot is
(a) Spike (b) Panicle
(c) Umbel (d) Catkin

935. Type of inflorescence in cassava is

(a) Spike (b) Panicle
(c) Umble (d) Catkin

936. Self pollination is a form of (JRF-2008)

(a) Outbreeding (b) Inbreeding
(c) Random mating (d) None of the above

937. Which of the following is sexually sterile diploid among all ?

(a) Onion (b) Garlic
(c) Leek (d) Beet

938. Palak is a

(a) Self-pollinated (b) Often cross-pollinated
(c) Cross-pollinated (d) None of these

939. Which of the following is wind pollinated ?

(a) Onion (b) Radish
(c) Carrot (d) None of these

940. Which of the following mechanisms in brinjal favours cross pollination ?

(a) Heterostyly (b) Protoandry
(c) Protogyny (d) None of these

941. Which of the following mechanisms exists in cole crops

(a) Heterostyly (b) Protoandry
(c) Protogyny (d) None of these

942. Which of the following mechanisms exists in carrot ?

(a) Heterostyly (b) Protoandry
(c) Protogyny (d) None of these

943. The edible part of rutabaga is

(a) Root (b) Spear
(c) Stem (d) Corm

944. Which of the following is winter crop ?

(a) Broad bean (b) Lima bean
(c) French bean (d) Cluster bean

945. ***Cyamposis tetragonolobus*** **is botanical name of**
(a) Broad bean (b) Cluster bean
(c) Lima bean (d) Scarlet bean

946. Which of the following is Summer season crop ?
(a) Broad bean (b) Lima bean
(c) Pea (d) None of these

947. Edible part of most of the summer season vegetable crops is
(a) Fruit (b) Stem
(c) Leaf (d) Root

948. ***Xanthosoma sagittifolium*** **is botanical name of**
(a) Giant Taro (b) Tannia
(c) Taro (d) Arrowroot

949. Cultivation of two crops together in the same field by mixing their seeds is called (JRF-2008)
(a) Mixed farming (b) Intercropping
(c) Mixed cropping (d) Double cropping

950. Which of the following genus belong to cucurbitaceae ?
(a) *Benincasa* (b) *Trichosanthes*
(c) *Legenaria* (d) All of these

951. ***Sesbania grandiflora*****, agathi belongs to family**
(a) Leguminosae (b) Labiateae
(c) Portulacae (d) Moringaceae

952. Which of the following vegetable belong to C4 plant ?
(a) Carrot (b) Potato
(c) Amaranths (d) Bean

953. "Potherbs" are generally
(a) Cool season crops (b) Rainy season crops
(c) Warm season crops (d) None of these

954. All the vegetable crops belongs to the Plant – Division
(a) Angiosperm (b) Both a and c
(c) Gymnosperm (d) None of these

955. **Which one of the following crops is hardy ?**
(a) Asparagus (b) Sweet potato
(c) Potato (d) Tapioca

956. **Which of the following crops has the highest respiration rate ?**
(a) Broccoli (b) Cabbage
(c) Onion (d) Leek

957. **Which of the following has the lowest respiration rate ?**
(a) Onion (b) Pea
(c) Sweet corn (d) Parsley

958. **Which of the following is an hexaploid ?**
(a) Sweet Potato (b) Watermelon
(c) Potato (d) Muskmelon

959. **Which of the following is an autotetraploid ?**
(a) Potato (b) Leek
(c) Okra (d) None of these

960. **Seed rate of cloves (kg/ha.) as garlic is**
(a) 100 - 200 (b) 400 - 500
(c) 600 - 800 (d) 900 - 1000

961. **Garlic contains amino acid mainly**
(a) Allicin (b) Alliin
(c) Allinage (d) Diaalyl disulphide

962. **Vine crops referred to which group of crop ?**
(a) Cucurbits (b) Bulb crop
(c) Solanaceous (d) Cole crop

963. **Which of the following is shallow rooted crop ?**
(a) Asparagus (b) Sweet potato
(c) Artichoke (d) Onion

964. **Cultivated single seeded cucurbits is**
(a) Chayote (b) Kakrol
(c) Kartoli (d) Pointed gourd

965. **The non-preferance insect resistant mechanism is known as (JRF-2008)**
(a) Antichlorosis (b) Antibiosis
(c) Antixenosis (d) Antipheromonosis

966. Which of the following is deep rooted crop ?
(a) Tomato (Wolf apple) (b) Onion
(c) Cabbage (d) Garlic

967. *Mamordica cochinchinensis* is commonly known as
(a) Bitter gourd (b) Kakrol
(c) Sweet gourd (d) Round melon

968. Anthesis in which of the following cucurbits takes place in the evening ?
(a) Cucumber (b) Muskmelon
(c) Sponge gourd (d) Bottle gourd

969. In which of the following crops, cleistogamy is observed?
(a) Lettuce (b) Capsicum
(c) Tomato (d) Onion

970. Satputia variety of ridge gourd have sex form
(a) Monoecious (b) Dioecious
(c) Hermaphrodite (d) Gynoecious

971. Herkogamy is observed in
(a) Lima bean (b) Capsicum
(c) Tomato (d) All of these

972. In which of the following vegetable crops, maximum varieties are released in India ?
(a) Tomato (b) Pepper
(c) Brinjal (d) Potato

973. Genetic male sterility is observed in
(a) Cole crops (b) Watermelon
(c) Muskmelon (d) All of these

974. Cytoplasmic male sterility is observed in
(a) Carrot (b) Cucumber
(c) Sweet pepper (d) All of these

975. "Solasodine" is present in
(a) Tomato (b) Onion
(c) Brinjal (d) Garlic

976. Which of following variety of carrot is excellent for canning ?
(a) Chantaney (b) Pusa Kesar
(c) Pusa Lal (d) Pusa Meghali

977. White brinjal is recommended for
(a) Diabetes (b) Cancer
(c) Heart disease (d) Skin problems

978. Which of the following acts as an anti-oxidant ?
(a) Selenium (b) Iodine
(c) Boron (d) Calcium

979. Bitterness in carrot is due to
(a) Sulphoraphane (b) Isopentanol
(c) Isocoumarin (d) Carotene

980. Tenderometer is used to measure the maturity of
(a) Cowpea (b) Cluster bean
(c) Pea pods (d) French bean

981. The per capita consumption of vegetable/availability of vegetable in India is
(a) 300 g (b) 285 g
(c) 145-175 g (d) 225 g

982. Which of the following are self blanched varieties of cauliflower ?
(a) Pusa Snowball-K (b) Pusa Himjyoti
(c) Pusa Sharadh (d) Pant Shubra

983. Which of the following countries has the highest per capita intake of vegetables ?
(a) Greece (b) Republic of Korea
(c) Turkey (d) India

984. Which of the following is the first resistant variety released in vegetable in India?
(a) Pusa Sawani in okra (b) Pusa Jawala in chilli
(c) Sel. 120 in tomato (d) Annamalai in brinjal

985. Pusa Manjari is an improved variety of
(a) Bittergourd (b) Ridgegourd
(c) Bottlegourd (d) Cucumber

986. Which of the following is first hybrid released in vegetable crop in public sector ?

(a) Pusa Meghdoot in bottle gourd
(b) Pusa Sanyog in cucumber
(c) Pusa Alankar in summer squash
(d) Arka Navneet in brinjal

987. Scooping is a practice that facilitates bolting in

(a) Cabbage (b) Brussels sprout
(c) Cauliflower (d) Broccoli

988. Which of the following are thermo-insensitive varieties of cauliflower ?

(a) Pusa Katki (b) Pusa Synthetic
(c) Pusa Deepali (d) All of these

989. Blanching is practiced in

(a) Okara (b) Asparagus
(c) Tomato (d) Onion

990. Which of the followings are self blanched varieties of cauliflower ?

(a) Pusa Deepali (b) Pusa Snowball – 16
(c) Pusa Snowball K-1 (d) All of these

991. Lincon is an exotic variety of

(a) Cowpea (b) Pea
(c) French bean (d) Broad bean

992. Which of the following radish variety is resistant to rust?

(a) Scarlet Globe (b) White Icicle
(c) French Breakfast (d) Arka Nishant

993. Which of the following is late maturing variety of cauliflower? (JRF-2008)

(a) Pusa Him Jyoti (b) Pusa Synthetic
(c) Pusa Katki (d) Snowball-16

994. Which of the following vegetable require mulching

(a) Potato (b) Tomato
(c) Lettuce (d) Brinjal

995. **Pusa Bedana, Durgapura Meetha, Durgapura Kesar are improved varieties of**
(a) Watermelon (b) Muskmelon
(c) Bottlegourd (d) Bittergourd

996. **Which of the following onion variety is suitable for dehydration ?**
(a) Udaipur – 101 (b) Jobner Green
(c) Udaipur – 102 (d) Pusa Jyoti

997. **Which of the following palak variety is development through spontaneous mutation ?**
(a) Pusa Palak (b) Jobner Green
(c) Pusa Harit (d) Pusa Jyoti

998. **Pusa Sumeet is an improved variety of**
(a) Broad bean (b) Cluster bean
(c) Lima bean (d) Peas

999. **All Green is a variety of** **(JRF-2008)**
(a) Beet leaf (b) Spinach
(c) Methi (d) Broccoli

1000. **Which of the following French Bean varieties is developed through mutation ?**
(a) Pusa Parvathi (b) Arka Komal
(c) Phule Surekha (d) Pant Anupma

1001. **Which of the following physiological disorder is not associated with nutrient deficiency/excess ?**
(a) Catface (b) Blossom end rot
(c) Fruit cracking (d) Flower drop/poor fruit set

1002. **In which of the following crop, TIBA is exogenously applied ?**
(a) Watermelon (b) Cucumber
(c) Musk melon (d) Bottle gourd

1003. **Which of the following is present in cluseter bean ?**
(a) Asparagine (b) Mannogalacton
(c) Glutamine (d) All of these

1004. Which of the following are the methods of seed extraction in tomato ?

(a) Fermentation (b) Alkali treatment
(c) Acid treatment (d) All of these

1005. Which of the following is pole type variety of French bean ?

(a) Kentucky Wonder (b) Giant Stringless
(c) Contender (d) Pant Anupama

1006. Favisim, an illness caused due to an allergy of which plant part of broad bean ?

(a) Green pod (b) Pollen
(c) Root (d) Both (a) and (b)

1007. An important ingredient of Panchkutta vegetable is

(a) Tomato (b) Beans
(c) Cucurbits (d) Brinjal

1008. Brown Heart of turnip is cause by the deficiency of

(a) B (b) Water stress
(c) Ca (d) High temperature

1009. About 95 to 99.5 portion of the plant tissue are made of (JRF-2008)

(a) C, H and O (b) N, P and K
(c) Ca, Mg and C (d) Cu, Zn and Fe

1010. Each beet root fruit (Capsule) contains 2-6 seeds hence requires

(a) Thinning (b) Weeding
(c) Hoeing (d) None of these

1011. Black leg in potato is caused by

(a) Fungus (b) Virus
(c) Bacteria (d) MLOs

1012. Majority of the cucubits are seed propagated except

(a) Pointed gourd (b) Kartoli and Kakrol
(c) Little gourd (d) All of these

1013. *Macrophomina phaseoli*, in potato causes

(a) Charcoal rot (b) Soft rot
(c) Brown rot (d) Wart

1014. Arka Sheetal is an improved variety of
(a) Long melon (b) Snap melon
(c) Water melon (d) Musk melon

1015. Virginia Savoy is a variety of
(a) Spinach (b) Beet
(c) Palak (d) None of these

1016. "Cinkle of potato" is caused by
(a) Fungus (b) Virus
(c) Bacteria (d) MLOs

1017. Headquarter of 'AGMARK' is located at (JRF-2008)
(a) Jaipur (b) Nagpur
(c) Lucknow (d) New Delhi

1018. Which of the following is a serious pest of sweet potato?
(a) Hairy caterpillar (b) Horned caterpillar
(c) Sweet potato weevil (d) Leaf eating catterpillar

1019. Golden nematode (Globodera *Heterodera rostochinenesis*) is serious pest of
(a) Potato (b) Chilli
(c) Tomato (d) MLOs

1020. *Ditylenchus dipsaci* is a serious nematode of
(a) Onion (b) Brinjal
(c) Tomato (d) Chilli

1021. Majority of the cucurbits are
(a) Annual and monoecious (b) Parennial dioecious
(c) Annual and dioecious (d) Perennial and monoecious

1022. "Potato Withes Broom" is caused by
(a) MLOs (b) Bacteria
(c) Fungus (d) Virus

1023. Which of the following is a temperate turnip ?
(a) Pusa Kanchan (b) Pusa Chandrima
(c) Pusa Swati (d) Pusa Swarnima

1024. Splitting in carrot is basically due to
(a) B deficiency (b) Fluctuating moisture imbalance
(c) Ca deficiency (d) High temperature

1025. **Which of the following is an early dwarf but wrinkled seeded variety of pea ?**
(a) Alaska (b) Meteor
(c) Asauji (d) Arkel

1026. **London Flag is improved variety of**
(a) Leek (b) Garlic
(c) Onion (d) Welsh Onion

1027. **Onion variety most suitable for export is**
(a) N-53 (b) Arka Bindu
(c) Arka Kalyan (d) Pusa Red

1028. **Pusa Rohini is the variety of**
(a) Tomato (b) Chilli
(c) Cabbage (d) Brinjal

1029. **Red pumpkin beetle (*Aulacophora foveicollis*) is less damaging in which of the following cucurbits ?**
(a) Pumpkin (b) Bottle gourd
(c) Muskmelon (d) Bitter gourd

1030. **Which of the following cucurbits is less affected by powdery mildew ?**
(a) *Luffa* (b) *Cucumis*
(c) *Lagenaria* (d) *Momordica*

1031. **Which of the following nutrient element effect sex expression in pumpkin ?**
(a) B (b) Ca
(c) Mo (d) Both a and c

1032. **Common methi bears flowers is of which colour ?**
(a) White (b) Red
(c) Pink (d) Yellow

1033. **Which of the following is non-climacteric ?**
(a) Cucumber (b) Tomato
(c) Muskmelon (d) Watermelon

1034. **Maleness in gynoecious line of cucumber is induced by**
(a) Silver nitrate (b) ABA
(c) NAA (d) IAA

1035. Self blanching is the characteristics of which group of cauliflower variety ?

(a) Late (b) Mid season
(c) Early (d) None of these

1036. Optimum temperature for tuber production is

(a) 15 - 20 °C (b) 20 - 25 °C
(c) 10 - 15 °C (d) > 30 °C

1037. Which of the following enzymes converts starch to sugar in potato ?

(a) Starch phosphorylase (b) Both (a) and (c)
(c) Glucose-1-Phosphate (d) None of these

1038. Which of the following is a physiological disorder of potato ?

(a) Phyllody (b) Leaf roll
(c) Purple top wilt (d) Hollow or black heart

1039. Starch content % of sweet potato tuber is

(a) 16 (b) 6
(c) 26 (d) 0.6

1040. Yerrusseri product is prepared from which vegetable ?

(a) Cucumber (b) Pumpkin
(c) Carrot (d) Radish

1041. "Allinase" is an enzyme that converts alliin in garlic into

(a) Alicin (b) Allicine
(c) Allinin (d) Allinine

1042. In sweet potato anti nutritional factor is

(a) Trypsin Inhibitor (b) Cynoglucoside
(c) Coumarin (d) Quercitin

1043. "Blossom end rot" is a physiological disorder of

(a) Tomato (b) Watermelon
(c) Chilli (d) All of these

1044. Flavour of cabbage leaves is due to the glucoside

(a) Cucurbitacin (b) Oleoresin
(c) Sinigrin (d) Eruic acid

1045. Allin is an odourless and colourless

(a) Acid (b) Pigment
(c) Alcohol (d) Amino acid

1046. Frog eye rot is a physiological disorder of

(a) Chilli (b) Potato
(c) Brinjal (d) Tomato

1047. The predominant pigment, responsible for red colour of tomatoes is due to (JRF-2008)

(a) Beta-carotene (b) Anthocyanin
(c) Lycopene (d) Xanthophyll

1048. Optimum seed rate (g/ha) of cauliflower is

(a) 200 - 300 (b) 300 - 500
(c) 700 - 800 (d) 800 - 900

1049. "Pillow" is a physiological disorder of .... caused due to calcium deficiency

(a) Cucumber (b) Radish
(c) Carrot (d) Turnip

1050. Buttoning, Riceyness, Fuziness and Leafy stem are the physiological disorder of

(a) Onion (b) Garlic
(c) Cabbage (d) Cauliflower

1051. "Metsubre" is a disorder of .... due to calcium deficiency

(a) Colocasia (b) Amorphophallus
(c) Yam (d) Potato

1052. Which of the following carrot variety is introduced from Germany and suitable for Nilgiri hills ?

(a) Imperator (b) Zeno
(c) Chanteny (d) Danvers

1053. Which of the following is a flower type observed in brinjal?

(a) Long styled (b) Short styled
(c) Pseudo short styled (d) All of these

1054. Which organ of the chilli fruit is richest in pungency ?

(a) Seed (b) Pericarp
(c) Placental (d) Leaves

1055. Which of the following cauliflower group is highly self incompatible ?

(a) Early maturing Group – I

(b) Mid season Group – II

(c) Late maturing Group – III

(d) None of these

1056. *Solanum acaule*, a potato species is resistant to

(a) Frost (b) Heat

(c) Nematode (d) Water stagnation

1057. Red colour development in tomato is favoured by

(a) Red light (b) Both (a) and (c)

(c) Far red light (d) None of these

1058. "Thermoperiodicity" refers to response of the plant to high day temperature and low temperature rather than constant temperature. It is commonly observed in

(a) Tomato (b) Brinjal

(c) Chilli (d) Okra

1059. Pungency in horse radish is due to

(a) Allyl iso thiocyanates (b) Butyle thiocyanates

(c) Both (a) and (b) (d) None of these

1060. Select the element constituent of amino acid which is essential for protein synthesis (JRF-2008)

(a) Calcium (b) Magnesium

(c) Mangnese (d) Sulphur

1061. Sponge gourd and ridge gourd are

(a) Intercrossable (b) Intersterile

(c) Both a and b (d) None of these

1062. Which of the following source is used for YVMV resistance transfer in Arka Anamika and Arka Abhay ?

(a) *A. tetraphyllus* (b) *A.esculentus*

(c) *A.manihot* (d) *A.manihot* spp.manihot

1063. Which of the following cucurbit groups have the highest chromosome number ?

(a) Pumpkin and squash (b) Cucumber

(c) Gourds (d) Melons

1064. Calcium deficiency causes blossom end rot in (JRF-2008)
(a) Potato (b) Tomato
(c) Cauliflower (d) Cabbage

1065. Which of the following cucurbit is dioecious ?
(a) Pointed gourd (b) Chayote
(c) Little gourd (d) All of these

1066. Which of the following cucurbits exhibits vivipary ?
(a) Cucumber (b) Chow Chow
(c) Pumpkin (d) Muskmelon

1067. The dioecious crop among the following is
(a) Brinjal (b) Sweet potato
(c) Asparagus (d) Ridge gourd

1068. Absence of axillary bud is the typical feature of
(a) Cauliflower (b) Cabbage
(c) Brussels sprout (d) Sprout broccoli

.1069. A nullisomic individual is represented by (JRF-2008)
(a) $2n - 1$ (b) $2n + 2$
(c) $2n - 2$ (d) $2n + 1$

1070. Which of the following cucurbits have the lowest chromosome numbers ?
(a) Cucumber (b) Muskmelon
(c) Watermelon (d) Longmelon

1071. Linkagę between gene affect
(a) Independent assortment (b) Fertilization
(c) Vernalization (d) Anaphase

1072. Seed borne disease of cauliflower is managed by
(a) Formaldehyde (b) Hot water treatment
(c) Methyl bromide (d) Soil sterilization

1073. Arka Jyoti is watermelon while Pusa Jyoti is
(a) Palak (b) Sugarbeet
(c) Amaranth (d) Beetroot

1074. Powdery mildew of pea is caused by
(a) Fungi (b) Bacteria
(c) Virus (d) None

1075. Arka Jyoti is an improved variety of
(a) Muskmelon (b) Bottle gourd
(c) Watermelon (d) Snapmelon

1076. Crossing over during meiosis results in
(a) Promoting linkage (b) Breaking linkage
(c) Help in mutation (d) None

1077. Pusa Harit is a palak while Arka Harit is a
(a) Beetroot (b) Bittergourd
(c) Spinach (d) Lettuce

1078. Pusa Rasraj is a F1 hybrid of **(JRF-2008)**
(a) Tomato (b) Watermelon
(c) Muskmelon (d) Cucumber

1079. Back cross method of breeding means
(a) Mutation breeding
(b) Tissue culture
(c) Polyploidy breeding
(d) Crossing F1 to desirable parents

1080. Arka Suryamukhi is an improved variety of
(a) Summer squash (b) Pumpkin
(c) Winter squash (d) Musk melon

1081. Mass selection is essentially a method of
(a) Hybridization technique (b) Clonal selection
(c) Population improvement (d) Mutation breeding

1082. In heterosis breeding one should capitalize on
(a) Inbreeding depression (b) Homozygosity
(c) Over dominance (d) High seed cost

1083. Optimum night temperature range for fruit set in tomato is : **(JRF-2008)**
(a) 10 - 20 °C (b) 15 - 20 °C
(c) 15 - 25 °C (d) 25 - 30 °C

1084. Which vegetable has high calorific value ?
(a) Fenugreek (b) Potato
(c) Amaranthus (d) Tomato

1085. Arka Vikas is a tomato whereas Pusa Vikas is a
(a) Cucumber (b) Pumpkin
(c) Ridgegourd (d) Snakegourd

1086. Transgenic plants can be produced under
(a) Open field condition (b) Controlled condition
(c) In a small plot (d) None

1087. India is the largest producer, consumer as well as exporter of :
(a) Spices (b) Fruits
(c) Milk (d) Rice

1088. Naveen is a tomato hybrid while Pusa Naveen is a
(a) Bottle gourd (b) Ridge gourd
(c) Sponge gourd (d) Bitter gourd

1089. Which of the following tomato variety has the indeterminate growth habit? (JRF-2008)
(a) Sioux (b) Roma
(c) Pusa Gaurav (d) Pusa Early Dwarf

1090. In cabbage which of the gollowing variety is first tropical variety which can set seed in plains
(a) Pusa Sambandh (b) Pusa Ageti
(c) Pusa Drum Head (d) Golden Acre

1091. Cavity spot in carrot is due to deficiency of
(a) Calcium (b) Boron
(c) Iron (d) Magnesium

1092. The yellow colour in tomato is due to pigment
(a) Anthocyanin (b) Carotene
(c) Lycopene (d) Pro-lycopene

1093. Arka Jeet is a popular cultivar of
(a) Muskmelon (b) Watermelon
(c) Watermelon (d) Bitter gourd

1094. Pusa Beedana, a triploid watermelon was released in
(a) 1970 (b) 1969
(c) 1989 (d) 1972

1095. Symptoms of 'Ca' deficiency in plants is first noticed in (JRF-2008)

(a) Middle leaves (b) Lower leaves
(c) Terminal leaves (d) All leaves simultaneously

1096. Arka Abhay is a variety of

(a) Brinjal (b) Chilli
(c) Tomato (d) Bhindi

1097. Pusa Sheetal is tomato while Arka Sheetal is

(a) Long melon (b) Water melon
(c) Musk melon (d) Snap melon

1098. 'Arkel' Garden pea is the most suitable for

(a) Early season (b) Late season
(c) Mid season (d) None of the above

1099. Bitter taste in brinjal fruit is due to

(a) Anti vitamina E factor (b) Solasodine
(c) CN glycosides (d) Trypsine inhibitors

1100. Pugency in garlic is due to

(a) Diallyl disulphide (b) Isothiocynates
(c) Allyl propyl disulphide (d) Erucic acid

1101. Many fold effects of a gene refers to (JRF-2008)

(a) Penetrance (b) Epistasis
(c) Pleiotrophy (d) None of these

1102. Black heart of potato develops upon storage of tubers at temperature (°C) of

(a) 5 (b) -5
(c) 0 (d) None of the above

1103. Black rot of cauliflower is

(a) Fungal disease (b) Viral disease
(c) Bacterial disease (d) None of the above

1104. Black rot of cabbage is transmitted by

(a) Mechanical means (b) Seed
(c) Vectors (d) None of the above

1105. While Pusa Gaurav is a tomato, Arka Gaurav is

(a) Tomato (b) Capsicum
(c) Brinjal (d) Chili

1106. Brown heart is a physiological disorder of

(a) Radish (b) Turnip
(c) Carrot (d) Beet

1107. Dormancy in potato can be break by use of

(a) Thiourea (b) Ethylene chlorohydrins
(c) Potassium thiocynate (d) All of the above

1108. Cabbage is a cross pollinated primarily owing to

(a) Floral morphology (b) Protogyny
(c) Protandry (d) Self-incompatibility

1109. Favism is a term associated with pollen allergy of

(a) Pea (b) Cow pea
(c) Broad bean (d) French bean

1110. Gherkin, *Cucumis anguria* is a mutant of

(a) *C. moschata* (b) *C. pepo*
(c) *C. longipes* (d) *C. sativus*

1111. Ploidy level of cultivated potato is

(a) Diploid (b) Tetraploid
(c) Triploid (d) Hexaploid

1112. "Sui generis" is a latin phrase which means (JRF-2008)

(a) A new system
(b) A secure system
(c) A system of it's own
(d) A conventional system

1113. Tuberization in potato is favoured by

(a) Short day low temperature
(b) Short day high temperature
(c) Long day low temperature
(d) Long day high temperature

1114. Cluster bean contains mucilaginous substance called

(a) Allantoin (b) Mannogalacton
(c) Glycoprotein (d) None of these

1115. Pant Samrat is brinjal while Samrat is

(a) Potato
(b) Sweet potato
(c) Tapioca
(d) Colocasia

1116. Colour of Pusa Swati variety of turnip is

(a) Red
(b) White
(c) Golden
(d) None of the above

1117. Large sized potato tubers are more prone to

(a) Black heart
(b) Hollow heart
(c) Brown heart
(d) All of the above

1118. Which of the following vegetable is known as 'God Sent vegetable' ? (JRF-2008)

(a) Pea
(b) Tomato
(c) Winged bean
(d) Brinjal

1119. The recommended daily consumption of leafy vegetable is

(a) 50 g
(b) 116 g
(c) 250 g
(d) None of the above

1120. Fruit setting in brinjal usually occurs in the flower having

(a) Long and medium style
(b) Medium style
(c) Short and medium style
(d) Short style

1121. What is the optimum temperature for seed germination in case of pea ?

(a) 8 °C
(b) 10 °C
(c) 20 °C
(d) 30 °C

1122. While Pusa Lal is Sweet potato; Pusa Red is a

(a) Tomato
(b) Onion
(c) Carrot
(d) Tapioca

1123. Variety resistant to cyst nematode in potato is

(a) Kufri Swarna
(b) Kufri Sutlej
(c) Kufri Giriraj
(d) Kufri Ashoka

1124. Garden pea belongs to the sub species

(a) *arvense*
(b) *elatius*
(c) *hortense*
(d) None of the above

1125. Golden nematodes is a serious pest of

(a) Tomato (b) Brinjal
(c) Potato (d) Cabbage

1126. While Arka Komal is French bean, Pusa Komal is

(a) Pea (b) Cowpea
(c) Cluster bean (d) Lima bean

1127. Yellow anther type variety in chilli is

(a) Bhagyalakshmi (b) Jawala Mukhi
(c) Andhra Jyoti (d) Bhaskar

1128. Haploid plants have been produced by anther culture in tomato by

(a) Greshof and Doy (b) Mehta and Upadhaya
(c) Tyagi and Prasad (d) H.B. Singh

1129. Haulms in potato seed crop are cut during

(a) End of December or first week of January
(b) First fortnight of January
(c) Second fortnight of January
(d) None of the above

1130. Hermophrodite cultivar of ridge gourd is

(a) Pusa Nasdar (b) CO1
(c) Satputia (d) Jhinga Torai

1131. Pusa Bindu is brinjal while Arka Bindu is

(a) Tomato (b) Onion
(c) Garlic (d) Cucumber

1132. The variety suitable for both cold and hot areas of the country in tomato is

(a) Pusa Sheetal (b) Pusa Sadhabahar
(c) Pusa Hybrid-1 (d) Arka Alok

1133. Hybrid variety of brinjal most suitable for Bartha making is :

(a) BH - 2 (b) Pusa Anmol
(c) BH - 1 (d) None of the above

1134. Seed rate (kg/ha) of fenugreek is

(a) 5 - 10 (b) 15 - 20
(c) 10 - 15 (d) 20 - 25

1135. Pinching of male buds before anthesis in the female parent in cucurbits was first suggested by

(a) Choudhary & Singh (b) Seshadhri and More
(c) Tyagi and Singh (d) V. Swaroop

1136. In green vegetables, the limiting amino acids is

(a) Methionine (b) Lysine
(c) Arginine (d) Tryptophan

1137. Netted varieties of muskmelon are known as

(a) Honey Dews (b) Casabas
(c) Cantaloupes (d) None of the above

1138. Seed (tuber) rate per hectare of potato is

(a) 1 quintal (b) 5 quintal
(c) 10 quintal (d) 15 quintal

1139. While Arka Pragati is an Onion; Pusa Pragati is

(a) Cluster bean (b) Garden pea
(c) Broad bean (d) Lima bean

1140. In tomato, which chemical is sprayed for fruit setting in adverse condition ?

(a) 2, 4-D (b) PCPA
(c) Etheral (d) Cycocel

1141. Phenolic factor present in onion which has antifungal properties is

(a) Sinigrin (b) Quercetin
(c) Catechol (d) Allylpropyl

1142. In watermelon, the most important indicator of maturity is

(a) Days after pollination
(b) Dull sound when fruit is thumped
(c) Metallic sound when fruit is thumped
(d) None of the above

1143. Which one of the following vegetable crop can tolerate acid soils ? **(JRF-2008)**

(a) Cucumber (b) Bean
(c) Okra (d) Watermelon

1144. Isolation distance for certified seed production in radish is
(a) 1000 m (b) 1600 m
(c) 800 m (d) 200 m

1145. In garden pea, the flower colour is
(a) White (b) Purple
(c) yellow (d) Pink

1146. *Kasuri methi* belongs to the following species
(a) *Foenum-graecum* (b) *Corniculata*
(c) *Oleracea* (d) None of the above

1147. Kufri Badshah variety of potato is resistant to
(a) Viral disease (b) Late blight
(c) Wart (d) None of the above

1148. Leaf curl virus in nature is spread by
(a) Aphids (b) Leafhopper
(c) Whitefly (d) None of the above

1149. Little leaf disease of brinjal is caused by
(a) MLOs (b) Fungi
(c) Bacteria (d) None of the above

1150. Much hoeing is not advisable in carrot as the smell of the leaves attracts
(a) Wire worm (b) Eel worm
(c) Carrot fly (d) Aphids

1151. Onion seeds remain viable for germination for following years ?
(a) 1-2 (b) 2-3
(c) 3-4 (d) 4-5

1152. Little leaf disease of brinjal is spread by
(a) Aphid (b) Leaf hopper
(c) Whitefly (d) None of the above

1153. Optimum time for planting onion bulbs for seed production is :
(a) Septermber (b) October
(c) December-January (d) None of the above

1154. Potato tuber production is maximum at following temperature

(a) 15 °C (b) 20 °C
(c) 25 °C (d) 30 °C

1155. Pea edible podded cultivar is

(a) Arkel (b) Perfection New Line
(c) Bonneville (d) Sylvia

1156. Indian Spinach/Poi/Basella belongs to the family

(a) Chenopodiaceae (b) Leguminosae
(c) Basellaceae (d) None of the above

1157. Poinsette is the most popular variety of

(a) Muskmelon (b) Longmelon
(c) Cucumber (d) None of the above

1158. Pollination in beetroot is by

(a) Insect (b) Water
(c) Wind (d) All of the above

1159. Potato is a .... plant for its tuber production

(a) Short-day (b) Day-neutral
(c) Long-day (d) None

1160. Optimum temperature for curd initiation and development in Snowball cauliflower is

(a) 20 - 27 °C (b) 12 - 16 °C
(c) 20 - 25 °C (d) 10 - 16 °C

1161. Puffiness in tomato is caused by

(a) Viral attack (b) Bacterial attack
(c) Physiological disorder (d) None of the above

1162. Pusa Chikni is a variety of

(a) Sponge gourd (b) Bottle gourd
(c) Ridge gourd (d) All season

1163. Quality of pea is ascertained by

(a) Refractometer (b) Potentiometer
(c) Tenderometer (d) None of the above

1164. Pusa Rasraj is

(a) A pure line variety of muskmelon
(b) A hybrid variety of longmelon

(c) A hybrid variety of muskmelon
(d) A seedless variety of watermelon

1165. Rat-tail radish is grown for its
(a) Long thin pod (b) Shoots
(c) Tender roots (d) None of the above

1166. Round head type of cabbage mature
(a) Early (b) Mid
(c) Late (d) None

1167. Salt tolerant variety of onion
(a) Punjab Selection (b) Nasik Red
(c) No. 404 (d) N - 53

1168. In turnip pithiness resistant variety is
(a) Pusa Kanchan (b) Pusa Swarnima
(c) Pusa Swati (d) Pusa Chandrima

1169. Single fruit of palak contains seeds
(a) 2 – 3 (b) 6 – 7
(c) 4 – 5 (d) 8 – 9

1170. Multigerm seed is found in case of
(a) Onion (b) Carrot
(c) Tomato (d) Beetroot

1171. The most suitable temperature for colour development in carrot is
(a) 10 - 15 °C (b) 15 - 20 °C
(c) 20 - 25 °C (d) None of the above

1172. The bitterness / acridity in colocasia corm is due to
(a) Calcium oxalate (b) Calcium carbonate
(c) Calcium chloride (d) Potassium oxalate

1173. The country where maximum genetic diversity in brinjal occurs is
(a) India (b) Peru and Bolivia
(c) Iran (d) China

1174. The crop with which turnip crosses easily is
(a) Radish (b) Cabbage
(c) Mustard (d) None of these

1175. The immediate precursor for ethylene synthesis is (JRF-2008)

(a) SAM (b) ACC
(c) Polyamine (d) Tryptophan

1176. The most ancient type of garden is

(a) Kitchen garden (b) Vegetable forcing
(c) Market garden (d) None of the above

1177. The most common cucurbit grown under glasshouses

(a) Bottle gourd (b) Cucumber
(c) Watermelon (d) None of the above

1178. The most commonly grown garden beet variety is

(a) Detroit Dard Red (b) Pusa Chandrima
(c) Pusa Kanchan (d) None of the above

1179. Which type of broccoli is most nutritive ?

(a) Purple coloured (b) White coloured
(c) Green coloured (d) None of the above

1180. Quarantine inspection of Agricultural and Horticultural crops is carried out by (JRF-2008)

(a) Botanical syrvey of India - Kolkatta
(b) Forest Research Institute - Dehradun
(c) Indian Agicultural Research Institute - New Delhi
(d) National Bureau of Plant Genetic Resources - New Delhi

1181. The most serious disease of okra

(a) Yellow vein mosaic (b) Powdery mildew
(c) Root rot (d) None of the above

1182. The pigment responsible for red colour in carrot is

(a) Carotene (b) Lycopene
(c) Anthocyanin (d) Quercetin

1183. Tuberization in sweet potato is promoted under (JRF-2005)

(a) 7 hr. light (b) 9 hr. light
(c) 11 hr. light (d) 13 hr. light

1184. Recommended seed rate (g/ha) of cabbage

(a) 300 - 400 g (b) 400 - 500 g
(c) 100 - 200 g (d) 500 - 600 g

1185. Delayed harvesting in radish may cause

(a) Forking (b) Deformed roots
(c) Pithiness (d) All of the above

1186. The primary objective of green house cultivation of vegetables is :

(a) To get early crop
(b) To protect the crop from disease
(c) To get off season crop
(d) To protect the crop from excessive rains

1187. *Trigonella corniculata* (Kasuri methi) produces flowers of following colour

(a) White (b) Purple
(c) Red (d) Orange to yellow

1188. Triploid watermelon variety is

(a) New Hampshire Midget
(b) Charleston Grey
(c) Pusa Bedana
(d) None of the above

1189. Major pollinating agent in cucurbits is (JRF-2005)

(a) Honey bee (b) House fly
(c) Ants (d) Bumble bee

1190. In case of tomato blotchy ripening is due to deficiency of

(a) N (b) P
(c) K (d) Ca

1191. Early cauliflower is ready for harvesting in the period of

(a) Mid-September to mid-November
(b) Mid-January to April
(c) Mid-November to mid-January
(d) None of the above

1192. Type of vegetable gardening followed on Dal lake of the Kashmir Valley is

(a) Floating garden (b) Kitchen garden
(c) Hydroponics (d) None of the above

1193. Variety of bottle gourd without 'crook neck' is
(a) Pusa Summer Prolific Long
(b) Pusa Summer Prolific Round
(c) Arka Bahar
(d) Pusa Meghdoot

1194. Which of the following is used to check sprouting of onion under storage ?
(a) NAA (b) GA
(c) MH (d) PCPA

1195. Which of the following seed rate of TPS is recommended for planting one ha area ?
(a) 150 g (b) 250 g
(c) 350 g (d) 450 g

1196. Which one ot tne following is a major disease of onion in India ?
(a) Leaf blight (b) Purple blotch
(c) Smut (d) White rot

1197. Which one of the following soil types is best for vegetable cultivation ?
(a) Sandy (b) Clay
(c) Sandy loam (d) Clay loam

1198. Which one of the following vegetables produce maximum seeds per fruit ?
(a) Tomato (b) Chilli
(c) Brinjal (d) Potato

1199. Curing of harvested sweet potato tubers takes place if the tubers are kept for 10 days at
(a) 20°C and 90% RH (b) 40°C and 70% RH
(c) 80°C and 30% RH (d) None of the above

1200. Pumpkin grown throughout India is (JRF-2005)
(a) *C. moshata* (b) *C. maxima*
(c) *C. ficifolia* (d) *C. paradisi*

1201. In .... the Fruit Product Order (FPO) was passed.
(a) 1950 (b) 1955
(c) 1960 (d) 1965

1202. A phenomenon in which a gene has more than one phenotypic effect is **(JRF, 2002)**

(a) Polymorphism (b) Penetrance
(c) Expressively (d) Pleiotropism

1203. Test cross is crossing F1 with **(JRF-2006)**

(a) Dominant parent (b) Recessive parent
(c) Either of the parents (d) None of these

1204. Single gene male sterility is transferred in .... years

(a) 3-4 (b) 5-6
(c) 7-8 (d) 4-5

1205. Self pollination in tomato is due to **(JRF-2009)**

(a) Allogamy (b) Cleistogamy
(c) Chasmogamy (d) Autogamy

1206. Which type of tomato are referred as "self topping" ?

(a) Determinate (b) Intermediate
(c) Indeterminate (d) None of these

1207. Which growth regular is useful for fruit set at high temperature in tomato ?

(a) Auxin (b) Ethylene
(c) Gibberllins (d) ABA

1208. Japanese White is a variety of **(JRF-2005)**

(a) Radish (b) Carrot
(c) Turnip (d) None

1209. Which crops have staminal sterility ?

(a) Tomato (b) Brinjal
(c) Cabbage (d) Cucumber (gynoecious line)

1210. Growth regulator which improves fruit set in tomato is **(JRF-2005)**

(a) 2, 4-D (b) GA
(c) MH (d) Ethephon

❑❑❑❑❑

ANSWERS

1	2	3	4	5	6	7	8	9	10
(b)	(b)	(b)	(a)	(a)	(b)	(a)	(a)	(c)	(d)
11	12	13	14	15	16	17	18	19	20
(b)	(c)	(a)	(d)	(b)	(b)	(c)	(c)	(c)	(d)
21	22	23	24	25	26	27	28	29	30
(b)	(d)	(a)	(a)	(d)	(b)	(a)	(b)	(d)	(b)
31	32	33	34	35	36	37	38	39	40
(d)	(c)	(c)	(a)	(d)	(c)	(d)	(c)	(b)	(c)
41	42	43	44	45	46	47	48	49	50
(c)	(b)	(a)	(b)	(c)	(c)	(c)	(a)	(c)	(b)
51	52	53	54	55	56	57	58	59	60
(c)	(d)	(a)	(a)	(c)	(a)	(b)	(b)	(a)	(b)
61	62	63	64	65	66	67	68	69	70
(a)	(c)	(c)	(b)	(c)	(*)	(c)	(a)	(d)	(b)
71	72	73	74	75	76	77	78	79	80
(b)	(d)	(a)	(b)	(a)	(a)	(c)	(a)	(d)	(b)
81	82	83	84	85	86	87	88	89	90
(c)	(a)	(b)	(d)	(b)	(b)	(b)	(a)	(d)	(a)
91	92	93	94	95	96	97	98	99	100
(d)	(a)	(a)	(b)	(c)	(a)	(c)	(b)	(b)	(a)
101	102	103	104	105	106	107	108	109	110
(c)	(b)	(c)	(b)	(b)	(a)	(c)	(a)	(c)	(d)
111	112	113	114	115	116	117	118	119	120
(b)	(b)	(c)	(a)	(b)	(a)	(b)	(a)	(a)	(a)
121	122	123	124	125	126	127	128	129	130
(a)	(d)	(c)	(b)	(a)	(d)	(a)	(c)	(c)	(d)
131	132	133	134	135	136	137	138	139	140
(a)	(d)	(c)	(b)	(c)	(a)	(a)	(a)	(a)	(b)
141	142	143	144	145	146	147	148	149	150
(c)	(d)	(c)	(a)	(d)	(a)	(b)	(a)	(a)	(a)

151	152	153	154	155	156	157	158	159	160
(c)	(b)	(a)	(c)	(d)	(c)	(d)	(a)	(c)	(c)
161	162	163	164	165	166	167	168	169	170
(b)	(b)	(d)	(b)	(a)	(c)	(a)	(a)	(b)	(a)
171	172	173	174	175	176	177	178	179	180
(c)	(c)	(b)	(c)	(a)	(c)	(a)	(c)	(a)	(c)
181	182	183	184	185	186	187	188	189	190
(a)	(b)	(a)	(b)	(c)	(c)	(a)	(a)	(b)	(c)
191	192	193	194	195	196	197	198	199	200
(a)	(a)	(a)	(d)	(b)	(d)	(d)	(a)	(a)	(b)
201	202	203	204	205	206	207	208	209	210
(c)	(a)	(b)	(c)	(a)	(a)	(a)	(a)	(c)	(b)
211	212	213	214	215	216	217	218	219	220
(b)	(b)	(d)	(b)	(b)	(a)	(a)	(b)	(a)	(a)
221	222	223	224	225	226	227	228	229	230
(d)	(a)	(b)	(a)	(b)	(d)	(a)	(c)	(b)	(d)
231	232	233	234	235	236	237	238	239	240
(a)	(b)	(a)	(a)	(b)	(a)	(c)	(c)	(b)	(c)
241	242	243	244	245	246	247	248	249	250
(b)	(b)	(b)	(d)	(b)	(c)	(c)	(b)	(a)	(d)
251	252	253	254	255	256	257	258	259	260
(c)	(a)	(b)	(b)	(a)	(d)	(a)	(b)	(d)	(a)
261	262	263	264	265	266	267	268	269	270
(b)	(c)	(a)	(c)	(b)	(c)	(d)	(b)	(d)	(a)
271	272	273	274	275	276	277	278	279	280
(b)	(b)	(d)	(a)	(b)	(a)	(a)	(c)	(b)	(b)
281	282	283	284	285	286	287	288	289	290
(a)	(c)	(d)	(b)	(b)	(b)	(a)	(d)	(b)	(a)
291	292	293	294	295	296	297	298	299	300
(b)	(d)	(b)	(d)	(b)	(b)	(a)	(a)	(a)	(a)
301	302	303	304	305	306	307	308	309	310
(a)	(d)	(b)	(b)	(b)	(c)	(a)	(d)	(d)	(d)

311	312	313	314	315	316	317	318	319	320
(e)	(b)	(d)	(a)	(a)	(a)	(a)	(a)	(a)	(a)
321	322	323	324	325	326	327	328	329	330
(a)	(b)	(b)	(d)	(a)	(b)	(b)	(a)	(c)	(b)
331	332	333	334	335	336	337	338	339	340
(a)	(a)	(d)	(b)	(a)	(b)	(a)	(c)	(b)	(b)
341	342	343	344	345	346	347	348	349	350
(a)	(b)	(a)	(b)	(c)	(a)	(a)	(d)	(c)	(a)
351	352	353	354	355	356	357	358	359	360
(a)	(a)	(a)	(b)	(b)	(c)	(a)	(d)	(c)	(d)
361	362	363	364	365	366	367	368	369	370
(d)	(d)	(a)	(c)	(c)	(b)	(a)	(c)	(d)	(a)
371	372	373	374	375	376	377	378	379	380
(a)	(d)	(c)	(c)	(d)	(c)	(a)	(d)	(a)	(a)
381	382	383	384	385	386	387	388	389	390
(c)	(c)	(d)	(d)	(d)	(d)	(b)	(a)	(c)	(b)
391	392	393	394	395	396	397	398	399	400
(b)	(b)	(b)	(b)	(b)	(a)	(a)	(c)	(a)	(d)
401	402	403	404	405	406	407	408	409	410
(a)	(b)	(d)	(d)	(d)	(d)	(d)	(c)	(d)	(c)
411	412	413	414	415	416	417	418	419	420
(a)	(d)	(d)	(c)	(b)	(b)	(b)	(c)	(a)	(a)
421	422	423	424	425	426	427	428	429	430
(b)	(a)	(d)	(b)	(a)	(b)	(b)	(b)	(a)	(b)
431	432	433	434	435	436	437	438	439	440
(a)	(a)	(a)	(b)	(a)	(c)	(a)	(b)	(a)	(d)
441	442	443	444	445	446	447	448	449	450
(b)	(d)	(a)	(b)	(a)	(c)	(c)	(a)	(a)	(d)
451	452	453	454	455	456	457	458	459	460
(c)	(d)	(d)	(b)	(a)	(c)	(a)	(a)	(c)	(d)
461	462	463	464	465	466	467	468	469	470
(d)	(a)	(b)	(c)	(c)	(d)	(c)	(d)	(d)	(c)

471	472	473	474	475	476	477	478	479	480
(d)	(b)	(d)	(a)	(b)	(c)	(b)	(b)	(b)	(b)
481	482	483	484	485	486	487	488	489	490
(b)	(b)	(c)	(c)	(c)	(c)	(c)	(a)	(b)	(d)
491	492	493	494	495	496	497	498	499	500
(a)	(d)	(a)	(b)	(b)	(c)	(b)	(c)	(a)	(c)
501	502	503	504	505	506	507	508	509	510
(c)	(d)	(b)	(d)	(b)	(b)	(a)	(b)	(a)	(b)
511	512	513	514	515	516	517	518	519	520
(d)	(d)	(c)	(d)	(a)	(d)	(a)	(a)	(b)	(a)
521	522	523	524	525	526	527	528	529	530
(c)	(b)	(d)	(b)	(d)	(c)	(d)	(a)	(b)	(d)
531	532	533	534	535	536	537	538	539	540
(d)	(a)	(c)	(a)	(c)	(d)	(c)	(b)	(b)	(a)
541	542	543	544	545	546	547	548	549	550
(a)	(d)	(d)	(d)	(a)	(b)	(a)	(a)	(d)	(b)
551	552	553	554	555	556	557	558	559	560
(d)	(b)	(d)	(a)	(b)	(c)	(b)	(d)	(a)	(d)
561	562	563	564	565	566	567	568	569	570
(d)	(b)	(d)	(a)	(b)	(b)	(d)	(a)	(c)	(a)
571	572	573	574	575	576	577	578	579	580
(a)	(b)	(c)	(a)	(a)	(c)	(d)	(a)	(c)	(a)
581	582	583	584	585	586	587	588	589	590
(c)	(d)	(c)	(c)	(b)	(d)	(b)	(c)	(a)	(a)
591	592	593	594	595	596	597	598	599	600
(c)	(b)	(d)	(d)	(d)	(a)	(a)	(b)	(b)	(a)
601	602	603	604	605	606	607	608	609	610
(c)	(b)	(c)	(d)	(c)	(a)	(b)	(b)	(c)	(b)
611	612	613	614	615	616	617	618	619	620
(a)	(b)	(d)	(a)	(b)	(d)	(b)	(a)	(d)	(d)
621	622	623	624	625	626	627	628	629	630
(d)	(b)	(c)	(a)	(a)	(a)	(b)	(b)	(a)	(b)

631	632	633	634	635	636	637	638	639	640
(b)	(c)	(b)	(b)	(b)	(c)	(a)	(a)	(a)	(c)
641	642	643	644	645	646	647	648	649	650
(c)	(a)	(b)	(b)	(a)	(b)	(c)	(b)	(a)	(a)
651	652	653	654	655	656	657	658	659	660
(b)	(c)	(b)	(c)	(a)	(c)	(a)	(d)	(b)	(b)
661	662	663	664	665	666	667	668	669	670
(a)	(c)	(a)	(b)	(a)	(c)	(b)	(d)	(d)	(d)
671	672	673	674	675	676	677	678	679	680
(a)	(d)	(b)	(b)	(b)	(d)	(b)	(c)	(d)	(c)
681	682	683	684	685	686	687	688	689	690
(b)	(a)	(d)	(b)	(b)	(a)	(a)	(c)	(c)	(b)
691	692	693	694	695	696	697	698	699	700
(d)	(c)	(c)	(c)	(b)	(a)	(b)	(b)	(c)	(c)
701	702	703	704	705	706	707	708	709	710
(a)	(a)	(a)	(c)	(c)	(d)	(a)	(c)	(b)	(c)
711	712	713	714	715	716	717	718	719	720
(d)	(b)	(c)	(c)	(a)	(b)	(d)	(b)	(c)	(c)
721	722	723	724	725	726	727	728	729	730
(d)	(a)	(d)	(c)	(d)	(a)	(d)	(d)	(b)	(c)
731	732	733	734	735	736	737	738	739	740
(b)	(b)	(b)	(b)	(c)	(b)	(b)	(d)	(b)	(d)
741	742	743	744	745	746	747	748	749	750
(b)	(a)	(a)	(b)	(c)	(c)	(a)	(c)	(d)	(b)
751	752	753	754	755	756	757	758	759	760
(d)	(b)	(a)	(c)	(c)	(b)	(b)	(b)	(c)	(b)
761	762	763	764	765	766	767	768	769	770
(b)	(c)	(a)	(d)	(a)	(b)	(b)	(b)	(d)	(d)
771	772	773	774	775	776	777	778	779	780
(b)	(a)	(b)	(c)	(d)	(a)	(c)	(c)	(c)	(a)
781	782	783	784	785	786	787	788	789	790
(a)	(b)	(a)	(d)	(a)	(a)	(d)	(c)	(d)	(a)

791	792	793	794	795	796	797	798	799	800
(c)	(a)	(a)	(a)	(a)	(b)	(a)	(a)	(b)	(a)
801	802	803	804	805	806	807	808	809	810
(d)	(a)	(d)	(d)	(c)	(b)	(d)	(c)	(d)	(b)
811	812	813	814	815	816	817	818	819	820
(b)	(c)	(d)	(a)	(c)	(b)	(d)	(b)	(a)	(a)
821	822	823	824	825	826	827	828	829	830
(a)	(a)	(b)	(b)	(a)	(c)	(d)	(b)	(b)	(d)
831	832	833	834	835	836	837	838	839	840
(b)	(b)	(c)	(b)	(b)	(b)	(b)	(b)	(d)	(a)
841	842	843	844	845	846	847	848	849	850
(a)	(d)	(b)	(b)	(c)	(a)	(a)	(b)	(b)	(a)
851	852	853	854	855	856	857	858	859	860
(c)	(c)	(c)	(d)	(a)	(c)	(c)	(b)	(a)	(a)
861	862	863	864	865	866	867	868	869	870
(d)	(c)	(c)	(d)	(c)	(b)	(c)	(d)	(b)	(c)
871	872	873	874	875	876	877	878	879	880
(a)	(b)	(a)	(c)	(c)	(b)	(a)	(d)	(a)	(a)
881	882	883	884	885	886	887	888	889	890
(b)	(a)	(a)	(a)	(a)	(a)	(a)	(d)	(b)	(d)
891	892	893	894	895	896	897	898	899	900
(c)	(b)	(d)	(a)	(c)	(a)	(c)	(b)	(a)	(b)
901	902	903	904	905	906	907	908	909	910
(d)	(a)	(a)	(d)	(d)	(a)	(a)	(b)	(d)	(b)
911	912	913	914	915	916	917	918	919	920
(a)	(b)	(d)	(d)	(c)	(b)	(a)	(d)	(d)	(d)
921	922	923	924	925	926	927	928	929	930
(a)	(b)	(c)	(d)	(a)	(d)	(b)	(d)	(d)	(d)
931	932	933	934	935	936	937	938	939	940
(d)	(b)	(a)	(c)	(b)	(b)	(b)	(c)	(d)	(a)
941	942	943	944	945	946	947	948	949	950
(c)	(b)	(a)	(a)	(b)	(b)	(a)	(b)	(c)	(d)

951	952	953	954	955	956	957	958	959	960
(a)	(c)	(a)	(a)	(a)	(a)	(a)	(a)	(a)	(b)
961	962	963	964	965	966	967	968	969	970
(a)	(a)	(d)	(a)	(c)	(a)	(b)	(d)	(a)	(c)
971	972	973	974	975	976	977	978	979	980
(a)	(c)	(d)	(d)	(c)	(a)	(a)	(a)	(c)	(c)
981	982	983	984	985	986	987	988	989	990
(c)	(a)	(a)	(a)	(c)	(a)	(c)	(d)	(b)	(d)
991	992	993	994	995	996	997	998	999	1000
(b)	(d)	(d)	(c)	(a)	(c)	(b)	(a)	(a)	(a)
1001	1002	1003	1004	1005	1006	1007	1008	1009	1010
(a)	(a)	(d)	(d)	(d)	(d)	(c)	(a)	(a)	(a)
1011	1012	1013	1014	1015	1016	1017	1018	1019	1020
(c)	(d)	(a)	(a)	(a)	(b)	(b)	(c)	(a)	(a)
1021	1022	1023	1024	1025	1026	1027	1028	1029	1030
(a)	(a)	(d)	(b)	(d)	(a)	(b)	(a)	(d)	(d)
1031	1032	1033	1034	1035	1036	1037	1038	1039	1040
(d)	(a)	(a)	(a)	(a)	(a)	(c)	(d)	(a)	(b)
1041	1042	1043	1044	1045	1046	1047	1048	1049	1050
(a)	(a)	(a)	(c)	(d)	(a)	(c)	(b)	(a)	(d)
1051	1052	1053	1054	1055	1056	1057	1058	1059	1060
(a)	(b)	(d)	(c)	(a)	(a)	(a)	(a)	(a)	(b)
1061	1062	1063	1064	1065	1066	1067	1068	1069	1070
(a)	(a)	(a)	(b)	(d)	(b)	(c)	(a)	(c)	(a)
1071	1072	1073	1074	1075	1076	1077	1078	1079	1080
(a)	(b)	(a)	(a)	(c)	(b)	(b)	(c)	(d)	(c)
1081	1082	1083	1084	1085	1086	1087	1088	1089	1090
(c)	(c)	(b)	(b)	(b)	(b)	(a)	(a)	(a)	(b)
1091	1092	1093	1094	1095	1096	1097	1098	1099	1100
(a)	(b)	(a)	(d)	(b)	(d)	(a)	(a)	(b)	(a)
1101	1102	1103	1104	1105	1106	1107	1108	1109	1110
(c)	(c)	(c)	(b)	(b)	(b)	(d)	(d)	(c)	(c)

1111	1112	1113	1114	1115	1116	1117	1118	1119	1120
(b)	(d)	(a)	(b)	(b)	(b)	(b)	(c)	(b)	(a)
1121	1122	1123	1124	1125	1126	1127	1128	1129	1130
(c)	(b)	(a)	(c)	(c)	(b)	(d)	(a)	(a)	(c)
1131	1132	1133	1134	1135	1136	1137	1138	1139	1140
(b)	(b)	(a)	(d)	(a)	(a)	(c)	(c)	(b)	(b)
1141	1142	1143	1144	1145	1146	1147	1148	1149	1150
(c)	(b)	(d)	(a)	(a)	(b)	(b)	(c)	(a)	(c)
1151	1152	1153	1154	1155	1156	1157	1158	1159	1160
(c)	(b)	(b)	(b)	(d)	(c)	(c)	(c)	(a)	(a)
1161	1162	1163	1164	1165	1166	1167	1168	1169	1170
(c)	(a)	(c)	(c)	(a)	(c)	(a)	(a)	(a)	(d)
1171	1172	1173	1174	1175	1176	1177	1178	1179	1180
(b)	(a)	(a)	(c)	(*)	(a)	(b)	(a)	(c)	(d)
1181	1182	1183	1184	1185	1186	1187	1188	1189	1190
(a)	(c)	(c)	(b)	(c)	(c)	(a)	(c)	(a)	(c)
1191	1192	1193	1194	1195	1196	1197	1198	1199	1200
(a)	(a)	(c)	(c)	(a)	(b)	(c)	(a)	(c)	(a)
1201	1202	1203	1204	1205	1206	1207	1208	1209	1210
(b)	(d)	(b)	(b)	(c)	(a)	(a)	(a)	(d)	(a)

****

Nutritive Value of Different Vegetables (per 100g edible portion on fresh weight basis)

S. no.	Name of Vegetable	Moisture (g)	Carbohy-drates (g)	Protein (g)	Fat (g)	Calorie (energy)	Vit. A (IU)	Thiamine (mg)	Ribofl-avin (mg)	Ascorbic acid (mg)	Calcium (mg)	Iron (mg)	Phosph-orus (mg)
1.	Potato	74.7	22.6	1.6	0.1	97	40	0.40	0.04	17.0	10.0	0.7	35.0
2.	Tomato (green)	93.1	3.6	1.9	0.1	23	307	0.07	0.01	31.0	20.0	1.8	-
3.	Tomato (ripe)	94.0	3.6	1.2	0.1	20	302	0.12	0.06	27.0	48.0	0.4	26.0
4.	Chilli	85.7	3.0	2.9	0.6	29	292	0.19	0.39	111.0	30.0	1.2	80.0
5.	Sweet pepper	92.4	4.3	1.3	0.3	24	683	0.06	0.03	175.0	10.0	1.2	-
6.	Brinjal	92.7	4.0	1.4	0.3	24	118	0.04	0.11	12.0	18.0	0.9	47.0
7.	Cabbage	92.4	5.3	1.4	0.2	29	80	0.06	0.05	100.0	46.0	0.8	38.0
8.	Cauliflower	91.7	4.9	2.4	0.2	31	70	0.04	0.03	75.0	30.0	17.0	76.0
9.	Knolkhol	90.1	6.7	2.1	0.1	36	20	0.05	0.10	50.0	20.0	0.4	60.0
10.	Broccoli	89.9	5.5	3.3	0.2	37	3500	0.05	0.12	137.0	80.0	0.8	79.0
11.	Brussels sprouts	84.9	8.9	4.4	0.5	58	400	0.05	0.13	105.0	40.0	1.6	80.0
12.	Kale	86.6	7.2	3.9	0.6	50	7540	0.05	0.16	115.0	85.0	1.6	62.0
13.	Wax gourd	96.5	1.9	0.4	0.1	10	0	0.06	0.01	1.0	30.0	0.08	20.0
14.	Bitter gourd (large)	92.4	4.2	1.6	0.2	25	210	0.07	0.09	88.0	20.0	1.8	70.0
15.	Bitter gourd (small)	83.2	10.6	2.1	1.0	60	210	0.07	0.06	96.0	23.0	2.0	38.0
16.	Bottle gourd	96.1	2.5	0.2	0.1	12	0	0.03	0.01	6.0	20.0	0.7	10.0
17.	Cucumber	96.1	2.5	0.4	0.1	13	0	0.03	0.01	7.0	10.0	1.5	25.0
18.	Sponge gourd	93.2	2.9	0.5	0.1	18	120	0.02	0.06	0.0	36.0	1.1	19.0
19.	Ridge gourd	95.2	3.4	0.5	0.1	17	56	0.01	0.01	5.0	18.0	0.5	26.0
20.	Snake gourd	94.6	3.3	0.5	0.3	18	160	0.04	0.06	0.0	26.0	0.3	20.0

21.	Pointed gourd	92.0	2.2	2.0	0.3	20	255	0.05	0.06	29.0	30.0	1.7	40.0
22.	Giant spine gourd	84.1	7.7	3.1	1.0	52	2592	0.05	0.18	0.0	33.0	4.6	42.0
23.	Ivy gourd	93.5	3.1	1.2	0.1	18	249	0.07	0.08	15.0	40.0	1.4	30.0
24.	Pumpkin (ripe)	86.0	4.6	1.4	0.1	25	2180	0.06	0.04	2.0	10.0	0.7	30.0
25.	Summer squash	94.8	3.5	1.0	0.1	19	260	0.05	0.03	18.0	10.0	0.6	30.0
26.	Winter squash (ripe)	86.0	6.6	1.1	0.2	32	3300	0.03	0.05	6.0	14.2	0.4	20.9
27.	Muskmelon	95.2	3.5	0.3	0.2	17	3420	0.11	0.08	26.0	32.0	1.4	14.0
28.	Watermelon	95.8	3.3	0.2	0.2	16	590	0.02	0.04	1.0	11.0	7.9	12.0
29.	Roundmelon	93.5	3.4	1.4	0.2	21	23	0.04	0.08	18.0	25.0	0.9	24.0
30.	Snapmelon	95.7	3.0	0.3	0.1	14	265	-	-	10.0	-	-	-
31.	Chowchow	92.5	5.7	0.7	0.1	27	8	0.01	0.04	4.0	140.0	0.6	30.0
32.	Radish (white)	94.4	3.4	0.7	0.1	17	50	0.06	0.02	15.0	50.0	0.04	22.0
33.	Radish (pink)	90.8	6.8	0.6	0.3	32	50	0.06	0.02	17.0	50.0	0.5	20.0
34.	Carrot	82.2	10.6	0.9	0.2	48	12000	0.04	0.02	3.0	48.0	0.6	30.0
35.	Beet	87.7	8.8	1.7	0.1	43	18	0.04	0.09	10.0	28.0	1.0	55.0
36.	Parsnip	78.6	18.2	1.5	0.5	78	0	0.08	0.12	18.0	57.0	0.6	80.0
37.	Turnip	91.6	6.2	0.5	0.2	28	4	0.04	0.04	43.0	30.0	0.4	40.0
38.	Rutabaga	89.1	8.9	1.1	0.1	38	330	0.07	0.08	36.0	55.0	1.0	41.0
39.	Onion	86.8	11.0	1.2	0.2	50	35	0.08	0.01	11.0	180.0	0.7	50.0
40.	Garlic	62.8	29.0	6.3	0.1	142	10	0.16	0.23	13.0	30.0	1.3	310.0
41.	Leek	78.9	5.0	1.8	0.1	28	30	0.23	-	11.0	-	2.3	70.0
42.	Shallot	-	-	2.6	-	-	-	0.06	0.02	1.0	37.0	1.3	60.0
43.	Lettuce	93.4	2.5	2.1	0.3	21	540	0.09	0.13	10.0	50.0	2.4	28.0
44.	Celery	88.0	1.6	6.3	0.6	38	0	0.01	0.11	7.0	50.0	0.6	38.0

45.	Sorrel	95.2	1.4	1.6	0.3	14	-	0.03	0.06	12.0	63.0	8.7	-
46.	Sweet corn	89.0	7.9	1.4	0.3	41	150	0.06	0.05	4.4	3.5	0.2	46.0
47.	Okra	89.6	6.4	1.9	0.2	35	88	0.07	0.10	13.0	66.0	1.5	56.0
48.	Pea	72.0	15.8	7.2	0.1	93	300	0.25	0.01	19.0	20.0	1.5	139.0
49.	French bean	91.4	4.5	1.7	0.1	25	321	0.08	0.06	16.0	50.0	1.7	28.0
50.	Cowpea	84.6	8.0	4.3	0.2	51	941	0.07	0.09	13.0	80.0	2.5	74.0
51.	Hycinth bean	86.1	6.7	3.8	0.7	48	312	0.1	0.06	9.0	210.0	1.7	68.0
52.	Cluster bean	81.0	10.8	3.2	0.4	59	316	0.09	0.09	47.0	130.0	5.0	50.0
53.	Winged bean	91.0	3.8	2.9	0.1	27	595	0.06	0.12	37.0	300.0	1.7	69.0
54.	Broad bean	85.4	7.2	4.5	0.1	48	14	0.08	-	12.0	50.0	1.4	-
55.	Lima bean	66.5	23.5	7.5	0.8	128	280	0.21	0.11	32.0	63.0	2.3	158.0
56.	Sweet potato	70.0	27.0	1.5	0.2	115	2800	0.08	0.04	24.0	46.0	0.8	49.0
57.	Taro corm	93.1	21.1	3.0	0.1	97	166	0.09	0.03	0.0	40.0	1.7	140.0
58.	Giant taro corm	81.2	16.9	0.6	0.1	71	-	0.10	0.02	0.0	50.0	0.5	46.0
59.	Tannia corm	75.4	21.0	2.2	0.1	93	-	0.12	0.04	8.0	34.0	1.2	62.0
60.	Elephant-foot yam	78.5	18.4	2.0	0.1	82	434	0.06	0.02	6.0	38.0	2.4	38.0
61.	Yam root	73.0	22.0	2.7	0.3	109	-	0.09	0.03	5.0	40.0	1.7	-
62.	Cassava	65.5	32.4	1.2	0.2	135	80	0.05	0.04	34.0	26.0	0.9	32.0
63.	Chinese potato	77.6	19.7	1.3	0.1	85	-	-	-	-	-	-	-
64.	Asparagus	91.7	2.9	1.7	0.2	20	762	0.12	0.13	25.0	15.8	0.7	46.9
65.	Globe artichoke	83.7	11.9	2.9	0.4	63	390	0.18	0.05	10.0	47.0	1.9	94.0
66.	Drumstick pod	86.9	3.7	2.5	0.1	25	176	0.05	0.07	120.0	30.0	3.3	110.0
67.	Jackfruit (unripe)	84.0	9.4	2.6	0.3	50	0	0.05	0.04	14.0	30.0	1.7	-
68.	Green banana	83.2	14.0	1.4	0.2	63	48	0.05	0.02	24.0	10.0	0.6	29.0

69. Green papaya	92.0	5.7	0.7	0.2	27	0	0.01	0.01	12.0	28.0	0.9	-
70. Green mango	87.5	10.1	0.7	0.1	44	144	0.04	0.01	3.0	10.0	-	-
71. Amaranth	85.7	6.3	4.0	0.5	45	9200	0.03	0.10	99.0	397.0	25.5	83.0
72. Palak	86.4	6.5	3.4	0.8	46	9770	0.26	0.56	70.0	380.0	16.2	30.0
73. Spinach	92.1	2.9	2.0	0.7	27	9300	0.03	0.07	28.0	73.0	10.9	21.0
74. Indian spinach	90.8	4.2	2.8	0.4	31	3250	0.03	0.16	87.0	200.0	1.0	35.0
75. Water spinach	92.4	-	1.9	-	-	4800	-	-	58.0	90.0	4.8	-
76. Fenugreek leaves	86.1	6.0	4.4	0.9	49	3744	0.05	-	54.0	360.0	17.2	51.0
77. Mustard leaves	89.9	3.2	4.0	0.6	34	4195	0.03	-	33.0	155.0	16.3	26.0
78. Bengal gram leaves	73.4	14.1	7.0	1.4	97	1564	0.09	0.10	61.0	340.0	23.8	-
79. Bottle gourd leaves	87.9	6.1	2.3	0.7	40	-	-	-	-	80.0	-	59.0
80. Pumpkin leaves	81.9	7.9	4.6	0.8	57	5760	-	-	-	392.0	-	112.0
81. Pointed gourd leaves	80.5	5.8	5.4	1.1	54	-	-	-	-	531.0	-	73.0
82. Coriander leaves	86.3	6.3	3.3	0.6	44	11168	0.50	0.06	135.0	184.0	18.5	-
83. Pigweed	89.6	2.9	3.7	0.4	30	2784	0.01	0.14	35.0	150.0	4.2	-
84. Portulaca	90.5	2.7	2.4	0.6	26	3664	0.10	0.22	29.0	111.0	14.8	-
85. Curry leaves	63.8	18.7	6.1	1.0	36	12096	0.08	0.01	4.0	830.0	7.0	57.0
86. Radish top	90.3	3.4	2.7	0.6	30	18	0.03	0.16	103.0	310.0	16.1	60.0
87. Carrot top	83.9	8.3	5.1	0.5	58	-	-	-	-	340.0	8.3	110.0
88. Beet top	90.4	5.6	2.0	0.3	27	6700	0.08	0.18	35.0	118.0	3.1	45.0
89. Agathi leaves	73.1	11.8	8.4	1.4	93	8640	0.21	0.09	58.0	404.0	5.0	-
90. Mint	84.9	5.8	4.8	0.6	48	2592	0.05	0.20	27.0	200.0	15.6	-
91. Ceylon spinach	93.6	4.0	1.6	-	22	5600	-	-	86.0	106.0	1.6	43.2
92. Swiss chard	91.8	4.4	1.6	0.2	21	3110	0.03	0.09	34.0	110.0	3.6	29.0

93. Cowpea leaves	89.0	4.1	3.4	0.7	20	9715	0.05	0.18	4.0	290.0	20.1	-
94. Drumstick leaves	75.9	12.5	6.7	1.7	92	10848	0.06	0.05	220.0	440.0	7.0	70.0
95. Chekkurmanis	73.6	11.6	6.8	3.2	102	9510	0.48	0.32	247.0	570.0	28.0	200.0
96. Onion scape	87.6	8.9	0.9	0.2	41	952	0.00	0.03	17.0	50.0	7.5	-
97. Green onion	87.6	10.6	1.0	0.2	45	50	0.03	0.04	24.0	135.0	0.9	24.0
98. Taro petioles	93.8	4.6	0.2	0.2	20	335	0.01	0.02	8.0	57.0	1.4	23.0
99. Chicory	94.2	2.9	1.6	0.3	21	10000	0.05	0.20	15.0	18.0	0.7	21.0
100. Chives	86.0	7.8	3.8	0.6	52	500	0.12	-	70.0	48.0	8.4	57.0
101. Water cress	93.6	3.3	1.7	0.3	18	4720	0.08	0.16	77.0	195.0	2.0	46.0
102. Endive	93.3	4.0	1.6	0.2	20	3000	0.07	0.12	11.0	79.0	1.7	56.0
103. Parsley	83.9	9.0	3.7	1.0	50	8230	0.11	0.28	193.0	193.0	4.3	84.0
104. Banana flower	89.9	5.1	1.7	0.7	33	43	0.05	0.02	16.0	32.0	1.6	-
105. Pumpkin flower	89.1	5.8	2.2	0.8	39	-	-	-	-	120.0	-	60.0
106. Agathi flower	87.4	-	1.8	0.6	-	-	0.13	-	41.0	-	-	-
107. Lotus root	85.9	11.3	1.7	0.1	53	-	0.10	-	22.0	21.0	0.4	-
108. Mushrooms	91.1	4.0	2.4	0.3	16	0	0.10	0.44	5.0	9.0	1.0	115.0

Note:

0.6 ug B carotene = 1 IU vitamin A

\- = data not available

1 µg = 10^{-6} gram

Eight independent Vavilovian centre of origin

S. No.	Centre of origin	Region
1.	The Chinese Centre	Mountainous regions of central and western China together with adjacent low lands.
2.	The Indian (Hindustani) centre	Indian subcontinent (dominated by India, the area also includes Bangladesh, Nepal, Sri Lanka and Bhutan) including Burma (now Myanmar).
2a.	The Indo-Malay centre	All of Malay archipelago (part of Malaysia which hangs off the corner of mainland southeast Asia stretching from the foot hills of the Himalayas in the north to the equator in the south), the Philippines and Indo-China. It is supplementary to Indian centre of origin.
3.	The Central Asiatic centre	Western Punjab (Now in Pakistan), northern part of Pakistan, all of Afghanistan, the Tadjik and Uzbk of former Soviet Republics (Now separate country Tajikistan, Uzbekistan) and Western Tien shan (China).
4.	The Near East Centre	Interior of Asia Minor (Syria, Lebanon, Turkey), high land of Turkmenistan, Iran and the whole Transcaucasia (a region to the south of the Caucasus mountains between the black sea and the Caspian, and comprising the present-day republics of Georgia, Armenia and Azerbaijan).
5.	The Mediterranean Centre	Israel, Jordon, Syria, Cyprus, Turkey, Lybia, Lebanon
6.	The Abyssinian centre	Ethiopia of Africa
7.	South Meican and Central American centre	Southern part of Mexico, small countries of central America which lie within a neck of land that joins north and south America (Guatemala, El Salvador, Costa Rica, Panama, Nicaragua, Houdurus, Haiti, Trinidad and Tobago, Cuba, etc.)
8.	The South American Centre	Mountainous areas and plateaus of Columbia, Ecuador, Peru, Paraguay, Bolivia and Chilli

Centres of Diversity of Major Vegetable Crops

Gene centre	Primary centre of origin & domestication	Secondary centre
Chinese-Japanese	Eggplant, wax gourd, Chinese cabbage, welsh onion	Watermelon, amaranth
Indo-Chinese	Sponge ground, ridge gourd, bottle gourd, sword bean, winged bean, taro, chayote, cucumber, yams, bitter gourd	Chinese cabbage, bottle gourd, yam bean, amaranth, cucumber, yard long bean, shallot
Hindustani Centre	Eggplant, wax gourd, cucumber, ridge gourd, sponge gourd, bitter gourd, hyacinth bean, drumstick	Watermelon, melon, bottle gourd, amaranth, okra, vegetable cowpea
Central Asia	Onion, garlic, carot, spinach, Brassica spp.	Eggplant, watermelon, melon, cauliflower
Near Eatst	Onion, garlic, leek, Brassica spp.	Okra
Mediterranean	Cabbage, cauliflower, broccoli, radish, beans	Sweet pepper, garlic, okra, pepper, onion, lima bean
African	Eggplant, watermelon, melon, bottle gourd, cowpea, okra	Amaranth, shallot
European	Lettuce	Onion, cauliflower, cabbage, carrot, common bean, spinach
Central-American & Mexican region	Tomato, hot pepper, pumpkin, squashes, sweet potato, amaranth	
Sourth American	Tomato, hot pepper, cassava, Xanthosoma, pumpkin and squashes, lima bean chayote.	Common bean Tomato, eggplant, melon, watermelon, squashes, onion, lettuce, lima bean

Variability of major vegetable crops in different agro-ecological regions of India

S. No.	Agro-ecological region	Geographical ranges	Variability in major crops
1	Humid Western Himalayan Region	Jammu and Kashmir, Himachal Pradesh and hills of Uttar Pradesh	Cucurbits, radish, carrot, turnip, peas, cowpea, brinjal, okra, spinach, fenugreek, amaranth, *Solanum khasianum, S. hirsutum, Sechium edule, Basella rubra.*
2	Humid Bengal/ Assam Basin	West Bengal and Assam	Cucurbits, radish, cowpea, brinjal, okra, spinach beet, amaranth, *Abelmoschus manihot, S. indicum, S. indicum, S. khasianum, S. surattense, Cucumis sativus var. sikkimensis, Edgeria darjelingensis, Melothria assamica, Momordica cochinchinensis, Sechium edule, Tuladiantha cardifolia, Basella rubra.*
3	Humid Eastern Himalayan Region	Arunachal Pradesh, Nagaland, Manipur, Mizoram, Tripura, Meghalaya and Andaman & Nicobar Islands	Cucurbits, radish, peas, cowpea, brinjal, okra, spinach, beet, amaranth, *A. manihot spp. Tetraphyllus, S. khasianum, S. torvum, S. sisymbriifolium, S. forex, S. vervascifolium, Cucumis histrix, Luffa echinata, Sechium edule.*
4	Sub-humid Sutlej Ganga Alluvial Plants	Punjab, Plains of Uttar Pradesh and Bihar	Cucurbits, radish, carrot, peas brinjal, okra, spinach, beet, fenugreek, onion, garlic, *A. manihot spp. Tetraphyllus var. pungens, A. tuberculatus, S. indicum, S. torvum, S. khasianum, S. surrattense, S. hispidum, Cucumis hardwickii, C. trigonus.*

5	Humid Eastern and Sourth Eastern Uplands	East Madhya Pradesh, Orissa and Andhra Pradeshj	Cucurbits, radish, carrot, cowpea, chillies, brinjal, okra spinach, amaranth, garlic, *A. manihot* var. *manihot, Solanum surattense, S. torvum.*
6	Arid Western Plains	Haryana, Rajasthan and Gujarat	Cucurbits, cauliflower, radish, carrot, peas, cowpea, chillies, brinjal, okra, spinach beet, fenugreek, onion, garlic, *A. tuberculatus, A. ficulneus, A. manihot ssp. Tetraphyllus, A. tuberculatus, Solanum surattense, S. nigrum, Citrullus colocynthis*
7	Semi-Arid Lava Plateau and Central Highlands	Maharashtra and West Madhya Pradesh	Cucurbits, cauliflower, radish, carrot, cowpea, chillies, brinjal, okra, spinach beet, fenugreek, amaranth, onion, *S. surattense, S. torvum, S. nigrum, S. khasianum, Cucumis setosus, Luffa acutangula* var. *acutangula*
8	Humid to Semi-Arid Western Ghats and Kanataka Plateau	Karnataka, Tamil Nadu, Kerala and Lakshadweep Islands	Cucurbits, chillies, brinjal, okra, amaranth, *A. crunitus, A. angulosus, A. ficulneus, A. moschatus, A. manihot* var. *tetraphyllus, Solanum trilobatum, S. indicum, S. insamum, S. pubescens, S. surattense, S. torvum, Luffa acutangula var. acutangula, Melothria angulata, Basella rubra.*

Classification of vegetables based on nutrient content

Nutrients	Sources
Calories	Immature seeds of lima bean, broad bean and peas, tapioca, sweet potato, yams, colocasia (corms), potato, garlic, brussel's sprouts and onion
Protein	Peas, double bean, winged bean, garlic, brussel's sprouts, cowpea, lima bean, lima bean seeds, amaranth leaves, drumstick leaves, and mathi.
Precursor of Vitamin-A (β-Carotene)	Carrot, spinach, turnip green, palak, mustard green, amaranth, coriander, colocasia leaves, sweet potato (yellow), pumpkin (yellow) and tomato
Vitamin-B complex	Peas, broad bean, lima bean, garlic, asparagus, colocasia corms and tomato
Vitamin-C (Ascorbic acid)	Turnip green, green chillies, brussel's sprouts, mustard green and amranths, coriander, drumstick leaves, cauliflower, knoll khol, spinach, cabbage, bitter gourd and radish leaves.
Calcium	Curry leaves, amaranth leaves, drumstick leaves, methi, turnip green, mustard green, coraider and palak.
Iron	Drumstick leaves and fruits, amaranth, methi, mint, coriander, spinach, palak and musgard green leaves.
Roughages	Spinach, lettuce, cabbage, amaranth and root vegetables.
Vegetable milk	Pea pods and cabbage leaves.

Classification of vegetables based on tenderness

Tender Crops	Very tender crops	
Cowpea	Cucumber	Bitter gourd
New Zealand spinach	Brinjal	Watermelon
French bean	Limba bean	Ridge gourd
Soybean	Cluster bean	Muskmelon
Sweet corn	Winged bean	Cassava
Tomato	Amaranth	Pointed gourd
Sweet potato	Hyacinth bean	Indian spinach
Sweet pepper	Okra	Elephant – foot yam
Summer squash	Chilli	
Pumpkin	Bottle gourd	

Classification of vegetables based on Respiratory activity

Respiratory activity of the produce	Crop
(i) Very low	Potato, onion
(ii) Low	Cabbage, sweet potato, turnip, cucumber
(iii) Moderate	Chilli, carrot, tomato, beet, sweet pepper
(iv) High	Hyacinth bean, French bean, peas, lettuce, lima bean, radish
(v) Very high	Green onion, muskmelon, watermelon, cauliflower, okra, leafy greens, broccoli, asparagus, parsley

Classification of vegetable crops based on their Temperature Requirement

Monthly mean temperature (^{0}C) (day and night)			Vegetable crops
Min.	**Max.**	**Optimum**	
Group: Hot			
18.0	35.0	25.0-27.0	Okra, watermelon, muskmelon, chilli, sweet potato, martynia, yam, cassava, amaranth, winged bean, cluster bean
Group: Warm			
10.0	35.0	20.0-25.0	Tomato, sweet pepper, brinjal, lima bean, French bean, hyacinth bean, cowpea, sweet corn, palak, Newzealand spinach, cucumber, pumpkin, pointed gourd, bottle gourd, bitter gourd, ridge gourd, snake gourd, summer squash, drumstick, Indian spinach, portulaca, elephant foot yam
Group: Cool-hot			
5.0	30.0	20.0-25.0	Onion, leek, garlic, shallot, chicory, chives, salsify, scorzonera, pak-choi, globe artichoke, taro, Indian cauliflower.
Group: Cool-warm			
5.0	25.0	18.0-25.0	Pea, broad bean, cauliflower, Chinese cabbage, cabbage, broccoli, Brussels sprouts, Kale, Knolkhol, turnip, radish, carrot, beet, rutabaga, horse radish, parsnip, potato, parsley, fennel, dil, Swiss chard, spinach, lettuce, celery, celeriac, asparagus, garden cress, mustard, cardoon.

Soil Temperature Requirement for Seed Germination

Crop	Soil Temperature			
	Minimum	Optimum	Optimum range	Maximum
Tomato	10.0	15.5-25.0	25.0	30.0
Brinjal	15.5	24.0-32.0	28.0	35.0
Chilli	15.5	18.0-35.0	29.0	35.0
Potato	10.0	15.0-25.0	25.0	30.0
Cabbage	4.0	10.0-30.0	20.0	37.8
Cauliflower	4.0	10.0-30.0	20.0	37.8
Radish	4.0	7.0-32.0	28.0	35.0
Carrot	4.0	7.0-26. 0	24.0	35.0
Turnip	4.0	15.0-35.0	29.0	40.0
Beet	4.0	10.0-29.0	26.0	35.0
Parsnip	1.6	10.0-21.0	18.0	29.0
Asparagus	10.0	25.0-30.0	30.0	35.0
Muskmelon	15.5	24.0-35.0	30.0	37.8
Cucumber	15.5	15.5-35.0	30.0	40.5
Watermelon	15.5	21.0-35.0	30.0	40.5
Summer squash	15.5	21.0-35.0	30.0	37.8
Pumpkin	15.5	21.0-32.0	30.0	37.8
Bitter gourd	15.5	20.0-35.5	32.0	37.8
Ridge gourd	15.5	21.0-35.5	30.0	40.5
Sweet corn	10.0	15.5-35.0	35.0	40.5
Okra	17.0	21.0-35.0	29.0	40.5
Pea	4.4	4.4-24.0	22.0	29.4
French bean	15.5	15.5-29.0	26.0	35.0
Lima bean	15.5	18.0-29.0	26.0	29.0
Cowpea	15.5	15.5-29.0	28.5	40.5
Hyacinth bean	15.5	15.5-29.0	28.0	35.0
Cluster bean	15.5	15.5-35.0	29.0	35.0
Palak	10.0	15.0-30.0	25.0	35.0
Amaranth	10.0	15.0-35.0	32.0	35.0
Spinach	1.6	7.2-23.9	21.1	29.4
Celery	4.4	15.5-21.1	21.1	29.
Swiss chard	4.4	10.0-29.4	29.4	35.0
Lettuce	1.6	4.4-26.6	23.9	29.4
Parsley	4.4	10.0-29.4	23.9	32.2
Onion	1.6	10.0-35.0	23.9	35.0
Drumstick	15.0	20.0-35.0	2?.0	40.0

Classification of Cool Season Vegetable Crops

Hardy Crops		Half hardy crops	
Broad bean	Brussels sprouts	Salsify	Asparagus
Broccoli	Chives	Carrot	Rutabaga
Cabbage	Garlic	Cauliflower	Beet
Collards	Kale	Celery	
Horse radish	Leek	Chard	
Knolkhol	Onion	Chicory	
Mustard	Peas	Chinese cabbage	
Parsley	Rhubarb	Globe artichoke	
Radish	Turnip	Endive	
Spinach	Pak-choi	Parsnip	
Sorrel		Potato	

Classification of vegetable crops based on their Relative Tolerance to Soil Acidity

Less tolerant (pH 6.8-6.0)	Moderately tolerant (pH 6.8-5.5)	Very tolerant (pH 6.8-5.0)
Asparagus	French bean	Chicory
Beet	Lima bean	Dendelion
Broccoli	Brussels sprouts	Endive
Cabbage	Carrot	Fennel
Cauliflower	Collard	Potato
Celery	Cucumber	Rhubarb
Palak	Brinjal	Shallot
Swiss chard	Garlic	Sorrel
Cress	Horse radish	Sweet potato
Chinese cabbage	Kale	Watermelon
Leek	Knolkhol	
Onion	Parsley	
Lettuce	Pea	
Muskmelon	Chilli	
New Zealand spinach	Pumpkin	
Okra	Radish	
Onion	Rutabaga	
Orach	Summer squash	
Parsnip	Winter squash	
Salsify	Tomato	
Spinach	Turnip	
Water cress		

Classification of vegetable crops based on their water requirement

High	Moderate	Low	Very Low
Palak Amaranth Lettuce Celery Other leafy greens Sweet pepper Cabbage Cauliflower Asparagus Broccoli Rhubarb Brussels sprouts Horse radish Radish Beet Turnip Ridge gourd Rutabaga Green onion	Onion Cucumber Chilli Brinjal Tomato Carrot Potato	Pea French bean Cowpea Broad bean Cluster bean Winged bean Hyacinth bean	Watermelon Muskmelon Pumpkin Wax gourd

*Note: The water requirement is expressed as the units of absorbed water required for the production of one unit of dry matter.

Classification of vegetable crops based on seed longevity in normal storage condition

Longevity	Vegetable Seeds
Up to 1 year	Onion, French bean
Up to 2 years	Okra, carrot, chilli, sweet pepper, cauliflower, cabbage, broccoli
Up to 3 years	Radish, beet, peas, ridge gourd, turnip, palak
Up to 4 years	Tomato, brinjal, pumpkin, summer squash, bottle gourd, bitter gourd, watermelon, muskmelon, cucumber, spinach

Critical Stages of Irrigation in Some Vegetable Crops

Crop	Stages of growth
Tomato	Flower development, fruitset and after each harvest
Brinjal	Same
Chilli	Tenth leaf to flowering, to fruiting and after periodical harvests
Potato	Stolon formation, tuberization and tuber enlargement
Pumpkin, Cucumber, Watermelon, Gourds, Muskmelon, Summer Squash, etc.	Flower bud development and early fruit development
Cabbage	Head formation and enlargement
Cauliflower	Throughout the whole vegetation period
Broccoli	Early stages of development
Onion	Bulb formation and enlargement
Garlic	Bulbing
Radish, Carrot, Beet, Tumip	Constant water supply during the whole vegetation period
Pea	Beginning of flower bud development and pod filling
French bean, Hyacinth bean	Flowering and pod development
Cowpea	Prior to flowering and after podset
Lettuce (head type)	Head formation
Celery	Throughout the whole vegetation period
Sweet potato	40-50 days after planting at tuber formation
Okra	After fruitset
Leafy vegetables	During the whole vegetation period
Elephant-foot yam	Shoot emergence

* Growth stage(s) when water stress reduces the yield most.

Classification of vegetable crops based on their Root Depth

Very shallow rooted (15-30 cm)	Shallow-rooted (30-60 cm)	Moderately deep-rooted (60-90 cm)	Deep-rooted (90-120 cm)	Very deep-rooted (120-180 cm)
Onion	Broccoli	French bean	Pea	Artichoke
Lettuce	Brussels sprouts	Beet	Chilli	Asparagus
Small radish	Cabbage	Carrot	Rutabaga	Lima bean
	Cauliflower	Chard	Summer squash	Parsnip
	Celery	Cucumber	Turnip	Pumpkin
	Chinese cabbage	Brinjal		Winter squash
	Endive	Muskmelon		Sweet potato
	Garlic			Tomato
	Leek			
	Parsley			
	Potato			
	Palak			
	Spinach			
	Radish			
	Cowpea			

Types for variation in chromosome number of vegetables

Type	Number	Description
Euploids		
Monoploids	n	Genetically hemizygous
Diploids	2n	Two homologous sets
Polyploids		
Triploids	3n	Three homologous sets
Tetraploids	4n	Four homologous sets
Hexaploid	6n	Sex homologous sets
Aneuploids		
Hypoploids	2n – x	Loss of one or more chromosomes
Nullisomics	2n – 2	Loss of homologous pair
Monosomic	2n – 1	Loss one member of pair
Double monosomic	2n – 1 – 1	Loss of two non-homologous chromosomes
Hyperploids	2n + x	Addition of one or more chromosomes
Trisomic	2n + 1	Addition of one chromosome of a pair
Tetrasomic	2n + 2	Addition of two chromosomes of a pair
Double tetrasomic	2n + 2 + 2	Addition of two chromosomes of two sepa rate pair

Symptoms of Nutrient Deficiency

Nutrient	Plant symptoms	Occurrence
Nitrogen	The deficiency symptoms are mostly generalized over the whole plant; the deficiency results in decrease in carbohydrate synthesis and loss in chlorophyll; slow and stunted growth; stem thin, erect and hard; leaves smaller than the normal; premature yellowing of lower leaves and new leaves are light green in colôur; in acute deficiency flowering is greatly reduced. Some specific nitrogen deficiency symptoms are yellow lower leaves with purple veins and light green young leaves, (tomato); yellow lower leaves, light green young leaves and slender upright stem (potato); older leaves orange to purple pigmented and young leaves pale green (cole crops); leaves pale green and turning reddish to purple (beet); leaves light green followed by yellowing (carrot and radish); stem hard and leaves yellowish green to yellow (cucurbits); stiff, upright, short and light green leaves (onion); small, sickly and pale green leaves (okra); yellowish green leaves with reddish tinged areas (sweet potato); yellow older leaves and pale green young leaves (celery and lettuce).	Occur in light soil where leaching is more due to heavy rainfall.
Phosphorus	The deficiency symptoms are mostly generalized over the whole plant; the deficiency causes delay in starch development, accumulation of sugars and consequent development of anthocyanin pigment; early growth gets restricted with poor root development; stem thin and shortened; mature leaves have characteristic dark to blue-green colouration and often purple colouration is developed; the purple colouration first appears on underside of leaves and later throughout; lower leaves sometimes turn yellow or greenish brown in colour; delayed flowering and maturity. Some specific phosphorus deficiency symptoms are olive-green leaves with purple colouration in the undersurface (tomato); forward rolling of leaves with marginal scorch (potato); dull green leaves with purple colouration in the undersurface (cole crops); small dull purple	Occur in acid soil (pH below 6.0) due to fixation of soluble phosphates. It also occurs in cold and moist soils.

	leaves (beet); dull green leaves with purple colouration in lower surface (carrot, radish); dark green leaves with dropping of older leaves (pea, bean); tip die back with wilting of older leaves (onion); dull green leaves with purple tinge in the lower surface (lettuce).	
Potassium	Symptoms of potassium deficiency progressively appear on younger leaves; slow and stunted growth with weak stem; in older leaves interveinal chlorosis near the margins occur followed by the development of grey or tan areas near the leaf margin and eventually a scorch appears around the entire leaf margin; leaf edges may become twisted or curled; chlorotic areas may develop throughout the leaf; necrotic areas may develop at the tip of the plant; potassium deficiency interferes the absorption of water and nutrient from soil, so the plant looks withered. Some specific potassium deficiency symptoms are bronze or grayish green lower leaves with tip necrosis starting from tip (potato); yellowing of older leaves with browning at margin (cabbage, cauliflower); leaves bluish green with surface crinkling (beet); leaves bluish green with pale yellow to brown margin (radish); dull green leaves followed by browning (carrot); leaves bluish green near veins followed by browning and necrosis at margins (cucurbits); chlorotic leaflet with necrotic brown areas at margins (beans); mottled leaflet followed by necrosis (cowpea); young leaves at margins (pea); young leaves brown-yellow followed by scorching (okra); yellowing and drying of older leaves from tip (onion); chlorosis followed by necrosis at margin in older leaves (sweet potato); dark green leaves followed by curling and browning (celery).	Occur on light soil where leaching is high due to heavy rainfall.
Calcium	The young leaves are affected first which often become small and get distorted and twisted; tip of the innermost unfolding leaves get twisted, gelatinized and finally die back at tips and margin; hooking appearance of the dry leaves of terminal bud is the characteristic symptom; leaf blades may be whitish green and sometimes brown to light brown spots appear on the leaves; stem becomes weak and brittle; root tip dies and root growth is	Occur on acidic and highly leached sandy soil. It also occurs on very dry soil and soil with very high potassium and nitrogen level.

	restricted; specific deficiency disorder includes blossom end rot of tomato, sweet pepper and watermelon; cavity in root phloem immediately below the epidermis in carrot and parsnip; black heart of celery; brown heart of endive; tip burn of lettuce and internal breakdown of cabbage. Other symptoms manifested in the foliage are: young leaves turn yellow, brown or purple, roll upwards at margins and become necrotic (tomato); small, pale green leaves which roll upwards (potato); leaves rolled up at margins and become discoloured (cabbage, cauliflower); leaves turn pale green around margins which curl upward (beet); leaves chlorotic and scorched (carrot); narrow white band at leaf margin (radish); chlorotic older leaves and curled young leaves (pea); young leaves light green and older leaves show some reddish areas (sweet potato).	
Magnesium	The symptom begins with a loss of green colour at the tips of the lower leaves between the veins; as the deficiency develop, larger portions of the lower leaves are affected, and the leaves further up in the plant become bleached out and turn whitish in colour; leaf becomes lifeless in texture, small in size and tints of orange, red and purple colour may develop unevenly; leaves may curve upwards at margin; older leaves may fall; stem become thin and brittle; maturity gets delayed. Some specific magnesium deficiency symptoms are interveinal chlorosis and necrosis in older leaves (tomato, brinjal, chilli); interveinal chlorosis and puckering of older leves (cabbage); chlorotic leaves with reddish tint between veins (beet); pale green leaves with light yellow or brown spots at tips (carrot); interveinal chlorosis in older leaves (radish); interveinal yellow spots followed by chlorosis in leaves (okra); leaves die back at tips (onion); browning of leaf tips (pea).	Occur on acidic soils, on highly leached light soils and on soils having high potassium levels.
Sulphur	Characteristic uniform yellowish green appearance of the younger leaves that resembles nitrogen starvation; in severe cases, the older leaves may turn pale green; the stem and petiole become brittle and collapse; shoot growth gets restricted; stem becomes stiff, woody and smaller in diameter; the terminal bud	Occur in the soil having low organic matter; also occur due to excessive leaching on light soils.

	commonly remains alive and necrotic spots may or may not be present over the leaf.	
Iron	Typical symptoms are distinctive green vein colour and yellow intrveinal areas on the young leaves; points and margins of the leaves keep their green colour for long time; in severe cases, the entire leaf, veins and interveinal area turn yellow which sometimes become bleached.	Occur on soil with pH above 6.8; on soil which contains considerable sodium and calcium; commonly occur on calcareous soil; due to excess phosphorus application.
Copper	Yellowing of young leaves; leaves may become elongated and tips may curl; leaf edges may become ragged; plant top may wither; in onion, bulbs become soft with thin pale yellow scales.	Occur on soils with 5 per cent or higher organic matter content and on peat soil; it may also occur on heavily cropped sandy soil with acidic reaction and on alkaline soil.
Zine	Deficiency symptoms mostly appear on the 2nd or 3rd fully mature leaves from the top of the plants; leaves become mottled or chlorotic and thick: rosette growth due to shortened internodes. Characteristic symptoms include small reddish brown spots on cotyledonary leaves of bean; green and yellow broad striping at the base of the leaves of corn; interveinal yellowing with marginal burning in beet; 'fernleaf' disorder in potato; tomato leaves become smaller, chlorotic and inwardly curved; onion leaves turn yellow; whole plant turns yellow in garlic.	Occur in neutral to alkaline sandy soils; also occur due to heavy phosphorus fertilization; occur in calcareous soils after leaching and erosin and acid-leached soils.
Manganese	Chlorosis between the veins and appearance of chlorotic and necrotic spots in the interveinal areas; grayish areas may appear near the base of the younger leaves; leave become smaller; typical symptoms include densely red foliage in beet; narrow, yellow striping in the leaves of onion and corn; chlorosis followed by necrosis and yellowing in the leaves of bean: small, yellowish leaves with interveinal yellow mottling in cabbage; interveinal chlorosis in pea and chlorotic and forwardly rolled leaves in tomato.	Occur on soil with pH above 6.7; it may also occur in sandy, peat and muck soil; often occur in calcareous soils.

Boron	Breakdown of the growing tip resulting in death of shoot tip; rosette appearance of the plant due to shortening of terminal growth; young leaves of terminal buds are light green, thick, curled and brittle; root growth gets restricted; flowers become barren. Some typical symptoms include browning of cauliflower; hollow stem in cole crops, particularly in cabbage; black heart of beet and internal browning of turnip due to beakdown of internal tissues; cracked stem of celery; stem resetting and open fruit locule in tomato.	Occur on soil with pH above 6.8; it may occur due to high soil phosphate levels; occur in crops having high boron requirement; occur on acidic soil where native boron has been leached; also occur on coarse textured soil, peats and mucks; water stressed conditions and overlimed acid soil.
Molybdenum	Impairment of growth; deformation of shoots, older leaves show interveinal yellowing with necrotic margins; leaves may crinkle as found in cole crops; in legume crops, the symptoms are indentical with those of nitrogen deficiency. Typical symptoms include whiptail leaf and small, open, loose curd in cauliflower; bleached, scorched and mottled older leaves in cabbage; pale, upwardly rolled leaves with diffused marginal and interveinal in tomato; decomposition and granulation in veins of chilli; leaf chlorosis with prominent red vein in beet; marginal yellowing followed by inward red colouration in carrot; leaf cupping; chlorosis, mottling and whiptail in turnip; discoloured leaf and death of growing tip in radish; yellow or white veins in leaves of pea; pale green leaves with interveinal mottling and scorching in bean; deep blue leaves with conspicuous yellow and green mottling in onion.	Occur on very acid and sandy soil having pH below 5.2 and on highly leached sandy soil.

Names and Structure of Some Plant Growth Regulators Used in Vegetable Production

Abbreviation or Common Name	Chemical Name or Structure	Major uses
ABA	Abscisic acid	Defoliant
Chiormequat (CCC)	2-chioroethyl-trimethyl ammonium chloride	LP, SR
3-CPA	2-(3-chlorophenoxy) propionic acid	FT
4-CPA	4-chlorophenoxy-acetic acid	FT
2,4-D	(2,4-dichlorophenoxy) acetic acid	Herbs.
Daminozide (B 9, Alar)	Succinic acid-2, 2-demethyl hydrazine	GR
Ethephon	2-chloroethyl phosphonic acid	SR
GA_3	Gibberellic acid	GE
Kinetin	6-furfuryl aminopurine	DB
IAA	Indole-3-acetic acid	Enlarge.
IBA	Indole-3-butyric acid	Enlarge.
Maleic hydrazide	1,2-dihydro-3, 6-pyridazinedione	Herb., GR
Mepiquat chloride	N,N-dinethyl piperidinium chloride	GR
NAA	1-naphthaleneacetic acid	FT, GR
NOXA BNOA	2-naphthalenyloxyacetic acid	GS
Paclobutrazol	(2RS, 3RS)-1-(4-chlorophenyl)-4, 4-dimethyl-2-(1 ,2,4-triazol-1-yl)-pentan-3-ol	GR
2,4,5-T	(2,4,5-trichiorophenoxy) acetic acid	Herb.
TIBA	2,3,5-triiodo-benzoic acid	GR
Uniconazol	(E)-(4-chlorophenyl)-4, 4-dimethyl-2-(1,2,4-triazol-1 -yl)-1-penten-3-ol	GR, Fung.

Abbreviations: DB-Dormancy breaker; Enlarg-PIant cell enlarger; FT-Fruit thinner; Fung-Fungicide; GR-Growth retardant; GS-Growth stimulant; Herb -Herbicide; LP-Lodging preventor; GE-Grape enlarger; SR-Sugarcane ripener.

Mode of application and concentration of plant growth regulators in vegetable production.

S. No.	Growth Regulators	Conc-(mg/L)	Method of applica--tions	Crops	Attributes affected
1	Cycocel (CCC)	250-500	Foliar Sprays	Cucurbits Tomato, Okra	Flowering, sex-expression, fruit yield.
		250	Seed treatment	Okra	Resistant to salt. Fruiting and yield.
		250	Seedling Treatment	Tomato	Resistant to virus, Fruiting and yield.
2.	Isopropl-N	5000	Dust or tuber dip	Potato	Storage at room temperature.
3	p-Chlophenoxy acetic acid	50	Foliar spray	Tomato	Fruit set, fruiting and yield.
4	2.4-Dichloro-phenoxy acetic acid	0.5	Foliar spray	Tomato	Fruit set, fruiting and yield
5	Ethephon (CEPA)	100-500	Foliar spray	Cucurbit, Okra, Tomato	Flowering , Fruiting, sex-expression and yield.
		1000	Pre-harvest	Tomato, Chillies	Fruit ripening , earliness and yield
		2000	Post-harvest	Tomato, Chillies	Fruit ripening
6	Ethylene chlorohydrin	(11/20g)	Vapour treatment	Potato	Breaking dormancy
7	Gibberellic acid (GA)	1	Tuber dip	Potato	Breaking dormancy
		10	Foliar sprays	Watermelon, tomato	Sex expression, fruiting and yield
		40-100	Seed or foliar	Okra, Tomato Brinjal	Seed germination fruit set and yield.
8	Indole acetic acid (IAA)	10-15	Seed or Foliar	Spinach, Fenugreek, Okra, Tomato, Brinjal, Cowpea, Onion	Seed germination fruit set and yield
9	Indolebutyric acid (IBA)	0.2	Seedling roots dip	Cabbage, Cauliflower	Seedlings growth and yield.

		25-100	Foliar sprays	Lettuce, Chinese Cabbage	Growth and yield
		250	Foliar Sprays	Cow Pea	Fruit set and yield
10	Maleic hydrazide (MH)	25	Foliar sprays	Peas	Growth and yield
		50-150	Foliar sprays	Cucurbits	Flowering sex-expression and yield
		2500-3000	Foliar Sprays	Onion, Garlic	Reducing storage losses
11	Methyl ester of NAA	5000	Foliar Sprays	Potato	Improving shelf life
12	Naphthalene acetic acid (NAA)	0.2	Seedling root dip	Cauliflower Cabbage, Tomato, Brinjal Onion, Chillies	Seedling set, growth and yield
		10-20	Foliar sprays	Tomato, Chillies	Flower drop, fruit set and yield
		25-30	Seed/Foliar	Okra, Tomato Brinjal, Onion Cucurbits	Seed germination growth and yield
		100	Foliar sprays	Cabbage, Cauliflower	Plant growth and yield
14	Napthoxyacetic acid (NOA)	25-100	Seed/Foliar	Tomato Okra	Germination , growth and yield
15	2,4,5-Trichloro-phenoxyacetic acid (2,4,5-T)	75-125	Foliar sprays	Potato	Improving shelf life
16	Silver nitrate	500	Foliar sprays	Cucumber	Induction of male flowers in gynoeceous lines
17	Silver thio Sulphate	400	Foliar sprays	Muskmelon	Induction of male flowers in gynoeceous lines
18	2,3,5-Tri-iodobenzoic acid (TIBA)	25-50	Foliar sprays	Cucurbits	Flowering, sex-expression and yield
		700	Foliar sprays	Soybean	Fruit set and yield
19	Triacontantol	2	Foliar sprays	Chillies, Peas	Fruit set and yield
20	Thiourea	100	Tuber dip	Potato	Breaking dormancy

Plant parts to be sampled and Growth stage of the crop for Plant Tissue Nutrient Analysis

Crop	Plant parts to be sampled	Stage of growth
Potato	Fourth to sixth leaf from growing tip	Early growth (35-40 days after planting)
Tomato	Fourth to sixth leaf from growing tip	Early bloom
Chilli and Sweet pepper	Young mature leaves	Early fruit set
Cauliflower	Young mature outside leaves	Button stage of curd
Cabbage	First mature leaf from central whorl	Prior to heading
Broccoli	Young mature leaves	First bud formation stage
Brussels sprouts	Young mature leaves	Midgrowth
Watermelon	Mature leaves near the base	Early growth
Muskmelon, Cucumber, Pumpkin, etc.	Portion of plant on main stem	Prior to fruit set
Beans (French bean, Cowpea, Lima bean, hyacinth bean, etc.)	Two to three fully develop leaves at top of the plant	Initial flowering
Pea	Leaves from third node down from top of the plant	Initial flowering
Root crops (Carrot, Radish, Beet, Turnip)	Young mature leaves from central rings	Prior to root enlargement
Bulb crops (Onion, Garlic etc.)	Young mature leaves from central	Prior to bulbing
Celery	Petiole of youngest fully elongated leaf	Midgrowth (30-35 cm tall)
Lettuce (head type)	Middrib of wrapper leaf	Heading
Lettuce (leaf type)	Youngest mature leaf	Midgrowth
Leafy greens (Palak, Spinach, etc.)	Youngest mature leaf	Midgrowth
Sweet potato	Fourth to sixth leaf from the growing tip	Prior to root enlargement
Asparagus	Top 10 cm of the new fern branch	Midgrowth of the fern
Sweet corn	Entire leaf at the ear node	Tasseling

Recommended Storage life for Different Vegetables

Vegetables	Temperature (°C)	Relative humidity (%)	Storage life (weeks)
Brinjal	10.0-11.1	92	2-3
Tomato (unripe)	8.9-10.0	85-90	4-5
Tomato (ripe)	7.2	90	1.0
Chilli (green)	7.2	85-90	3-5
Chilli (ripe)	5.6-7.2	90-95	2.0
Potato	3.0-4.4	85	34.0
Cabbage (early)	0.0-1.7	92-95	4-6
Cabbage (late)	1.0-1.7	92-95	12.0
Cauliflower (late)	0.0-1.7	85-95	7.0
Bitter gourd	0.6-1.7	85-90	4.0
Cucumber	10.0-11.7	92	2.0
Muskmelon	1.7-3.3	85-90	1.5
Pumpkin	10.0-12.0	70-75	24-36
Winter squash	12.8-15.6	70-75	24-36
Watermelon	7.2-15.6	80-90	2.0
Beet (topped)	0.0-1.7	90-95	8-14
Beet (bunched)	0.0	90	1.5
Carrot (topped)	0.0	95	20-24
Radish (topped)	0.0	88-92	3-5
Turnip	0.0	90-95	8-16
Hyacinth bean	0.6-1.7	90	3.0
Lima bean	4.4-7.2	90-95	1.5-2.0
Pea	0.0	88-92	2-3
Garlic	0.0	65	28-36
Onion (bulb)	0.0	70-65	20-24
Onion (leaves)	0.0	90-95	2.0
Taro	11.1-12.8	85-90	21
Cassava	0.0-1.7	85	23
Sweet potato	10.0-12.8	80-90	13-20
Yam	26.7	66-70	3-5
Ginger	7.2-10.0	75	16-24
Okra	8.9	90	2.0
Sweetcorn	0.6-1.7	90-95	1.0
Celery	0.0-0.6	92-95	8.0
Lettuce (head type)	0.0	90-95	3.0
Lettuce (leaf type)	0.0	95	1.0
Coriander leaves	0.0-1.7	90	5.0
Asparagus	0.0	95	3-4

Potential Vegetables for export and quality specifications:

Item	Specifications
Onion	
Big	Light to dark red, globose, 4-6 cm in size and strongly pungent (for Bangladesh, South-East Asia and Gulf countries): yellow or brown, globose and 4-6 cm in size (for European and Japan markets).
Multiplier	Bigger cloved bulb of bright red colour
Small	Dark red, round, 2-3 cm in size
Okra	Green, tender, 6-8 cm, 4-5 ridges, slender and roundish with pointed tips.
Potato	White-skinned long and extra long tubers with yellow to light yellow flesh (for Iraq and Iran); both white and red types of oval shape and 4-5 cm in diameter size (for Bangladesh).
Bitter gourd	Complete green, lustrous, more prickly, medium long and spindle shaped.
Chillies	Slender, dark green, lustrous, 7-8 cm, smooth without wrinkles and bends.
Garlic	White, round, 3.5-4 cm having 15-20 big-sized cloves.
Gherkin	Green, small sized (7.5 cm x 2.5 cm), 160-300 fruits/kg is considered as premium grade.
Brinjal	Round, thornless, shiny purple with dark green calyx.
Baby corn	Yellow seeded, thick walled, predominantly lobed, firm and shining green.
Lettuce	Round, medium to large head (750g to 1.0 kg) with wrinkled and green non-wrapper leaves, brittle textured and tight folded.
Cabbage	Round to flat-round head with dark green, smooth outer leaves, solid, short cone.
Bottle gourd	Green, tender, cylindrical, 20-25 cm with soft hairs.
French bean	Tender, green, 15 cm in length
Tomato	Round, medium-sized and red in colour
Cluster bean	Green, tender, 7-10 cm in length

Suitable varieties for export

Crop	Variety
Onion: Big	N-53, Nasik Red, Agrifound Dark Red, Agrifound Light Red, Arka Niketan (for Gulf and other countries); Arka Pitambar, Granex-33 and Spanish Brown (for Japan and Europe); Arka Bindu (for dried onion).
Okra	Parbhani Kranti, Arka Anamika, Arka Abhay, Varsha Uphar, Pusa Sawani, Kufri Sinduri, Kufri Jyoti (for Bangladesh)
Bitter Gourd	Arka Harit, Pusa Vishesh, MBTH-101, Hirkani, Konkan Tara, Prethi
Chilli	Bhagyalakshmi (G-4), Andhra Jyoti (G-5), G-3, Pant C-1, Arka Lohit, X-235, Panvel, Ansari, Musalwadi
Garlic	G-41, Jamnagar Ladwa, Malek. Agrifound Parvati, Yamuna Safed.
Bottle gourd	Arka Bahar, Pusa Naveen, Varad, Samrat (Hybrid)
Tomato	Round-fruited hybrids
Brinjal	Pant Rituraj, Pusa Uttam, Pusa Upkar, Pusa Kranti, Arka Navneet and Pusa Anmol (hybrid)
Gherkin	Ajax, Calypsoplus, BSS-168, BSS-169 (Hybrid)
Baby corn	VL42 (OP), MEH 114 (Hybrid), COBC-1 (Synthetic).
Sweet pepper	Bharath, Green Gold, Indira, Hira, Lario (Hybrid).
Watermelon	Sugar Baby, Arka Manik
Cluster bean	Pusa Sadabahar, Pusa Musami, Pusa Naubahar
French bean	Contender, Masterpiece, Premier, Arka Komal
Elephant-foot yam	Kovour, Gajendra

Effect of seed moisture content on seeds in storage

Seed moisture content	Effect on seed
33-60%	Germination occurs on imbibation, quick loss of viability
18-20%	High respiratory rate, high heating with inadequate ventilation, high activity of moulds and storage pests, seeds susceptible to mechanical damage in handling
13-17%	Still high respiratory rate, moulds and insects can be damaging, fumigation can affect seed viability, resistant to mechanical damage.
10-12%	Seeds can be stored reasonably well for 6-18 months in open storage in temperate climate, insects can be a problem, fumigation can safely be done.
8-10%	Seed can safely be stored for 1-3 years in open storage, very little insect activity, seeds susceptible to mechanical damage.
4-8%	Safe moisture content for sealed storage
0-4%	Excessive desiccation is damaging to seeds, seeds may lose viability.

Temperature range for multiplication of the organisms associated with seed storage

Organism	Temperature range (°C)		Relative humidity %
	For multiplication	Optimum	
Insects	21-42	27-37	30-95
Mites	8-31	19-31	60-100
Fungi	8-80	20-40	60-100
Bacteria	8-80	26-28	91-100

Approximate moisture content of vegetable seeds in equilibrium in air having different relative humidities at room temperature (20°C)

Seeds of the vegetable	Per cent relative humidity in air					
	10	20	30	45	60	75
Pea	5.4	7.3	8.6	10.1	11.9	15.0
Broad bean	4.2	5.8	7.2	9.3	11.1	14.5
Lima bean	4.6	6.6	7.7	9.2	11.0	13.8
French bean	3.0	4.8	6.8	9.4	12.0	15.0
Cabbage	3.2	4.6	5.4	6.4	7.6	9.6
Chinese cabbage	2.4	3.4	4.6	6.3	7.8	9.4
Radish	2.6	3.8	5.1	6.8	8.3	10.2
Carrot	4.5	5.9	6.8	7.9	9.2	11.6
Beet	2.1	4.0	5.8	7.6	9.4	11.2
Turnip	2.6	4.0	5.1	6.3	7.4	9.0
Tomato	3.2	5.0	6.3	7.8	9.2	11.1
Brinjal	3.1	4.9	6.3	8.0	9.8	11.9
Chillies	2.8	4.5	6.0	7.8	9.2	11.0
Cucumber	2.6	4.3	5.6	7.1	8.4	10.1
Winter squash	3.0	4.3	5.6	7.4	9.0	10.8
Watermelon	3.0	4.8	6.1	7.6	8.8	10.4
Okra	3.8	7.2	8.3	10.0	11.2	13.1
Onion	4.6	6.8	8.0	9.5	11.2	13.4
Welsh onion	3.4	5.1	6.9	9.4	11.8	14.0
Parsnip	5.0	6.1	7.0	8.2	9.5	11.2
Celery	5.8	7.0	7.8	9.0	10.4	12.4
Lettuce	2.8	4.2	5.1	5.9	7.1	9.6
Spinach	4.6	6.5	7.8	9.5	11.1	13.2
Mustard	1.8	3.2	4.6	6.3	7.8	9.4
Sweet corn	3.8	5.8	7.0	9.0	10.6	12.8

Methods of germination test of the seeds of some vegetable crops

Crop	Substrate	Temp(°C)	First count (days)	Final count (days)	Additional recommendation to break dormancy
Okra	TP; BP	20-30	4	21	-
Onion	BP; TP	20	6	12	Pre-chill at 5°C
Beet	TP; BP	20-30	4	14	Pre-wash 2-4 hours in running water
Beet	TP; BP	20-30	4	14	Pre-wash 2-4 hours in running water
Cabbage	TP	20-30	5	10	Pre-chill, soaking the paper in 0.2% KNO_3 solution
Cauliflower	TP	20-30	5	10	Do
Chilli	TP, BP	20-30	7	14	Soaking paper in 0.2% KNO_3 solution
Muskmelon	BP	20-30	4	8	-
Cucumber	TP;BP	20-30	4	8	-
Winter squash	BP	20-30	4	8	-
Pumpkin	BP	20-30	4	8	-
Summer squash	BP	20-30	4	8	-
Carrot	TP; BP	20-30	7	14	-
Lettuce	TP; BP	20	4	7	Pre-chill at 5°C
Tomato	TP; BP	20-30	5	14	Soaking the paper in 0.2% KNO_3 solution
Pea	BP	20	5	8	-
Brinjal	TP; BP	20-30	7	14	-
Broad bean	BP	20	4	14	-
Cowpea	BP	20-30	5	8	-

BP: Between two or more layers of germination or filter paper

TP: Top of the one or more layer of filter or germination paper

Note: Potassium nitrate (KNO_3) solution at 0.2% concentration is to be applied to the papers (substrate) at the point of saturation only at the beginning of the test. Subsequent moistening is to be done normally with water.

Radiation Dose for different vegetables

Crop	Dose (Krd)	Crop	Dose (Krd)	Crop	Dose (Krd)
Potato	5-6	Peas	15-25	Carrot	30-50
Tomato	15-25	French bean	10-25	Brassica sp.	60-80
Brinjal	20-30	Broad bean	10-25	Radish	70-80
Chilli	10-20	Onion	10-15	Okra	30-60
				Sweet potato	3-15

Prescribed Field Standards (Isolation Requirements, Minimum Field Inspections and Maximum Permitted off-Types) for Vegetables Seed Production

Crops	Minimum isolation distance (m		
	F	C	To be isolated from
Potato	5	5	Other varieties
Tomato	50	100	Other varieties
Brinjal	200	100	Other varieties
Chilli	400	200	Other varieties & chillies from sweet pepper
Sweet pepper	400	200	Other varieties & sweet pepper from chillies
Okra	400	200	Other varieties and wild *Abelmoschus spp.*
Pea	20	10	Other varieties
Other beans like cowpea, French bean, cluster bean, hyacinth bean, etc.	50	25	Other varieties of the crop
Cabbage	1600	1000	Other varieties of the crop and other crops like, cauliflower, Brussels sprouts, knolkhol, kale, collards
Cauliflower	1600	1000	Other varieties of the crop and other crops like, cabbage, sprouting broccoli, Brussels sprouts, knolkhol, collard, kale

Knolkhol	1600	1000	Other varieties of the crop and other crops like cabbage, cauliflower, Brussels sprouts, broccoli, collards, kale
Chinese cabbage	1600	1000	Same
Onion	1000	500	Other varieties
Radish	1600	1000	Other varieties of the crop
Turnip	1600	1000	Other varieties of the crop and other crops, like Chinese cabbage, repeseed, mustard and rutabaga
Carrot	1000	800	Other varieties
Beet	1600	1000	Other varieties of the crop and other crops, like palak, sugar beet, swiss chard
Amaranth	400	200	Other varieties and wild *Amaranthus* spp.
Palak	1600	1000	Other varieties of the crops and other crops like beet, sugar beet, Swiss chard
Fenugreek	50	25	Other varieties
Lettuce	50	25	Other varieties
Cucurbits	800	400	Other varieties and other crops Like long melon, snap melon
Muskmelon	-do-	-do-	Other varieties and other crops Like long melon, snap melon
Pumpkin	-do-	-do-	Other varieties and other crops like Summer squash and winter squash
Summer squash	-do-	-do-	Other varieties and other crops like pumpkin and winter squash
Winter squash	-do-	-do-	Other varieties and other crops like pumpkin and summer squash
Other cucurbits, like bitter gourd, bottle gourd, ridge gourd, sponge gourd, snake gourd, watermelon, cucumber, tinda etc.	-do-	-do-	Other varieties of the crop

Crops	Minimum field inspections		Maximum permitted off-types (%)	
	No.	Stages of crop growth	F	C
Potato	3	45 days after planting, just before haulm cutting, after haulm cutting	0.2	0.5
Tomato	3	Before flowering, flowering to immature fruit stage, mature fruit stage	0.1	0.1
Brinjal	3	Before flowering, flowering to immature fruiting, mature fruit stage	0.1	0.2
Chilli	3	Before flowering, flowering to fruiting, mature fruit stage	0.1	0.2
Sweet pepper	3	Before flowering, flowering to fruiting, mature fruit stage	0.1	0.2
Okra	3	Before flowering, flowering to fruiting, mature fruit stage	0.1	0.2
Pea	3	Before flowering, flowering, edible pod stage	0.1	0.2
Other beans like cowpea, French bean, cluster bean, hyacinth bean, etc.	2	Before flowering, flowering to fruiting	0.1	0.2
Cabbage	3	Head formation stage, fully developed head stage, flowering stage	0.1	0.2
Cauliflower	4	Before initiation of curds, curd initiative on stage, fully developed curd stage, flowering stage	0.1	0.2
Knolkhol	3	Before initiation of stem knob, fully developed knob stage, flowering stage	0.1	0.2
Chinese cabbage	3	Head formation stage, fully developed head stage, flowering stage	0.1	0.2
Onion	3*	20-30 days after sowing, fully developed bulb stage, flowering stage. (*seed-to-seed method)	0.1	0.2

	4**	20-30 days after sowing, after lifting of bulbs, at replanting of bulbs, flowering stage (*bulb-to-seed method)		
Radish	3	20-30 days after sowing, at lifting and replanting, flowering, (root-to-seed method)	0.1	0.2
Turnip	3	20-30 days after sowing, at lifting and replanting, flowering, fully developed umbel stage (root-to-seed method)	0.1	0.2
Carrot	4	20-30 days after sowing, at lifting and replanting, flowering, fully developed umbel stage (root-to-seed method)	0.1	0.2
Beet	3	20-30 days after sowing, at lifting and replanting, flowering	0.1	0.2
Amaranth	2	Fully developed stage before flowering, flowering stage	0.1	0.2
Palak	2	Fully developed stage before flowering, flowering stage	0.1	0.2
Fenugreek	2	Fully developed stage before flowering, flowering stage	0.1	0.2
Lettuce	3	Before head formation, fully developed head stage, flowering stage (head type lettuce) and before fully grown stage, at full grown stage, flowering stage (non-head type lettuce)	0.1	0.2
Cucurbits	3	Before flowering, flowering to immature fruiting, mature fruit stage	0.1	0.2
Muskmelon	-do-	-do-	-do-	-do-
Pumpkin	-do-	-do-	-do-	-do-
Summer squash	-do-	-do-	-do-	-do-
Winter squash	-do-	-do-	-do-	-do-
Other cucurbits, like bitter gourd, bottle gourd, ridge gourd, sponge gourd, snake gourd, watermelon, cucumber, tinda etc.	-do-	-do-	-do-	-do-

Specified Seed Standards for different Vegetable Crops

Crop	Minimum physical purity (%)		Minimum germination (%)		Maximum moisture (%) in ordinary container		Maximum moisture (%) in vapour-proof container	
	F	C	F	C	F	C	F	C
Tomato	98	98	70	70	8	8	6	6
Brinjal	98	98	70	70	8	8	6	6
Chilli	98	98	70	70	8	8	6	6
Sweet pepper	98	98	70	70	8	8	6	6
Okra	99	99	65	65	10	10	8	8
Pea	98	98	75	75	9	9	8	8
Cowpea	98	98	75	75	9	3	-	-
Hyacinth bean	98	98	75	75	9	9	7	7
French bean	98	98	75	75	9	9	7	7
Cluster bean	98	98	70	70	9	9	-	-
Cabbage	98	98	70	70	7	7	5	5
Cauliflower	98	98	65	65	7	7	5	5
Knolkhol	98	98	70	70	7	7	5	5
Chinese cabbage	98	98	70	70	7	7	5	5
Onion	98	98	70	70	8	8	5	5
Radish	98	98	70	70	6	6	5	5
Turnip	98	98	70	70	6	6	5	5
Carrot	95	95	60	60	8	8	7	7
Beet	96	96	60	60	9	9	8	8
Amaranth	95	95	70	70	8	8	6	6
Palak	96	96	60	60	9	9	8	8
Fenugreek	98	98	70	70	8	8	-	-
Lettuce	98	98	70	70	8	8	6	6
Cucumber	99	99	60	60	7	7	6	6
Muskmelon	99	99	-60	60	7	7	6	6
Longmelon	99	99	60	60	7	7	6	6
Watermelon	99	99	60	60	7	7	6	6
Pumpkin	99	99	60	60	7	7	6	6
Summer squash	99	99	60	60	7	7	6	6
Winter squash	99	99	60	60	7	7	6	6
Bitter gourd	99	99	60	60	7	7	6	6
Bottle gourd	99	99	60	60	7	7	6	6
Ridge gourd	99	99	60	60	7	7	6	6
Sponge gourd	99	99	60	60	7	7	6	6
Snake gourd	99	99	60	60	7	7	6	6
Round melon	99	99	60	60	7	7	6	6

F = Foundation and C = Certified seed

Structural chromosomal changes in vegetables

Terminology	Type of structural chromosomal aberrations
1. Deletion	Loss of a segment from chromosome
a. Terminal deletion	Lost segment contains telomere of affected chromosome (rare)
b. Interstitial deletion	A segment between telomere and centromere is lost (common)
2. Duplication	A chromosome segment is present in more than two copies in same nucleus
a. Tandem duplication	Additional chromosome segment is located just after the normal segment and gene sequence remains the same
b. Reverse duplication	Gene sequence of the extra segment is in reverse order, i.e., inverted (edc in place of cde)
c. Displace duplication	Additional segment is located in same chromosome but away from the normal segment
d. Translocation duplication	Additional segment is located in nonhomologous chromosome
3. Inversion	A chromosome segment contains gene in a sequence which is reverse of the normal gene sequence
a. Para-centric inversion	The inverted segment does not contains centromere
b. Peri-centric inversion	The inverted segment contains centromere
4. Translocation	Integration of a chromosome segment into a nonhomologous chromosome
a. Simple translocation (shift)	A segment of a chromosome integrated into a nonhomologous chromosome
b. Reciprocal translocation (exchange)	It is produced when two nonhomologous chromosome exchange segments from each other

Physical purity test:

Physical purity denotes the percentage (by weight) of seeds belonging to the variety under certification. The working sample from the seed lot is classified into the components like (a) pure seed i.e., seeds of the variety under certification, (b) seeds of other varieties of the same crop, (c) seeds of other crops, (d) seeds of weeds and (e) inert matters like sand, straw, stone, pebbles, soil particles, etc.

So, by purity analysis, the composition of the sample is ascertained and the results are expressed as weight percentage.

$$\textbf{Purity (\%)} = \frac{\text{Weight of pure seed}}{\text{Total weight of working sample}} \times 100$$

Similarly, weight percentage of the impurities can be determined separately.

Determination of real value of seed:

Real value of the seed ascertains the percentage of seed of the working sample of the seed lot that would produce normal seedlings of the variety under certification.

$$\text{Real value of seed (\%)} = \frac{\text{Purity (\%)} \times \text{Germination (\%)}}{100}$$

Storage of certified seeds:

The storage period ranges from 8-9 months. If the seeds are stored in moisture vapour permeable containers like, cloth bag, paper packet, multiwall paper bags, etc. the safe seed moisture content is around 9 per cent at ambient temperature. If the seeds are stored in moisture vapour-proof containers like, tin cans, low density polyethylene bags, rigid plastic containers, aluminium foil pouches, etc. the seed moisture content must be within 4-6 per cent. Short-lived seeds like onion, French bean are also stored in cold storage condition. So, adequate facilities should be available for seed drying so that such low seed moisture content is kept in the seeds. Generally, hybrid seeds are packed in moisture vapour-proof containers.

Storage of germplasm seeds:

Germplasm seeds are required to be kept for many years. Basic requirements of such long-term storage are to provide cold temperature and low relative humidity in the storage atmosphere. Properly dried seeds to perfect moisture level of below 8 per cent is preferably stored at 5° to 10°C and 30% relative humidity condition.

Storage of breeder or foundation stock seeds:

Such stock seeds need to be stored for 4-5 years to minimize genetic drift in the varietal population upon reproducing every year. Properly dried seeds having around 9 per cent moisture under moisture vapour permeable packaging should be stored under a storage atmosphere with 25% relative humidity at 30°C or less or 45% relative humidity at 20°C or less temperature. The required relative humidity in the storage room can be achieved by making the room moisture-proof and by using dehumidifier.

Self-pollinated crops:

In these crops, pollination involves the stigma of the same flower that produces the pollens or other flowers of the same plant, which effect natural self-fertilization. In these crops, natural cross-pollination is normally less than 10 per cent. Sexform of these crops are hermaphrodite e.g., potato, tomato, brinjal, chilli, sweet pepper, peas, cowpea, hyacinth bean, French bean, cluster bean, runner bean, fenugreek, lettuce, etc.

Cross-pollinated crops:

In these crops, pollens from the flowers of one plant is transferred by some agents (insects, wind, etc.) to the stigma of the flowers of another plant to effect pollination. Cross-pollination is indicative of high efficiency of reproduction systems based on cross-fertilization. In cross-pollinated vegetable crops, natural cross-pollination generally exceeds 60 per cent to the extent of even 100 per cent. Sexform, different adaptations and some inherent mechanisms favour cross-pollination. Different cross-pollinated vegetable crops are cabbage, cauliflower,

knolkhol, Brussels sprouts, sprouting broccoli, radish, beet, turnip, carrot, onion, leek, amaranth, palak, spinach, asparagus, pumpkin, bottle gourd, watermelon, muskmelon, ridge gourd, snake gourd, wax gourd, summer squash, bitter gourd, cucumber, sweet corn, pointed gourd, giant spine gourd, cassava, sweet potato, coriander, parsnip, etc.

Often-cross-pollinated crops:

These crops are predominantly self-pollinated but natural cross-pollination normally exceeds 25 %. Okra, lima bean, chilli, sweet pepper and potato are sometimes regarded as often-cross-pollinated crops due to high natural cross-pollination.

Vegetatively propagated crops:

Crop	Plant parts for regeneration
Potato	Underground stem tuber
Sweet potato	Underground root tuber and above ground vines
Elephant-foot yam	Corm, corm pieces and cormel
Taro	Cormel
Garlic	Clove (bulblet), bulbil
Pointed gourd	Fleshy root and vine cuttings
Giant spine gourd	Fleshy root
Cassava	Stem cutting
Drumstick	Limb cutting
Jerusalem artichoke	Root tuber
Globe artichoke	Root suckers or offshoots and division of old crown
Yam	Pieces of stem tuber and vine cuttings
Rhubarb	Division of crowns

Floral organization favouring self-pollination:

Bisexual flower structure and homogamy (stigmas and stamens maturing at the same time) are the main aspects favouring self-pollination. In vegetable crops normally homogamy occurs after flowers have opened (Chasmogamy). Homogamy in closed flowers (Cleistogamy) is not generally met with in vegetable crops. In peas and bean, stamens and pistil in opened flower remain enveloped inside a tube-like structure of jointed petals called keel.

Floral organism favouring cross-pollination:

(a) Sexform favouring cross-pollination:

Monoecy:

Male and female flowers are borne separately in the same plant. This unisexualism favours cross-pollination e.g., most of the cucurbitaceous vegetable crops, amaranth, sweet corn, cassava, etc. In muskmelon, most of the varieties are andromonoecious (separate male and hermaphrodite flowers in the same plant) while some are monoecious.

Dioecy:

Male and female flowers are borne on separate plants meaning that the male and female plants are separate e.g., pointed gourd, asparagus, spinach, giant spine gourd, yam, etc.

(b) Inherent mechanisms favouring cross-pollination:

Despite hermaphrodite sexform, self-fertilization is hindred favouring cross-pollination due to some genetically controlled mechanisms.

Self-incompatibility:

It is genetically controlled physiological hindrance to self-fertilization, which marks inability of the functional pollens to fertilize the vaule after self-pollination. It is found in cabbage, cauliflower, knolkhol, Brussels sprouts, radish, turnip and beet.

Male sterility:

In this phenomenon, male parts of the flower are non-functional

without affecting the female fertility. It is controlled by nuclear genes or sterile cytoplasm or by the interaction between nuclear gene and sterile cytoplasm. It is found in tomato, brinjal, chilli, musk-melon, onion, carrot, sprouting broccoli, Brussels sprouts, cabbage, cauliflower, sweet potato, etc.

Dichogamy

Protandry:

In protandrous flowers, pollens are shed before stigma becomes receptive e.g., onion, carrot, beet, parsley, leek, etc.

Protogyny:

In protogynous flowers, stigma becomes receptive before pollens are shed from the anthers of the same flower e.g., elephant-foot yam.

Classification of male sterility in vegetable crops

S.	Type of male sterility		Inheritance	Crops in which Identified
1.	**Pollen sterility**	(a)	Cytoplasmic	Carrot, sweet pepper, turnip, Chinese cabbage, cucumber (in cauliflower, cabbage and broccoli this male sterility has been introduced from sterile cytoplasm of radish)
		(b)	Genic (i) Single recessive gene	Tomato, brinjal, pea, muskmelon, watermelon, chilli, lima bean, pumpkin, cucumber, cabbage, cauliflower, Brussels sprouts, broccoli, Chinese cabbage
			(ii) Duplicate dominant genes	Cauliflower
		(c)	Cytoplasmic-genic	
			(i) Single recessive gene	Onion, radish, sweet pepper
			(ii) Two recessive gene	Beet
			(iii) Single dominant gene	Carrot

2.	**Staminal Sterility** Male sterility arises because stamens are either altogether absent or present in rudimentary form. So, pollens are not formed. The flowers are having normal corolla and style.	Genic (Single recessive gene)	Tomato
3.	**Functional male sterility**		
(a)	**Positional sterility** Male sterility arises because another fails to dehisce to liberate the normal and fertile pollens so that the pollens cannot reach the stigma. This male sterility is maintained easily by manual self-pollination. The male sterile plants can easily be identified by examining the anthers.	Gene (single recessive)	Tomato, brinjal, sweet pepper
(b)	**Exerted stigma sterility** Male sterility is caused due to long style where stigma protrudes through the anther cone and becomes out of the reach of the pollens of the same flower. However, long styly is very much influenced by environment	Gene (single recessive)	Tomato

Breeding of self-pollinated vegetable crops introduction of elite varieties

Crop	Variety	Introduced from
Tomato	Roma	USA
	Labonita	USA
	Sioux	USA
	Marvel	USA
	Best of All	USA
	Money Maker	USA
Sweet pepper	California Wonder	USA
	Yolo Wonder	USA
	World Beater	USA
	Chinese Giant	USA
	Golden Cal Wonder	USA
	Bullnose	USA
Okra	Perkins Long Green	Exotic
Pea	Early Superb	UK
	Meteor	UK
	Arkel	UK
	Little Marvel	UK
	Alaska	UK
	Early Badger	USA
	Bonneville	USA
	Lincon	USA
	Alderman	USA
	Perfection New Line	USA
	Sylvia	Sweden
French bean	Contender	USA
	Gaint Stringless	USA
	Ketucky Wonder	USA
	Bountiful	USA
	Master piece	USA
	Premier	USA
	Jampa	Mexico
Cowpea	Philippines Early	Philippines

Estimated seed requirement in India (hybrid and open pollinated (OP) varieties

Crops	Area under cultivation (ha)			Seed requirement (tones)		
	OP	Hybrid	Total	OP	Hybrid	Total
Tomato	199780	299670	499450	119.87	89.91	209.77
Brinjal	384390	108420	492810	307.52	43.37	350.88
Chilli	378886	42100	420985	378.89	21.05	399.94
Cabbage	156431	156430	312860	125.15	78.22	203.36
Cauliflower	265935	46930	312860	212.75	23.47	236.21
Gourds	273640	147345	420985	1368.20	736.73	2104.92
Melons	130650	43550	174197	522.59	130.65	653.24
Okra	286730	95575	282302	2867.27	955.76	3823.03
Total	2076435	940015	3016449	5902.22	2079.13	7981.34

The commonly utilized mechanisms/methods for developing commercial hybrids

Mechanism	Commercially exploited crops
Hand emasculation + MP	Tomato, eggplant, okra, sweet pepper
Bagging/protection of staminate and pistillate flowers + MP	Cucurbits (bottle gourd, pumpkin, watermelon)
Pinching of staminate flowers + MP/NP	Cucurbits (bitter gourd, bottle gourd, pumpkin, watermelon)
Male sterility + MP	Tomato, sweet pepper, chilli
Male sterility + NP chilli	Onion, cabbage, carrot, radish, muskmelon,
Self-incompatibility + NP	Cabbage, cauliflower, broccoli
Gynoecism + NP	Cucumber, muskmelon
PGR and pinching of staminate flowers + NP	Summer squash, winter squash, etc.

MP – Manual NP – Natural

Premium quality attributes of the hybrids

Crop/hybrid	Targeted quality attributes
Tomato	High temperature tolerance, better self-life, high TSS, acidity and high lycopene content.
Chilli	High oleoresin, specific fruit size, shape and pungency for various markets.
Sweet pepper	Adaptability to higher temperature condition
Cabbage	High temperature tolerance, longer staying capacity in field, narrow, short and soft core.
Cauliflower	High temperature tolerance, non ricy, compact, bract free and self blanching curd.
Okra	Attractive light green fruit with spineless, smooth skin.
Muskmelon	High TSS content, thick flesh.
Watermelon	High TSS content, less seed, pink flesh.
Bitter gourd	Spindle shaped green fruit.
Bottle gourd	Uniform shape, non ridged fruit surface.
Cucumber	Bunching fruit habit, finger size fruit (for gherkin), less seeds, spineless smooth skin.
Pumpkin	Fruits with high flesh recovery and beta carotene content.

Research priorities for developing multiple resistant hybrids against biotic stresses

Crops	Target for biotic stresses
Tomato	Tomato leaf curl virus + Early blight + Bacterial wilt + Root knot nematode
Brinjal	Phomopsis + Bacterial wilt + Fruit and shoot borer.
Chilli	Leaf curl + Thrips + Mites + Anthracnose.
Capsicum	Phytophthora + Thrips + Mites.
Okra	Yellow vein mosaic virus + Fruit borer.
Onion	Stemphyllium +Purple blotch + Thrips.
Cucumber	Downey mildew + Mosaic.
Muskmelon	Powdery mildew + Downey mildew + Anthracnose + Fusarium. + fruit fly.
Watermelon	Powdery mildew + Downey mildew + Anthracnose.
Cabbage	Black rot + Diamond back moth.
Cauliflower	Black rot + Diamond back moth.
Bittergourd	Mosaic + Fruit fly.
Bottlegourd	DM + PM
Cucumber	CMV + PM
Pumpkin	CMV.

Research Priorities for Hybrid Seed Production System and Specific Trait

Crops	F_1 seed production system*	Heterosis breeding aims
Tomato	Use of genetic cytoplasmic male sterile (cms) and genie male sterile lines High oleoresin	Tolerant to high temperature (off-season), high TSS and high lycopene.
Chilli	Use of cytoplasmic male sterile or genetic male lines.	High oleoresin.
Capsicum	Standardization of techniques for plains (use of protected infrastructure).	Adapted to the plains of North India.
Cauliflower	Use of strong self incompatible or CMS lines.	Heat tolerance (off-season).
Cabbage	Use of strong self incompatible or CMS lines.	Tolerant to high temperature (tropical cabbage).
Onion	Development and use of short day CMS lines.	Photoperiod insensitive (off-season).
Carrot	Use of cytoplasmic male sterile lines.	High carotene per cent.
Watermelon	Use of male sterile lines.	High TSS and seedlessness.
Muskmelon	Use of male sterile lines.	High TSS.
Cucumber	Use. of gynoecious line.	Pickling type (for export).

*Initiative may be taken to identify, characterize and utilize conditional mutants of male sterility in all the vegetables

Hybrids recommended for release and cultivation

Crop	Name of identified hybrids
Tomato	ARTH-4, MTH-6 ARTH-3, Pusa Hybrid-2, Arka Vardhan, NA-501, DTH-4, KT-4, NA-601, FMH-2, BSS-20, Avinash-2, HOE-303, Sun-496, DTH-8, Swarn Baibhav (CHTH-1), ARTH-128, KTH-2.
Brinjal	Arka Navneet, Pusa Hybrid-6, Pusa Hybrid-5, HRBH-201, NDBH-1, ABH-1, MHB-10, MHB-39, NDBH-6, ABH-2, Phule Hybrid-2, Pusa Hybrid-9, ARBH-541, PBH-6, JBH-1, BH-1, BH-2, Kashi Sandesh.
Chilli	HOE-888, ARCH-236, Sungro-86-235, ARCH-228.
Capsicum	KT-1, SEL-11, Lario, DARL-202.
Okra	DVR-1, DVR-2, DVR-3, DVR-4.
Cauliflower	Pusa Hybrid-1, DCH-541; **Synthetics**: Synthetic-1, Early Synthetic. Shri
Cabbage	Ganesh Gol, Nath-401, BSS-32, Nath-501, Quisto; **Synthetics**: Pusa Synthetic
Carrot	Hybrid-1.
Muskmelon	Hybrid M-3, MHY-5
Watermelon	Arka Jyoti
Bitter gourd	Pusa Hybrid-2
Cucumber	PCUCH-1
Bottle gourd	PBOG-2, PBOG-1, NDBH-4

Some popular vegetable hybrids in India

Crops	Name of Popular hybrids
Tomato	Rupali, Naveen, Rashmi, Tolstoi, Minakshi, NS-815, NS-2530, TH-2312, TH-802
Brinjal	Pusa Hybrid-5, Pusa Hybrid-6, Manju, Kalpataru
Chilli	CH-1, CH-3, NS-1101, ARCH-236, Tejashwani
Sweet pepper	Bharat, Kt-1, Indira, Lario
Cauliflower	NS-60, Punam, Priya, Summer King, Agati Himlata, Sweta, Pawas, NS-66
Cabbage	Sumit, Nav Kranti, Manisha, Sri Ganesh Gol, Bajrang, NS 25
Cucumber	Pant Sanker, Khira-1, Malini
Bitter gourd	Pusa hybrid-1
Bottle gourd	Pusa Hybrid-3, Varad (MGH-4)
Muskmelon	Punjab Hybrid-1
Watermelon	NS-295

Public sector hybrids in solanaceous vegetable crops

Name of hybrid/crop	Source
Tomato	
Pusa Hybrid-1, Pusa Hybrid-2, DTH-4, DTH-8, Pusa Hybrid-4,	IARI
KT-4, Arka Vishal, Arka Vardan, FMH-1, Arka Shrestha,	IIHR
Arka Abhijit	Pantnagar
Pant Hybrid-1, Pant Hybrid-2, Pant Hybrid-10,	Faizabad
Pant Hybrid-11, NDTH-1, NDTH-2, NDTH-6, NDTH-4,	Rahuri
Rajashree, Phule Hybrid-1, Hybrid-37,	PAU, Ludhiana
TH-2312	
Brinjal	
Arka Navneet,	IIHR
Pusa Anmol, PusaHybrid-6, Pusa Hybrid-5, Pusa Hybrid-9,	IARI
Vijay Hybrid, Azad Hybrid	Kanpur
NDBH—1, NDBH-6, NDTH-11, NDBH-7	Faizabad
Hybrid-2	Rahuri
Punjab Hybrid, BH-1	Ludhiana
ABH-1, ABH-2	Anand
Pant Hybrid-2	Pantnagar
Chilli	
CH-1	Ludhiana
CH-3	Ludhiana
Sweet pepper	
Solan Hybrid-1	Solan
KT-1 (Pusa Deepti), KT-2	Katrain
Sel-2	Srinagar
Okra	
DOH-3, DOH-4	IARI
JOH-5	Parvani
DVR-1, DVR-2	Varanasi

Public sector hybrids/synthetics of cole and bulb crops

Name of hybrid	Source
Cauliflower	
Pusa Hybrid-2 (F_1 hybrid)	IARI
Pusa Early Synthetic, Pusa Synthetic (synthetic)	IARI
Pant Gobi-3 (synthetic)	Pantnagar
Onion	
Arka Kirtiman, Arka Lalima	IIHR
Hybrid 63, Hybrid 35	IARI
Cabbage	
Pusa Synthetic (synthetic)	IARI
H-64 (Hybrid), BRH-5	Katrain

■ Mode of inheritance of resistance:

Oligogenic resistant:

Single or few major genes govern this disease resistance of the host plant, and both dominant and recessive genes may condition the resistance reaction. Such oligogenic resistance is specific to some pathotype of the pathogen.

Polygenic resistance:

This disease resistance is governed by many genes (polygenes) with small effect expressed in cumulative manner thus showing continuous variation for disease reaction. In this case several genes contribute to a quantitative trait loci (QTL) that show additive effect against different races of the pathogen

Oligogenic Resistance

Crops	Genetic control for resistance to disease
Tomato	Single dominant gene for Fusarium wilt (I), Verticillium wilt (Ve), early blight (Ad), late blight (Ph-1), leaf mould (Cf), leaf spot (Sm), mosaic virus (TM), leaf curl virus (Tlc); Two independent genes for Fusarium wilt.
Potato	Single dominant gene for late blight (R), potato virus X, potato virus Y and few dominant major genes for bacterial wilt.
Brinjal	Single dominant gene for bacterial wilt (Rps); A pair of dominant genes for phomopsis blight
Chilli	Single dominant gene for Phytophthora fruit rot, bacterial leaf spot, tobacco mosaic, single recessive gene for leaf curl virus (lc), two independent dominant genes for phytophthora root rot.
Okra	Two complementary dominant genes for yellow vein mosaic virus in A. manihot ssp. Manihot.
Pea	Single dominant gene for Fusarium wilt (Fnw) and ascohyta blight and single recessive gene for powdery mildew (er-1), pea-seed-born mosaic virus (sbm)
Cowpea	Single dominant gene for cercospora leaf spot (Cls-1) and bean yellow mosaic, single recessive gene for powdery mildew, bacterial blight and cercospora leaf spot.
Cucumber	Single dominant (P) for downey mildew; two pairs of duplicate genes and three complementary recessive genes for powdery mildew.
Muskmelon	Two incompletely dominant genes for downey mildew, single dominant gene for powdery mildew and watermelon mosaic, three recessive genes for cucumber mosaic.
Watermelon	Single dominant gene for fusarium wilt and anthracnose.

Polygenic Resistance

Crop	Polygenic control for resistance to disease
Tomato	Bacterial wilt, bacterial spot, bacterial canker, leaf curl in L. pimpinellifolium, L.peruvianum (3-5 recessive genes); Early blight (two or more recessive genes).
Potato	Potato leaf roll virus
Brinjal	Phomopsis blight
Cauliflower	Black rot (dominant polygenes)

Wild and allied species of the vegetable crops carrying resistance attribute

Crops	Disease	Pathogen	Species
Potato	Late blight	*Phytophthora infestans*	*Solanum tuberosum ssp. Andigena, S.demissum S.curtilobum, S.stoloniferum*
Tomato	Fusarium wilt	*Fusarium oxysporum f.sp.lycopersici*	*L. pimpinelifolium, L. hirsutum*
	Early blight	Alternatia solani	*L. pimpinelifolium, L. hirsutum f.glabratum*
	Bacterial wilt	*Pseudomonas solanacearum*	*L. pimpinellifolium*
	Leaf curl	Tomato leaf curl virus	*L. peruvianum, L. pimpinellifolium, L. hirsutum f. glabratum*
Brinjal	Bacterial wilt	*Pseudomonas solanacearum*	*S. sisymbrifolium, S. integrifolium*
	Phomopsis blight	*Phomopsis vexans*	*S.xanthocarpum, S.indicum, S. gilo, S.khasianum, S.nigrum, S.sisymbrifolium*
	Little leaf	Phytoplasma	*S.viarum, S.incanum*
Chilli	Tobacco mosaic	Tobacco mosaic virus	*C. chacoense*
	Phytophthora rot	*Phytophthora capsici*	*C. frutescens*
	Leaf curl	Leaf curl virus	*C. frutescens*
	Anthracnose	*Colletotrichum capsici*	*C. baccatum, C.chinense*
Okra	Yellow vein mosaic	Yellow vein mosaic virus	*A.manihot ssp. manihot A.manihot ssp. Tetraphylus*
Onion	Purple blotch	*Alternaria porri*	*Allium fistulosum*
Cabbage & cauliflower	Club root	*Plasmodiophora brassicae*	*B. napus,*
	White rust	*Albugo candida*	*B. campestris*
			B. carinata

Disease Resistant Varieties Developed/Identified (ICAR Institutes and State Agricultural Universities)

Crops	Diseases	Name of the resistant Varieties or breeding lines	Name of the centres
Tomato	Bacterial wilt (*Pseudomonas Solanacearum*)	LB-79 (Shakti)	KAU
		BWR-5 (Arka Alok)	IIHR
		BWR-1 (Arka Abha)	
		BT- 1 (Utkal Palilavi)	OUAT
		BT-2 (Utkal Deepti)	
	Late blight (*Phytophcra infestans*)	TRB-I and TRB-2	PAU
	Leaf curl virus	H-24 (Hisar Anmol)	CCSHAU
		H-36, H-86, H-88	IIVR
Brinjal	Bacterial wilt (*Pseudomonas solanacearum*)	BWR-12 (Arka Keshav)	IIHR
		Arka Nidhi	
		Arka Nellkanth	
		BB-7 (Utkal Tarini)	OUAT
	Phomopsis rot (*Phornopsis vexans*)	SM-6-6 (Soorya)	KAU
		JC- I and JC-2	AAU
		Pusa Bhairav	IARI
Okra	Yellow vein mosaic virus	Pusa Sawani & 91-4	IARI
		Sel-2	NBPGR
		Punjab Padmini and P-7	PAU
		PB-57, Parbhani Kranti	MAU
		Sel-4 (Arka Abhay)	IIHR
		Sel- 10 (Arka Anamika)	

		HRB-9-2 (Varsha Uphar) HRB55 (Hisar Barsati) BO-2 Utkal Gaurav KS-404	CCSHAU OUAT VRS
Chilli	Leaf curl virus CMV, TMV, & Leaf curl Mosaic, Wilt & Dieback Virus Complex (Bacterial wilt) (*Pseudomonas solanacearum*)	Pusa Jwala Pusa Sadabahar Pant C-4 Punjab Lal and Punjab Surkh LCA 235, LCA 305 BC-21-2 (Utkal Rashmi) AAUM-1 & AAUM-2	IARI GBPUAT PAU OUAT AAU
Capsicum	Bacterial wilt (*Pseudomonas solanacearum*)	Arka Gaurav	IIHR
Muskmelon	Downey mildew (*Pseudoperonospora cubensis*) Cucumber green mottle mosaic virus	MR-12 (Punjab Rasila) DVRM-I DVRM-2	PAU IARI
Water- -melon	Powdery mildew (*Erysiphae cichoracearuin*) and Anthracnose (*Colletotrichum lagenarium*)	Arka Manik	IIHR
Garden Pea	Powdery mildew (*Erysiphe polygoni*)	PRS-4, Rachna, KS-225, KS-245 JP-4, JP-83, JP-7l and JP-885 FC-1 (Arka Ajit) IP-3, Pant P-5 ,PMR-21 DPP-62 VP-9003 ,VP-8902	VRS JLNKVV IIHR GBPUAT HPKVV VPKAS

		DMR-7 (Alankar) HFP-4 & HFP-12 (Jayanti & Ultra) HUP-I	IARI CCSHAU BHU
Cauliflower	Black rot (*Xanthomonas Compestris*) & Curd & Inflorescence blight (*brassicicola*)	Pusa Shubhra	IARI
Cabbage	Black rot (*Xanthornonas compestris*)	Pusa Mukta	IARI RS
Cowpea	Bacterial blight (*Xanthomonas compestris cv. vignicola*)	Pusa Komal	IARI RS
French bean	Common mosaic virus and rust (Uromyces)	UPF 191 (Pant Anupama)	GBPUAT

KAU = Kerala Agriculthral University, Vellanikkara.
IIHR = Indian Institute of Horticultural Research, Hessarghatta.
GBPUAT = GB Pant University of Agriculture and Technology, Panthagar.
IARIRS = IARI, Regional Station, Katrain
IARI = Indian Agricultural Research Institute, New Delhi.
BHU = Bananaras Hindu University, Varanasi.
CCSHAU = CCS, Haryana Agricultural University, Hissar.
HPKVV = Himachal Pradesh Krishi Vishwa Vidyalaya, Palampur.
VPKAS = Vivekananda Parvatiya Krishi Anusaudhan Shala, Almora.
VRS = Vegetable Research Station, Kalyanpur,
JLNKVV = Jawahar Lal Nehru Krishi Vishwa Vidyalaya, Jabalpur.
PAU = Punjab Agricultural University, Ludhiana.
AAU = Assam Agricultural University, Jorhat.
OUAT = Orissa University of Agriculture & Technology, Bhubaneshwar.
MAU = Marathwada Agricultural University, Parbhani
NBPGR = NBPGR, New Delhi.
IIVR = Indian Institute of Vegetable Research Varanasi.

Pest Resistant Varieties Hybrids/Lines of Major Vegetable Crops

Crop	Pest	Varieties/hybrids
Brinjal	Shoot and fruit borer (Leucenodes orbonalis)	SM 17-4, PBr 129-5, Doli 5, ARU 2-C, Punjab Jamuni Gola, Punjab Sadbahar Punjab Barsati Pusa Purple Round Punjab Neelam, Kalyanpur-2 Punjab Chamkila.
	Jassids (Amraṣa undalis)	Black Beauty Ram Nagar Giant Vijay Hybrid.
	Nematode	Pusa Purple Cluster Pant Rituraj Arka Nidhi Arka Kesav Arka Neelkanth.
Chilli	Thrips (Thrips tabaci)	Bhaskar
Cabbage	Aphid (Brevicorync brassicae)	All Season, Red Drum Read, Sure Head, Express Mail.
Cauliflower	Stem borer (Hellula undcjljs)	EMS-3, KW-5, KW-8 Early Patna Kathmandu Local.
Okra	Jassid (Amrasa biguttula)	IC-7194, IC-139999, New Selection Punjab Padniani.

Onion	Thrips (Thrips tabaci)	PBR-2, PBR-6,PBR-4, PBR-5 , PBR-6. Arka Niketan Pusa Ratnar
Round gourd	Fruitfly (Dacus cucurbitae)	Arka Tinda.
Pumpkin	Fruitfly (Dacus cucurbitac)	Arka Suryamukhi.
Bitter gourd	Fruitfly (Dacus cucurbitac)	Hissar-II.

Vegetable varieties developed through various breeding methods

S. N.	Vegetable	Variety	Introduction /Selection: Country	Introduction /Selection: Pure line /SPS/ Clonal selection	Introduction /Selection: Mass selection	(Hybridization) Pedigree	Hybrids/ synthetics
1	Tomato	Sioux	USA	Meeruti		Pusa Ruby (Sioux x Improved Meeuti)	Pusa Hybrid-1 Pusa Hybrid-2
		Lebonita	USA	Pusa 120		Pusa Early Dwarf (Improved Meeruti x Red Cloud)	Arka Vardan
		Marglobe	USA			Punjab Chhuhara (EC 55005 x Punjab Tropic)	Arka Vishal
		Roma	USA			Arka Meghali (Arka Vikash x IHR 584)	Arka Shreshta
		Dwarf Money Maker	Israel			Hisar Lalima (Pusa Early Dwarf x HS 101)	Arka Abhijit
		Balkan	Bulgaria				DTH – 4
		Fireball	Canada				KT – 4
2	Pea	Arkel	England			Jawahar Matar 3 (T 9 x Early Badger)	
		Meteor	England			P 88 (Pusa 2 x Morasis 55)	
		Bonneville	USA			VL Ageti Matar 7 (Pant Uphar x Arkel)	
		Sylvia	Sweden			VL 3 (Old Sugar x Early Wrinkled Dwarf)	
		Little Marvel	England			Pant Sabji Matar 3 (Arkel x GC 141)	
3	French bean	Contender	USA	Pusa Himlata		Pant Bean 2 (Turkish Brown x Contender)	
		Bountiful	USA	Arka Komal			
4	Dolichos bean			CO 1		CO – 2 (CO 8 x CO 1)	
5	Cluster bean			Sharad Bahar IC 11388		Pusa Navbahar (Pusa Mausami x Pusa Sadabhar)	

6	Chilli			NP 46 A Patna Red		Pusa Jwala (NP 46 A x Puri Red) Pusa Sadabahar (Pusa Jwala x IC 31339)	Punjab Hybrid KT – 1
7	Brinjal			Pant Samrat	Pusa Purple Long Pusa Purple Round Arka Sheel	Pusa Kranti [(PPL x Hyderpur) x Wynad Giant] Arka Navneet (IIHR 21 x Supreme) Pant Rituraj (T 3 x PPL)	Pusa Hybrid - 5 Pusa Hybrid-6 Pusa Hybrid-9
8	Okra					Pusa Sawani (IC 1542 x Pusa Makhmali) Parbhani Kranti (Pusa Sawani x *A. manihot*) x Pusa Sawani	
9	Onion	Early Grano Red Grano Bermuda Yellow	USA USA Philippines	Pusa Red Hisar - 2 Punjab Selection	Arka Pragati Arka Bindu Arka Niketan		Arka Kirtiman Arka Lalima
10	Cauliflower	Snowball-16 Improved Japanese	Holland Israel	Dania Hisar–1		Pusa Snowball - 1 (EC 12013 x EC 12012) Pusa Subhra (MGS 23 x 15-1) x (D 96)	Pusa Synthetic Pusa Hybrid-2
11	Cabbage	September	Germany	Pusa Drum Head		Pusa Mukta (EC. 24855 x EC. 10109)	Pusa Synthetic
12	Watermelon	New Hamp-shire Midget Asahi Yamato Sugar baby Shipper	USA Japan USA USA			Arka Manik (IIHR 21 x Crimson Sweet)	

13	Cucumber	Poinsette	USA				Pusa Sanyog
14	Summer Squash	Patty Pen	USA				Pusa Alankar
15	Carrot	Jeno	Germany			Pusa Kesar (Local Red x Nantes) Pusa Meghali (Pusa Kesar x Nantes) Pusa Yamdagni (EC 9981 x Nantes)	Hybrid – 1
16	Muskmelon			Arka Jeet Arka Rajhans		Pusa Sharbati (Kutana x PMR 6) Punjab Sunehri (Hara Madhu x Edisto)	Punjab Hybrid Pusa Rasraj
17	Bottle Gourd			Pusa Summer Prolific Long		Pusa Summer Prolific Round	
18	Turnip					Pusa Chandrima (Japanese White x Snowball) (Asiatic x European) Pusa Swarnima (Japanese Whitev x Golden Ball) (Asiatic x European) Pusa Kanchan (local Red Round x Golden Ball)	
19	Radish					Punjab Safed (White 5 x Japanese White) Pusa Himani (Radish Black x Japanese White)	
20	Sweet Potato			Pusa Safed Pusa Lal Pusa Sinduri			
21	Potato			Kufri Red round Kufri Safed			

List of Contributors

1. Dr. A.K. Singh, Sr. Scientist (Vegetable), IARI, New Delhi.
2. Dr. A.K. Sureja, Sr. Scientist (Vegetable), IARI, New Delhi.
3. Dr. Aishwarya Dudi, Subject Matter Specialist (Home Science), CAZRI, Krishi Vigyan Kendra, Pali.
4. Dr. Amol Vasishth, Assistant Professor (Forestry), College of Horticulture and Forestry, MPUAT, Jhalawar.
5. Dr. Arjun Kumar Verma, SMS (Agronomy), KVK, MPUAT, Jhalawar.
6. Dr. Ashutosh Mishra, Associate Professor (Floriculture), College of Horticulture and Forestry, MPUAT, Jhalawar.
7. Dr. B.L. Dhaka, Assistant Professor (Extension Education), KVK, Bundi.
8. Dr. Balu Ram Choudhary, Scientist (Vegetables), CIAH, Bikaner.
9. Dr. Bhagirath Ram, Scientist (Sr. Scale) (Plant Breeding), CAZRI, Jodhpur.
10. Dr. Birbal, Scientist (Sr. Scale) (Horticulture), Regional Research Station, CAZRI, Bikaner.
11. Dr. Deepak Sarolia, Assistant Professor (Horticulture), Rajasthan College of Agriculture, MPUAT, Udaipur.
12. Dr. Dhirendra Singh, Associate Professor (Plant Breeding), SKN College of Agriculture, Jobner.

13. Dr. Gopal Lal, Principal Scientist, NRC on Seed Spices, Tabiji.
14. Dr. H.R. Mahla, Sr. Scientist (Plant Breeding), Regional Research Station, CAZRI, Jaisalmer.
15. Dr. H.S. Jat, Sr. Scientist (Agronomy), CSSRI, Karnal.
16. Dr. Haridayal, Subject Matter Specialist (Horticulture), CAZRI Krishi Vigyan Kendra, Pali.
17. Dr. I.B. Maurya, Associate Professor (Vegetable Science), College of Horticulture and Forestry, MPUAT, Jhalawar.
18. Dr. J.R. Manmohan, Assistant Professor (Forestry), College of Horticulture and Forestry, MPUAT, Jhalawar.
19. Dr. Jitendra Singn, Associate Professor (Pomology), College of Horticulture and Forestry, MPUAT, Jhalawar.
20. Dr. K.D. Ameta, Assistant Professor (Horticulture), Rajasthan College of Agriculture, MPUAT, Udaipur.
21. Dr. Kamal Meena, Assistant Professor (Pomology), KVK, Sawai Madhopur
22. Dr. Kavita Arvindakshan, Assistant Professor (Vegetables), College of Horticulture and Forestry, MPUAT, Jhalawar
23. Dr. Khem Chand, , Sr. Scientist (Agriculture Economics), Regional Research Station, CAZRI, Pali
24. Dr. L.K. Dashora, Professor (Horticulture) and Dean, College of Horticulture and Forestry, MPUAT, Jhalawar.
25. Dr. Lobsang Wangshu, Associate Professor (Pomology), CHF, Pasighat, Arunachal Pradesh.

26. Dr. M. K. Chaudhary, Subject Matter Specialist (Agronomy), CAZRI Krishi Vigyan Kendra, Pali.
27. Dr. M. L. Meena, Subject Matter Specialist (Agriculture Extension), CAZRI Krishi Vigyan Kendra, Pali.
28. Dr. M.C. Jain, Associate Professor (Pomology), College of Horticulture and Forestry, MPUAT, Jhalawar.
29. Dr. M.R. Meena, Assistant Professor (Vegetables), Ambedkar University, Lucknow.
30. Dr Madhu, Scientist (Microbiology), CSSRI, Karnal.
31. Dr. Mahesh Kumar, Scientist (Sr. Scale) (Agriculture Chemistry), CAZRI, Jodhpur.
32. Dr. Mahesh Punia, Assistant Professor (Vegetables), KVK, Bundi.
33. Dr. Mali Ram, Assistant Professor (Horticulture), SKN College of Agriculture, Jobner.
34. Dr. Manoj Kumar Sharma, Assistant Professor (Soil Science), College of Horticulture and Forestry, MPUAT, Jhalawar.
35. Dr. Narayan Singh Nathawat, Sr. Scientist (Plant Physiology), Regional Research Station, CAZRI, Bikaner.
36. Dr. O.P. Gadwal, Assistant Professor (Vegetables), KVK, Nagaur.
37. Dr. P.K. Singh, Sr. Scientist (Vegetables), IIVR Regional Station, Sargatia.
38. Dr. P.R. Meghwal, Sr. Scientist (Horticulture), CAZRI, Jodhpur.
39. Dr. P.S. Chauhan, Assistant Professor (Forestry), College of Horticulture and Forestry, MPUAT, Jhalawar.

40. Dr. Prabhakar Singh, Assistant Professor (Vegetable), SKN College of Agriculture, Jobner.
41. Dr. Prerak Bhatnagar, Assistant Professor (Pomology), College of Horticulture and Forestry, MPUAT, Jhalawar.
42. Dr. R. Paliwal, Assistant Professor (Vegetables), SKN College of Agriculture, Jobner.
43. Dr. R.A. Kaushik, Professor (Horticulture) & Head, Rajasthan College of Agriculture, MPUAT, Udaipur.
44. Dr. R.K. Bagri, SMS (Plant Pathology), KVK MPUAT, Jhalawar.
45. Dr. R.K. Singh, Sr. Scientist (Agriculture Extension), CSSRI, Karnal.
46. Dr. R.R. Meena, SMS (Horticulture), KVK MPUAT, Jhalawar.
47. Dr. R.S. Dhaka, Former Professor (Horticulture), SKN College of Agriculture, Jobner.
48. Dr. Ranjeet Kumar, Sr. Scientist (Agriculture Economics), CIAE, Bhopal.
49. Dr. S. C. Kachhawaha, Subject Matter Specialist (Veterinary Science), CAZRI Krishi Vigyan Kendra, Pali.
50. Dr. S.D. Dhakar, Programme Coordinator, KVK MPUAT, Jhalawar.
51. Dr. S.K. Jain, Associate Professor (Post Harvest Technology), College of Horticulture and Forestry, MPUAT, Jhalawar.
52. Dr. S.K. Moond, Assistant Professor (Floriculture), College of Horticulture and Forestry, MPUAT, Jhalawar.

53. Dr. S.K. Sanwal, Sr. Scientist (Vegetables), IIVR, Varansi.
54. Dr. S.P. Singh, Assistant Professor (Vegetables), KVK, Jalore.
55. Dr. S.P.S. Tanvar, Sr. Scientist (Agronomy), Regional Research Station, CAZRI, Pali.
56. Dr. S.S. Meena, Scientist (Sr. Scale) (Vegetable Science), NRC on Seed Spices, Tabiji.
57. Dr. S.S. Punia, Assistant Professor (Plant Breeding), ARS, MPUAT, Kota.
58. Dr. Shiv Dutt, Sr. Scientist (Plant Breeding), Regional Research Station, CAZRI, Pali.
59. Dr. Shiv Kumar Dhull, Sr. Scientist (Agriculture Economics), NCAP, New Delhi.
60. Dr. Sita Ram .Kantwa, Scientist (Agronomy), IARI regional station, Indore.
61. Dr. Sunil Pareek, Assistant Professor (Horticulture), MPUAT, Udaipur.
62. Dr. V.S. Rathore, Sr. Scientist (Agronomy), Regional Research Station, CAZRI, Bikaner.
63. Dr. Vikas Khandelwal, Assistant Professor (Plant Breeding), NAU, Navsari.
64. Dr Vishakha Bansal, SMS (Home Science), KVK, MPUAT, Jhalawar.
65. Sh. A. K. Maru, Programme Assistant (Agriculture), CAZRI Krishi Vigyan Kendra, Pali.
66. Sh. Bhim Singh, Assistant Professor (Forestry), College of Horticulture and Forestry, MPUAT, Jhalawar.

67. Sh. Deepak Raj Purohit, Assistant Professor (Biotechnology), College of Horticulture and Forestry, MPUAT, Jhalawar.
68. Sh. H.L. Bairwa, Assistant Professor (Horticulture), Rajasthan College of Agriculture, MPUAT, Udaipur.
69. Sh. M.S. Choudhary, Technical Officer (Horticulture), CAZRI, Krishi Vigyan Kendra, Pali.
70. Sh. P. K. Tomar, Programme Assistant (Computer), CAZRI Krishi Vigyan Kendra, Pali.
71. Sh. Pradeep Kumar, Scientist (Horticulture), CAZRI, Jodhpur.
72. Sh. S. K. Dashora, Technical Officer (Agriculture Economics, CAZRI Krishi Vigyan Kendra, Pali.
73. Sh. S.B.S. Pandey, Assistant Professor (Forestry), College of Horticulture and Forestry, MPUAT, Jhalawar.
74. Sh. S.C. Meena, SMS (Soil Science), KVK, MPUAT, Jhalawar.
75. Sh. V.C. Prahlad, Assistant Professor (Forestry), College of Horticulture and Forestry, MPUAT, Jhalawar.

****

Applied Entomology – Srinivasa Reddy

Aushadhiya Kharpatwar (Hindi Medium)
– Dr. Chandra Bhanu

Cotton Improvement in India – Dr. Phundan Singh

Crop Nutrition: Principles and Practices – Dr. Rajendra Prasad

Descriptive Entomology – Dr. Shravan Haldhar

Entomology Glossary – Dr. T.V.Prasad

Entomology Treatise – Dr. T.V.Prasad

General Agriculture for JRF – Pushpendra Khanna

General Agriculture Question Pool – Mangesh Y. Dudhe

Glaustas Horticulture – P. Muthukumar

Guide on Nematology – Sunanda

Handbook of Entomology – Dr. T.V.Prasad

Illustrated Dictionary of Plant Pathology – Mamta Singh

Indian Forest Service Question Papers

IPR and Plant Breeders Rights – Dr. Phundan Singh

IPR and Plant Breeders Rights: Objective – Dr. Phundan Singh

IPR and Plant Breeders Rights: At A Glance – Dr. Phundan Singh

Krishi Darpan – Dr. S.R.Kantwa

Krishi Vigyan – Dr. S.R.Kantwa

MCQ in Agriculture Extension – Sunil, V.G.

MCQs in Horticulture – Bharat Singh Hada

Objective Agriculture for A.R.S. Exam. – Dr. S.R.Kantwa

Objective Agriculture for J.R.F. Exam. – Dr. S.R.Kantwa

Objective Crop Protection – Dr. T.V.Prasad

Objective Entomology – Dr. T.V.Prasad

Objective Horticulture – Udal Singh

Objective Soil Science – T.Ramesh

Plant Pathology At a Glance – Dr. P. Srinivas

Plant Science: At A Glance – Mangesh Y. Dudhe

Ready Reckoner of Genetics – Dr. S.V.Sai Prasad

Ready Reckoner of Plant Breeding – Dr. S.V.Sai Prasad

Terminology of Environmental Science – Somen Acharya

Training - Process and Methods – Dr. B. L. Dhaka

Vegetable Science – D.Singh